PROCESS SYSTEMS ANALYSIS AND CONTROL

McGraw-Hill Chemical Engineering Series

Editorial Advisory Board

Building the Literature of a Profession

Fifteen prominent chemical engineers first met in New York more than 60 years ago to plan a continuing literature for their rapidly growing profession. From Industry came such pioneer practitioners as Leo H. Baekeland, Arthur D. Little, Charles L. Reese, John V. N. Dorr, M. C. Whitaker, and R. S. McBride. From the universities came such eminent educators as William H. Walker, Alfred H. White, D. D. Jackson, J. H. James, Warren K. Lewis, and Harry A. Curtis. H. C. Parmelee, then editor of *Chemical and Metallurgical Engineering,* served as chairman and was joined subsequently by S. D. Kirkpatrick as consulting editor.

After several meetings, this committee submitted its report to the McGraw-Hill Book Company in September 1925. In the report were detailed specifications for a correlated series of more than a dozen texts and reference books which have since become the McGraw-Hill Series in Chemical Engineering and which became the cornerstone of the chemical engineering curriculum.

From this beginning there has evolved a series of texts surpassing by far the scope and longevity envisioned by the founding Editorial Board. The McGraw-Hill Series in Chemical Engineering stands as a unique historical record of the development of chemical engineering education and practice. In the series one finds the milestones of the subject's evolution: industrial chemistry, stoichiometry, unit operations and processes, thermodynamics, kinetics, process control, and transfer operations.

Chemical engineering is a dynamic profession, and its literature continues to evolve. McGraw-Hill, with its editor, B.J. Clark and its consulting editors, remains committed to a publishing policy that will serve, and indeed lead, the needs of the chemical engineering profession during the years to come.

The Series

Bailey and Ollis: *Biochemical Engineering Fundamentals*
Bennett and Myers: *Momentum, Heat, and Mass Transfer*
Brodkey and Hershey: *Transport Phenomena: A Unified Approach*
Carberry: *Chemical and Catalytic Reaction Engineering*
Constantinides: *Applied Numerical Methods with Personal Computers*
Coughanowr: *Process Systems Analysis and Control*
Douglas: *Conceptual Design of Chemical Processes*
Edgar and Himmelblau: *Optimization of Chemical Processes*
Gates, Katzer, and Schuit: *Chemistry of Catalytic Processes*
Holland: *Fundamentals of Multicomponent Distillation*
Holland and Liapis: *Computer Methods for Solving Dynamic Separation Problems*
Katz and Lee: *Natural Gas Engineering: Production and Storage*
King: *Separation Processes*
Lee: *Fundamentals of Microelectronics Processing*
Luyben: *Process Modeling, Simulation, and Control for Chemical Engineers*
McCabe, Smith, J. C., and Harriott: *Unit Operations of Chemical Engineering*
Mickley, Sherwood, and Reed: *Applied Mathematics in Chemical Engineering*
Nelson: *Petroleum Refinery Engineering*
Perry and Chilton (Editors): *Perry's Chemical Engineers' Handbook*
Peters: *Elementary Chemical Engineering*
Peters and Timmerhaus: *Plant Design and Economics for Chemical Engineers*
Reid, Prausnitz, and Rolling: *Properties of Gases and Liquids*
Smith, J. M.: *Chemical Engineering Kinetics*
Smith, J. M., and Van Ness: *Introduction to Chemical Engineering Thermodynamics*
Treybal: *Mass Transfer Operations*
Valle-Riestra: *Project Evaluation in the Chemical Process Industries*
Wei, Russell, and Swartzlander: *The Structure of the Chemical Processing Industries*
Wentz: *Hazardous Waste Management*

PROCESS SYSTEMS ANALYSIS AND CONTROL

Second Edition

Donald R. Coughanowr

Department of Chemical Engineering
Drexel University

McGraw-Hill, Inc.

New York   St. Louis   San Francisco   Auckland   Bogotá   Caracas
Hamburg   Lisbon   London   Madrid   Mexico   Milan   Montreal   New Delhi
Paris   San Juan   São Paulo   Singapore   Sydney   Tokyo   Toronto

This book was set in Times Roman by Publication Services.
The editors were B.J. Clark and John M. Morriss;
the production supervisor was Louise Karam.
The cover was designed by Rafael Hernandez.
Project supervision was done by Publication Services.
R. R. Donnelley & Sons company was printer and binder.

PROCESS SYSTEMS ANALYSIS AND CONTROL

2 3 4 5 6 7 8 9 0 DOC DOC 9 0 9 8 7 6 5 4 3 2 1

ISBN 0-07-013212-7

Library of Congress Cataloging-in-Publication Data

Coughanowr, Donald R.
    Process systems analysis and control / by Donald R. Coughanowr. —
2nd ed.
      p.   cm. — (McGraw-Hill chemical engineering series)
    Includes index.
    ISBN 0-07-013212-7
    1. Chemical process control.   I. Title.   II. Series.
TP155.75.C68   1991
660'.02815—dc20   90-41740

PROCESS SYSTEMS ANALYSIS AND CONTROL

Second Edition

Donald R. Coughanowr

Department of Chemical Engineering
Drexel University

McGraw-Hill, Inc.

New York   St. Louis   San Francisco   Auckland   Bogotá   Caracas
Hamburg   Lisbon   London   Madrid   Mexico   Milan   Montreal   New Delhi
Paris   San Juan   São Paulo   Singapore   Sydney   Tokyo   Toronto

This book was set in Times Roman by Publication Services.
The editors were B.J. Clark and John M. Morriss;
the production supervisor was Louise Karam.
The cover was designed by Rafael Hernandez.
Project supervision was done by Publication Services.
R. R. Donnelley & Sons company was printer and binder.

PROCESS SYSTEMS ANALYSIS AND CONTROL

2 3 4 5 6 7 8 9 0 DOC DOC 9 0 9 8 7 6 5 4 3 2 1

ISBN 0-07-013212-7

Library of Congress Cataloging-in-Publication Data

Coughanowr, Donald R.
    Process systems analysis and control / by Donald R. Coughanowr. —
    2nd ed.
        p.   cm. — (McGraw-Hill chemical engineering series)
    Includes index.
    ISBN 0-07-013212-7
    1. Chemical process control.   I. Title.   II. Series.
TP155.75.C68   1991
660'.02815—dc20   90-41740

ABOUT THE AUTHOR

Donald R. Coughanowr is the Fletcher Professor of Chemical Engineering at Drexel University. He received a Ph.D. in chemical engineering from the University of Illinois in 1956, an M.S. degree in chemical engineering from the University of Pennsylvania in 1951, and a B.S. degree in chemical engineering from the Rose-Hulman Institute of Technology in 1949. He joined the faculty at Drexel University in 1967 as department head, a position he held until 1988. Before going to Drexel, he was a faculty member of the School of Chemical Engineering at Purdue University for eleven years.

At Drexel and Purdue he has taught a wide variety of courses, which include material and energy balances, thermodynamics, unit operations, transport phenomena, petroleum refinery engineering, environmental engineering, chemical engineering laboratory, applied mathematics, and process dynamics and control. At Purdue, he developed a new course and laboratory in process control and collaborated with Dr. Lowell B. Koppel on the writing of the first edition of *Process Systems Analysis and Control*.

His research interests include environmental engineering, diffusion with chemical reaction, and process dynamics and control. Much of his research in control has emphasized the development and evaluation of new control algorithms for processes that cannot be controlled easily by conventional control; some of the areas investigated are time-optimal control, adaptive pH control, direct digital control, and batch control of fermentors. He has reported on his research in numerous publications and has received support for research projects from the N.S.F. and industry. He has spent sabbatical leaves teaching and writing at Case-Western Reserve University, the Swiss Federal Institute, the University of Canterbury, the University of New South Wales, the University of Queensland, and Lehigh University.

Dr. Coughanowr's industrial experience includes process design and pilot plant at Standard Oil Co. (Indiana) and summer employment at Electronic Associates and Dow Chemical Company.

He is a member of the American Institute of Chemical Engineers, the Instrument Society of America, and the American Society for Engineering Education. He is also a delegate to the Council for Chemical Research. He has served the AIChE by participating in accreditation visits to departments of chemical engineering for ABET and by chairing sessions of the Department Heads Forum at the annual meetings of AIChE.

To
Effie, Corinne, Christine, and David

CONTENTS

Part IV   Frequency Response

Part V   Process Applications

Part VI   Sampled-Data Control Systems

Part VII   State-Space Methods

Part VIII Nonlinear Control

Part IX Computers in Process Control

PREFACE

Since the first edition of this book was published in 1965, many changes have taken place in process control. Nearly all undergraduate students in chemical engineering are now required to take a course in process dynamics and control. The purpose of this book is to take the student from the basic mathematics to a variety of design applications in a clear, concise manner.

The most significant change since the first edition is the use of the digital computer in complex problem-solving and in process control instrumentation. However, the fundamentals of process control, which remain the same, must be acquired before one can appreciate the advanced topics of control.

In its present form, this book represents a major revision of the first edition. The material for this book evolved from courses taught at Purdue University and Drexel University. The first 17 chapters on fundamentals are quite close to the first 20 chapters of the first edition. The remaining 18 chapters contain many new topics, which were considered very advanced when the first edition was published.

A knowledge of calculus, unit operations, and complex numbers is presumed on the part of the student. In certain later chapters, more advanced mathematical preparation is useful. Some examples would include partial differential equations in Chap. 21, linear algebra in Chaps. 28–30, and Fourier series in Chap. 33.

Analog computation and pneumatic controllers in the first edition have been replaced by digital computation and microprocessor-based controllers in Chaps. 34 and 35. The student should be assigned material from these chapters at the appropriate time in the development of the fundamentals. For example, obtaining the transient response for a system containing a transport lag can be obtained easily only with the use of computer simulation of transport lag. Some of the software now available for solving control problems should be available to the student; such software is described in Chap. 34. To understand the operation of modern microprocessor-based controllers, the student should have hands-on experience with these instruments in a laboratory.

Chapter 1 is intended to meet one of the problems consistently faced in presenting this material to chemical engineering students, that is, one of perspective. The methods of analysis used in the control area are so different from the previous experiences of students that the material comes to be regarded as a sequence of special mathematical techniques, rather than an integrated design approach to a class of real and practically significant industrial problems. Therefore, this chapter presents an overall, albeit superficial, look at a simple control-system design problem. The body of the text covers the following topics:

1. Laplace transforms, Chaps 2 to 4.
2. Transfer functions and responses of open-loop systems, Chaps. 5 to 8.
3. Basic techniques of closed-loop control, Chaps. 9 to 13.
4. Stability, Chap. 14.
5. Root-locus methods, Chap. 15.
6. Frequency-response methods and design, Chaps. 16 and 17.
7. Advanced control strategies (cascade, feedforward, Smith predictor, internal model control), Chap. 18.
8. Controller tuning and process identification, Chap. 19.
9. Control valves, Chap. 20.
10. Advanced process dynamics, Chap. 21.
11. Sampled-data control, Chaps. 22 to 27.
12. State-space methods and multivariable control, Chaps. 28 to 30.
13. Nonlinear control, Chaps. 31 to 33.
14. Digital computer simulation, Chap. 34.
15. Microprocessor-based controllers, Chap. 35.

It has been my experience that the book covers sufficient material for a one-semester (15-week) undergraduate course and an elective undergraduate course or part of a graduate course. In a lecture course meeting three hours per week during a 10-week term, I have covered the following Chapters: 1 to 10, 12 to 14, 16, 17, 20, 34, and 35.

After the first 14 chapters, the instructor may select the remaining chapters to fit a course of particular duration and scope. The chapters on the more advanced topics are written in a logical order; however, some can be skipped without creating a gap in understanding.

I gratefully acknowledge the support and encouragement of the Drexel University Department of Chemical Engineering for fostering the evolution of this text in its curriculum and for providing clerical staff and supplies for several editions of class notes. I want to acknowledge Dr. Lowell B. Koppel's important contribution as co-author of the first edition of this book. I also want to thank my colleague, Dr. Rajakannu Mutharasan, for his most helpful discussions and suggestions and for his sharing of some of the new problems. For her assistance in typing, I want to thank Dorothy Porter. Helpful suggestions were also provided by Drexel students, in particular Russell Anderson, Joseph Hahn, and Barbara Hayden. I also want to thank my wife Effie for helping me check the page proofs by reading to me the manuscript, the subject matter of which is far removed from her specialty of Greek and Latin.

McGraw-Hill and I would like to thank Ali Cinar, Illinois Institute of Technology; Joshua S. Dranoff, Northwestern University; H. R. Heichelheim, Texas Tech University; and James H. McMicking, Wayne State University, for their many helpful comments and suggestions in reviewing this second edition.

Donald R. Coughanowr

PROCESS SYSTEMS ANALYSIS AND CONTROL

CHAPTER 1

AN INTRODUCTORY EXAMPLE

In this chapter we consider an illustrative example of a control system. The goal is to introduce some of the basic principles and problems involved in process control and to give the reader an early look at an overall problem typical of those we shall face in later chapters.

The System

A liquid stream at temperature T_i is available at a constant flow rate of w in units of mass per time. It is desired to heat this stream to a higher temperature T_R. The proposed heating system is shown in Fig. 1.1. The fluid flows into a well-agitated tank equipped with a heating device. It is assumed that the agitation is sufficient to ensure that all fluid in the tank will be at the same temperature, T. Heated fluid is removed from the bottom of the tank at the flow rate w as the product of this heating process. Under these conditions, the mass of fluid retained in the tank remains constant in time, and the temperature of the effluent fluid is the same as that of the fluid in the tank. For a satisfactory design this temperature must be T_R. The specific heat of the fluid C is assumed to be constant, independent of temperature.

Steady-State Design

A process is said to be at steady state when none of the variables are changing with time. At the desired steady state, an energy balance around the heating process may be written as follows:

$$q_s = wC(T_s - T_{i_s}) \tag{1.1}$$

1

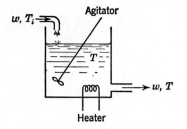

FIGURE 1-1
Agitated heating tank.

where q_s is the heat input to the tank and the subscript s is added to indicate a steady-state design value. Thus, for example, T_{i_s} is the normally anticipated inlet temperature to the tank. For a satisfactory design, the steady-state temperature of the effluent stream T_s must equal T_R. Hence

$$q_s = wC(T_R - T_{i_s}) \tag{1.2}$$

However, it is clear from the physical situation that, if the heater is set to deliver only the constant input q_s, then if process conditions change, the tank temperature will also change from T_R. A typical process condition that may change is the inlet temperature, T_i.

An obvious solution to the problem is to design the heater so that its energy input may be varied as required to maintain T at or near T_R.

Process Control

It is necessary to decide how much the heat input q is to be changed from q_s to correct any deviations of T from T_R. One solution would be to hire a process operator, who would be responsible for controlling the heating process. The operator would observe the temperature in the tank, presumably with a measuring instrument such as a thermocouple or thermometer, and compare this temperature with T_R. If T were less than T_R, he would increase the heat input and vice versa. As he became experienced at this task, he would learn just how much to change q for each situation. However, this relatively simple task can be easily and less expensively performed by a machine. The use of machines for this and similar purposes is known as *automatic process control*.

The Unsteady State

If a machine is to be used to control the process, it is necessary to decide in advance precisely what changes are to be made in the heat input q for every possible situation that might occur. We cannot rely on the judgment of the machine as we could on that of the operator. Machines do not think; they simply perform a predetermined task in a predetermined manner.

To be able to make these control decisions in advance, we must know how the tank temperature T changes in response to changes in T_i and q. This necessitates writing the *unsteady-state,* or *transient,* energy balance for the process. The input and output terms in this balance are the same as those used in the steady-state balance, Eq. (1.1). In addition, there is a transient accumulation of energy in the tank, which may be written

$$\text{Accumulation} = \rho V C \frac{dT}{dt} \qquad \text{energy units/time}^*$$

where ρ = fluid density
$\quad V$ = volume of fluid in the tank
$\quad t$ = independent variable, time

By the assumption of constant and equal inlet and outlet flow rates, the term ρV, which is the mass of fluid in the tank, is constant. Since

$$\text{Accumulation} = \text{input} - \text{output}$$

we have

$$\rho V C \frac{dT}{dt} = wC(T_i - T) + q \qquad (1.3)$$

Equation (1.1) is the steady-state solution of Eq. (1.3), obtained by setting the derivative to zero. We shall make use of Eq. (1.3) presently.

Feedback Control

As discussed above, the controller is to do the same job that the human operator was to do, except that the controller is told in advance *exactly* how to do it. This means that the controller will use the existing values of T and T_R to adjust the heat input according to a predetermined formula. Let the difference between these temperatures, $T_R - T$, be called *error*. Clearly, the larger this error, the less we are satisfied with the present state of affairs and vice versa. In fact, we are completely satisfied only when the error is exactly zero.

Based on these considerations, it is natural to suggest that the controller should change the heat input by an amount *proportional* to the error. Thus, a plausible formula for the controller to follow is

$$q(t) = wC(T_R - T_{i_s}) + K_c(T_R - T) \qquad (1.4)$$

where K_c is a (positive) constant of proportionality. This is called *proportional control*. In effect, the controller is instructed to maintain the heat input at the

---

*A rigorous application of the first law of thermodynamics would yield a term representing the transient change of internal energy with temperature at constant pressure. Use of the specific heat, at either constant pressure or constant volume, is an adequate engineering approximation for most liquids and will be applied extensively in this text.

steady-state design value q_s as long as T is equal to T_R [compare Eq. (1.2)], i.e., as long as the error is zero. If T deviates from T_R, causing an error, the controller is to use the magnitude of the error to change the heat input proportionally. (Readers should satisfy themselves that this change is in the right direction.) We shall reserve the right to vary the parameter K_c to suit our needs. This degree of freedom forms a part of our instructions to the controller.

The concept of using information about the deviation of the system from its desired state to control the system is called *feedback* control. Information about the state of the system is "fed back" to a controller, which utilizes this information to change the system in some way. In the present case, the information is the temperature T and the change is made in q. When the term $wC(T_R - T_{i_s})$ is abbreviated to q_s, Eq. (1.4) becomes

$$q = q_s + K_c(T_R - T) \tag{1.4a}$$

Transient Responses

Substituting Eq. (1.4a) into Eq. (1.3) and rearranging, we have

$$\tau_1 \frac{dT}{dt} + \left(\frac{K_c}{wC} + 1\right)T = T_i + \frac{K_c}{wC}T_R + \frac{q_s}{wC} \tag{1.5}$$

where

$$\tau_1 = \frac{\rho V}{w}$$

The term τ_1 has the dimensions of time and is known as the *time constant* of the tank. We shall study the significance of the time constant in more detail in Chap. 5. At present, it suffices to note that it is the time required to fill the tank at the flow rate, w. T_i is the inlet temperature, which we have assumed is a function of time. Its normal value is T_{i_s}, and q_s is based on this value. Equation (1.5) describes the way in which the tank temperature changes in response to changes in T_i and q.

Suppose that the process is proceeding smoothly at steady-state design conditions. At a time arbitrarily called zero, the inlet temperature, which was at T_{i_s}, suddenly undergoes a permanent rise of a few degrees to a new value $T_{i_s} + \Delta T_i$, as shown in Fig. 1.2. For mathematical convenience, this disturbance is idealized to

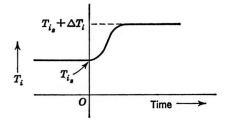

FIGURE 1-2
Inlet temperature versus time.

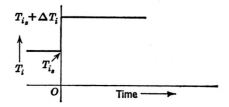

FIGURE 1-3
Idealized inlet temperature versus time.

the form shown in Fig. 1.3. The equation for the function $T_i(t)$ of Fig. 1.3 is

$$T_i(t) = \begin{cases} T_{i_s} & t < 0 \\ T_{i_s} + \Delta T_i & t > 0 \end{cases} \tag{1.6}$$

This type of function, known as a step function, is used extensively in the study of transient response because of the simplicity of Eq. (1.6). The justification for use of the step change is that the response of T to this function will not differ significantly from the response to the more realistic disturbance depicted in Fig. 1.2.

To determine the response of T to a step change in T_i, it is necessary to substitute Eq. (1.6) into (1.5) and solve the resulting differential equation for $T(t)$. Since the process is at steady state at (and before) time zero, the initial condition is

$$T(0) = T_R \tag{1.7}$$

The reader can easily verify (and should do so) that the solution to Eqs. (1.5), (1.6), and (1.7) is

$$T = T_R + \frac{\Delta T_i}{(K_c/wC) + 1}(1 - e^{-(K_c/wC + 1)t/\tau_1}) \tag{1.8}$$

This system *response,* or tank temperature versus time, to a step change in T_i is shown in Fig. 1.4 for various values of the adjustable control parameter K_c. The reader should compare these curves with Eq. (1.8), particularly in respect to the relative positions of the curves at the new steady states.

It may be seen that the higher K_c is made, the "better" will be the control, in the sense that the new steady-state value of T will be closer to T_R. At first

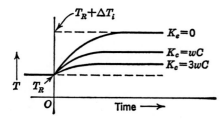

FIGURE 1-4
Tank temperature versus time for various values of K_c.

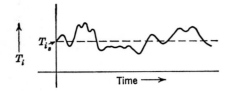

FIGURE 1-5
A fluctuating behavior of T_i.

glance, it would appear desirable to make K_c as large as possible, but a little reflection will show that large values of K_c are likely to cause other problems. For example, note that we have considered only one type of disturbance in T_i. Another possible behavior of T_i with time is shown in Fig. 1.5. Here, T_i is fluctuating about its steady-state value. A typical response of T to this type of disturbance in T_i, *without control action*, is shown in Fig. 1.6. The fluctuations in T_i are delayed and "smoothed" by the large volume of liquid in the tank, so that T does not fluctuate as much as T_i. Nevertheless, it should be clear from Eq. (1.4a) and Fig. 1.6 that a control system with a high value of K_c will have a tendency to overadjust. In other words, it will be too *sensitive* to disturbances that would tend to disappear in time *even without control action*. This will have the undesirable effect of *amplifying* the effects of these disturbances and causing excessive wear on the control system.

The dilemma may be summarized as follows: In order to obtain accurate control of T, despite "permanent" changes in T_i, we must make K_c larger (see Fig. 1.4). However, as K_c is increased, the system becomes oversensitive to spurious fluctuations in T_i. (These fluctuations, as depicted in Fig. 1.5, are called *noise*.) The reader is cautioned that there are additional effects produced by changing K_c that have not been discussed here for the sake of brevity, but which may be even more important. This will be one of the major subjects of interest in later chapters. The two effects mentioned are sufficient to illustrate the problem.

Integral Control

A considerable improvement may be obtained over the proportional control system by adding integral control. The controller is now instructed to change the heat input by an additional amount proportional to the time integral of the error. Quantitatively, the heat input function is to follow the relation

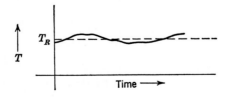

FIGURE 1-6
The response, without control action, to a fluctuating T_i.

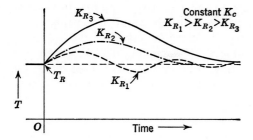

FIGURE 1-7
Tank temperature versus time: step input for proportional and integral control.

$$q(t) = q_s + K_c(T_R - T) + K_R \int_0^t (T_R - T)dt \qquad (1.9)$$

This control system is to have two adjustable parameters, K_c and K_R.

The response of the tank temperature T to a step change in T_i, using a control function described by (1.9), may be derived by solution of Eqs. (1.3), (1.6), (1.7), and (1.9). Curves representing this response, which the reader is asked to accept, are given for various values of K_R at a fixed value of K_c in Fig. 1.7. The value of K_c is a moderate one, and it may be seen that for all three values of K_R the steady-state temperature is T_R; that is, *the steady-state error is zero*. From this standpoint, the response is clearly superior to that of the system with proportional control only. It may be shown that the steady-state error is zero for *all* $K_R > 0$, thus eliminating the necessity for high values of K_c. (In subsequent chapters, methods will be given for rapidly constructing response curves such as those of Fig. 1.7.)

It is clear from Fig. 1.7 that the responses for $K_R = K_{R_2}$ and $K_R = K_{R_1}$ are better than the one for $K_R = K_{R_3}$ because T returns to T_R faster, but it may be difficult to choose between K_{R_2}, and K_{R_1}. The response for K_{R_2} "settles down" sooner, but it also has a higher maximum error. The choice might depend on the particular use for the heated stream. This and related questions form the study of *optimal* control systems. This important subject is mentioned in this book more to point out the existence of the problem than to solve it.

To recapitulate, the curves of Fig. 1.7 give the transient behavior of the tank temperature in response to a step change in T_i when the tank temperature is controlled according to Eq. (1.9). They show that the addition of integral control in this case eliminates steady-state error and allows use of moderate values of K_c.

More Complications

At this point, it would appear that the problem has been solved in some sense. A little further probing will shatter this illusion.

It has been assumed in writing Eqs. (1.4a) and (1.9) that the controller receives instantaneous information about the tank temperature, T. From a physical standpoint, some measuring device such as a thermocouple will be required to measure this temperature. The temperature of a thermocouple inserted in the tank may or may not be the same as the temperature of the fluid in the tank. This can be demonstrated by writing the energy balance for a typical thermocouple installation, such as the one depicted in Fig. 1.8. Assuming that the junction is at a uniform temperature T_m and neglecting any conduction of heat along the thermocouple lead wires, the net rate of input of energy to the thermocouple junction is

$$hA(T - T_m)$$

where h = heat-transfer coefficient between fluid and junction
     A = area of junction

The rate of accumulation of energy in the junction is

$$mC_m \frac{dT_m}{dt}$$

where C_m = specific heat of junction
     m = mass of junction

Combining these in an energy balance,

$$\tau_2 \frac{dT_m}{dt} + T_m = T \tag{1.10}$$

where $\tau_2 = mC_m/hA$ is the time constant of the thermocouple. Thus, changes in T are not instantaneously reproduced in T_m. A step change in T causes a response in T_m similar to the curve of Fig. 1.4 for $K_c = 0$ [see Eq. (1.5)]. This is analogous to the case of placing a mercury thermometer in a beaker of hot water. The thermometer does not instantaneously rise to the water temperature. Rather, it rises in the manner described.

Since the controller will receive values of T_m (possibly in the form of a thermoelectric voltage) and *not* values of T, Eq. (1.9) must be rewritten as

$$q = q_s + K_c(T_R - T_m) + K_R \int_0^t (T_R - T_m)dt \tag{1.9a}$$

The *apparent error* is given by $(T_R - T_m)$, and it is this quantity upon which the controller acts, rather than the true error $(T_R - T)$. The response of T to a step

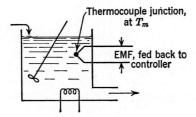

FIGURE 1-8
Thermocouple installation for heated-tank system.

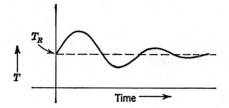

FIGURE 1-9

Tank temperature versus time with measuring lag.

change in T_i is now derived by simultaneous solution of (1.3), (1.6), (1.9a), and (1.10), with initial conditions

$$T(0) = T_m(0) = T_R \qquad (1.11)$$

Equation (1.11) implies that, at time zero, the system has been at rest at T_R for some time, so that the thermocouple junction is at the same temperature as the tank.

The solution to this system of equations is represented in Fig. 1.9 for a particular set of values of K_c and K_R. For this set of values, the effect of the thermocouple delay in transmission of the temperature to the controller is primarily to make the response somewhat more oscillatory than that shown in Fig. 1.7 for the same value of K_R. However, if K_R is increased somewhat over the value used in Fig. 1.9, the response is that shown in Fig. 1.10. The tank temperature oscillates with *increasing* amplitude and will continue to do so until the physical limitations of the heating system are reached. The control system has actually caused a deterioration in performance. Surely, the uncontrolled response for $K_c = 0$ in Fig. 1.4 is to be preferred over the *unstable* response of Fig. 1.10.

This problem of *stability* of response will be one of our major concerns in this text for obvious reasons. At present, it is sufficient to note that extreme care must be exercised in specifying control systems. In the case considered, the proportional and integral control mechanism described by Eq. (1.9a) will perform satisfactorily if K_R is kept lower than some particular value, as illustrated in Figs. 1.9 and 1.10. However, it is not difficult to construct examples of systems for which the addition of *any* amount of integral control will cause an unstable response. Since integral control usually has the desirable feature of eliminating steady-state error, as it did in Fig. 1.7, it is extremely important that we develop

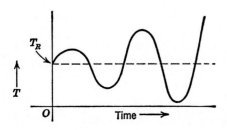

FIGURE 1-10

Tank temperature versus time for increased K_R.

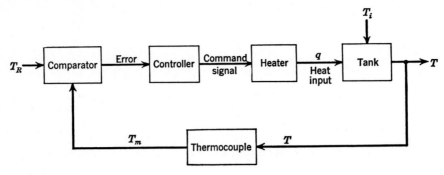

FIGURE 1-11
Block diagram for heated-tank system.

means for predicting the occurrence of unstable response in the design of any control system.

Block Diagram

A good overall picture of the relationships among variables in the heated-tank control system may be obtained by preparing a *block diagram*. This diagram, shown in Fig. 1.11, indicates the flow of information around the control system and the function of each part of the system. Much more will be said about block diagrams in Chap. 9, but the reader can undoubtedly form a good intuitive notion about them by comparing Fig. 1.11 with the physical description of the process given in the previous paragraphs. Particularly significant is the fact that each component of the system is represented by a block, with little regard for the actual physical characteristics of the represented component (e.g., the tank or controller). The major interest is in (1) the relationship between the signals entering and leaving the block and (2) the manner in which information flows around the system. For example, T_R and T_m enter the comparator. Their difference, the *error,* leaves the comparator and enters the controller.

SUMMARY

We have had an overall look at a typical control problem and some of its ramifications. At present, the reader has been asked to accept the mathematical results on faith and to concentrate on obtaining a physical understanding of the transient behavior of the heated tank. We shall in the forthcoming chapters develop tools for determining the response of such systems. As this new material is presented, the reader may find it helpful to refer back to this chapter in order to place the material in proper perspective to the overall control problem.

PROBLEMS

1.1. Draw a block diagram for the control system generated when a human being steers an automobile.

PART
I

THE LAPLACE
TRANSFORM

CHAPTER
2

THE LAPLACE TRANSFORM

Even from our brief look at the control problem of Chap. 1, it is evident that solution of differential equations will be one of our major tasks. The Laplace transform method provides an efficient way to solve linear, ordinary, differential equations with constant coefficients. Because an important class of control problems reduces to the solution of such equations, the next three chapters are devoted to a study of Laplace transforms before resuming our investigation of control problems.

Definition of the Transform

The Laplace transform of a function $f(t)$ is *defined* to be $f(s)$ according to the equation

$$f(s) = \int_0^\infty f(t)e^{-st}dt \tag{2.1}$$

We often abbreviate this notationally to

$$f(s) = L\{f(t)\}$$

where the operator L is defined by Eq. (2.1).*

---

*Many texts adopt some notational convention, such as capitalizing the transformed function as $F(s)$ or putting a bar over it as $\overline{f}(s)$. In general, the appearance of the variable s as the argument or in an equation involving f is sufficient to signify that the function has been transformed, and hence any such notation will seldom be required in this book.

Example 2.1. Find the Laplace transform of the function

$$f(t) = 1$$

According to Eq. (2.1),

$$f(s) = \int_0^\infty (1)e^{-st}\,dt = -\left.\frac{e^{-st}}{s}\right|_{t=0}^{t=\infty} = \frac{1}{s}$$

Thus,

$$L\{1\} = \frac{1}{s}$$

There are several facts worth noting at this point:

1. The Laplace transform $f(s)$ contains no information about the behavior of $f(t)$ for $t < 0$. This is not a limitation for control system study because t will represent the time variable and we shall be interested in the behavior of systems only for positive time. In fact, the variables and systems are usually defined so that $f(t) \equiv 0$ for $t < 0$. This will become clearer as we study specific examples.
2. Since the Laplace transform is defined in Eq. (2.1) by an improper integral, it will not exist for every function $f(t)$. A rigorous definition of the class of functions possessing Laplace transforms is beyond the scope of this book, but readers will note that every function of interest to us *does* satisfy the requirements for possession of a transform.*
3. The Laplace transform is linear. In mathematical notation, this means:

$$L\{af_1(t) + bf_2(t)\} = aL\{f_1(t)\} + bL\{f_2(t)\}$$

where a and b are constants, and f_1 and f_2 are two functions of t.

Proof. Using the definition,

$$L\{af_1(t) + bf_2(t)\} = \int_0^\infty [af_1(t) + bf_2(t)]e^{-st}\,dt$$

$$= a\int_0^\infty f_1(t)e^{-st}\,dt + b\int_0^\infty f_2(t)e^{-st}\,dt$$

$$= aL\{f_1(t)\} + bL\{f_2(t)\}$$

4. The Laplace transform operator transforms a function of the variable t to a function of the variable s. The t variable is eliminated by the integration.

Transforms of Simple Functions

We now proceed to derive the transforms of some simple and useful functions.

---

*For details on this and related mathematical topics, see Churchill (1972).

1. The step function

$$f(t) = \begin{cases} 0 & t < 0 \\ 1 & t > 0 \end{cases}$$

This important function is known as the unit-step function and will henceforth be denoted by $u(t)$. From Example 2.1, it is clear that

$$L\{u(t)\} = \frac{1}{s}$$

As expected, the behavior of the function for $t < 0$ has no effect on its Laplace transform. Note that as a consequence of linearity, the transform of any constant A, that is, $f(t) = Au(t)$, is just $f(s) = A/s$.

2. The exponential function

$$f(t) = \begin{cases} 0 & t < 0 \\ e^{-at} & t > 0 \end{cases} = u(t)e^{-at}$$

where $u(t)$ is the unit-step function. Again proceeding according to definition,

$$L\{u(t)e^{-at}\} = \int_0^\infty e^{-(s+a)t}\,dt = -\frac{1}{s+a}e^{-(s+a)t}\Big|_0^\infty = \frac{1}{s+a}$$

provided that $s + a > 0$, that is, $s > -a$. In this case, the convergence of the integral depends on a suitable choice of s. In case s is a complex number, it may be shown that this condition becomes

$$\text{Re}(s) > -a$$

For problems of interest to us it will always be possible to choose s so that these conditions are satisfied, and the reader uninterested in mathematical niceties can ignore this point.

3. The ramp function

$$f(t) = \begin{cases} 0 & t < 0 \\ t & t > 0 \end{cases} = tu(t)$$

$$L\{tu(t)\} = \int_0^\infty te^{-st}\,dt$$

Integration by parts yields

$$L\{tu(t)\} = -e^{-st}\left(\frac{t}{s} + \frac{1}{s^2}\right)\Big|_0^\infty = \frac{1}{s^2}$$

4. The sine function

$$f(t) = \begin{cases} 0 & t < 0 \\ \sin kt & t > 0 \end{cases} = u(t)\sin kt$$

$$L\{u(t)\sin kt\} = \int_0^\infty \sin kt\; e^{-st}\,dt$$

TABLE 2.1

Function	Graph	Transform
$u(t)$		$\dfrac{1}{s}$
$t\,u(t)$		$\dfrac{1}{s^2}$
$t^n u(t)$		$\dfrac{n!}{s^{n+1}}$
$e^{-at} u(t)$		$\dfrac{1}{s+a}$
$t^n e^{-at} u(t)$		$\dfrac{n!}{(s+a)^{n+1}}$
$\sin kt\, u(t)$		$\dfrac{k}{s^2+k^2}$

TABLE 2.1 *(Continued)*

Function	Graph	Transform
$\cos kt\, u(t)$		$\dfrac{s}{s^2 + k^2}$
$\sinh kt\, u(t)$		$\dfrac{k}{s^2 - k^2}$
$\cosh kt\, u(t)$	1	$\dfrac{s}{s^2 - k^2}$
$e^{-at} \sin kt\, u(t)$		$\dfrac{k}{(s + a)^2 + k^2}$
$e^{-at} \cos kt\, u(t)$		$\dfrac{s + a}{(s + a)^2 + k^2}$
$\delta(t)$, unit impulse	Area $= 1$	1

Integrating by parts,

$$L\{u(t)\sin kt\} = \left.\frac{-e^{-st}}{s^2 + k^2}(s \sin kt + k \cos kt)\right|_0^\infty$$

$$= \frac{k}{s^2 + k^2}$$

In a like manner, the transforms of other simple functions may be derived. Table 2.1 is a summary of transforms that will be of use to us. Those which have not been derived here can be easily established by direct integration, except for the transform of $\delta(t)$, which will be discussed in detail in Chap. 4.

Transforms of Derivatives

At this point, the reader may wonder what has been gained by introduction of the Laplace transform. The transform merely changes a function of t into a function of s. The functions of s look no simpler than those of t and, as in the case of $A \rightarrow A/s$, may actually be more complex. In the next few paragraphs, the motivation will become clear. It will be shown that the Laplace transform has the remarkable property of transforming the operation of differentiation with respect to t to that of multiplication by s. Thus, we claim that

$$L\left\{\frac{df(t)}{dt}\right\} = sf(s) - f(0) \tag{2.2}$$

where

$$f(s) = L\{f(t)\}$$

and $f(0)$ is $f(t)$ evaluated at $t = 0$. [It is essential not to interpret $f(0)$ as $f(s)$ with $s = 0$. This will be clear from the following proof.]*

  Proof.

$$L\left\{\frac{df(t)}{dt}\right\} = \int_0^\infty \frac{df}{dt}e^{-st}\,dt$$

To integrate this by parts, let

$$u = e^{-st} \qquad dv = \frac{df}{dt}\,dt$$

Then

$$du = -se^{-st}\,dt \qquad v = f(t)$$

---

* If $f(t)$ is discontinuous at $t = 0$, $f(0)$ should be evaluated at $t = 0^+$, i.e., just to the right of the origin. Since we shall seldom want to differentiate functions that are discontinuous at the origin, this detail is not of great importance. However, the reader is cautioned to watch carefully for situations in which such discontinuities occur.

Since

$$\int u \, dv = uv - \int v \, du$$

we have

$$\int_0^\infty \frac{df}{dt} e^{-st} dt = f(t)e^{-st}\Big|_0^\infty + s\int_0^\infty f(t)e^{-st} dt = -f(0) + sf(s)$$

The salient feature of this transformation is that whereas the function of t was to be differentiated with respect to t, the corresponding function of s is merely multiplied by s. We shall find this feature to be extremely useful in the solution of differential equations.

To find the transform of the second derivative we make use of the transform of the first derivative twice, as follows:

$$L\left\{\frac{d^2f}{dt^2}\right\} = L\left\{\frac{d}{dt}\left(\frac{df}{dt}\right)\right\} = sL\left\{\frac{df}{dt}\right\} - \frac{df(t)}{dt}\Big|_{t=0}$$
$$= s[sf(s) - f(0)] - f'(0)$$
$$= s^2f(s) - sf(0) - f'(0)$$

where we have abbreviated

$$\frac{df(t)}{dt}\Big|_{t=0} = f'(0)$$

In a similar manner, the reader can easily establish by induction that repeated application of Eq. (2.2) leads to

$$L\left\{\frac{d^n f}{dt^n}\right\} = s^n f(s) - s^{n-1}f(0) - s^{n-2}f^{(1)}(0) - \cdots - sf^{(n-2)}(0) - f^{(n-1)}(0)$$

where $f^{(i)}(0)$ indicates the ith derivative of $f(t)$ with respect to t, evaluated for $t = 0$.

Thus, the Laplace transform may be seen to change the operation of differentiation of the function to that of multiplication of the transform by s, the number of multiplications corresponding to the number of differentiations. In addition, some polynomial terms involving the initial values of $f(t)$ and its first $(n - 1)$ derivatives are involved. In later applications we shall usually define our variables so that these polynomial terms will vanish. Hence, they are of secondary concern here.

Example 2.2. Find the Laplace transform of the function $x(t)$ that satisfies the differential equation and initial conditions

$$\frac{d^3x}{dt^3} + 4\frac{d^2x}{dt^2} + 5\frac{dx}{dt} + 2x = 2$$

$$x(0) = \frac{dx(0)}{dt} = \frac{d^2x(0)}{dt^2} = 0$$

It is permissible mathematically to take the Laplace transforms of both sides of a differential equation and equate them, since equality of functions implies equality of their transforms. Doing this, there is obtained

$$s^3x(s) - s^2x(0) - sx'(0) - x''(0) + 4[s^2x(s) - sx(0) - x'(0)]$$
$$+ 5[sx(s) - x(0)] + 2x(s) = \frac{2}{s}$$

where $x(s) = L\{x(t)\}$. Use has been made of the linearity property and of the fact that only positive values of t are of interest. Inserting the initial conditions and solving for $x(s)$

$$x(s) = \frac{2}{s(s^3 + 4s^2 + 5s + 2)} \tag{2.3}$$

This is the required answer, the Laplace transform of $x(t)$.

Solution of Differential Equations

There are two important points to note regarding this last example. In the first place, application of the transformation resulted in an equation that was solved for the unknown function by *purely algebraic means*. Second, and most important, if the function $x(t)$, which has the Laplace transform $2/s(s^3 + 4s^2 + 5s + 2)$ were known, we would have the solution to the differential equation and boundary conditions. This suggests a procedure for solving differential equations that is analogous to that of using logarithms to multiply or divide. To use logarithms, one transforms the pertinent numbers to their logarithms and then adds or subtracts, which is much easier than multiplying or dividing. The result of the addition or subtraction is the logarithm of the desired answer. The answer is found by reference to a table to find the number having this logarithm. In the Laplace transform method for solution of differential equations, the functions are converted to their transforms and the resulting equations are solved for the unknown function *algebraically*. This is much easier than solving a differential equation. However, at the last step the analogy to logarithms is not complete. We obviously cannot hope to construct a table containing the Laplace transform of every function $f(t)$ that possesses a transform. Instead, we shall develop methods for reexpressing complicated transforms, such as $x(s)$ in Example 2.2, in terms of simple transforms that can be found in Table 2.1. For example, it is easily verified that the solution to the differential equation and boundary conditions of Example 2.2 is

$$x(t) = 1 - 2te^{-t} - e^{-2t} \tag{2.4}$$

The Laplace transform of x, using Eq. (2.4) and Table 2.1, is

$$x(s) = \frac{1}{s} - 2\frac{1}{(s+1)^2} - \frac{1}{s+2} \tag{2.5}$$

Equation (2.3) is actually the result of placing Eq. (2.5) over a common denominator. Although it is difficult to find $x(t)$ from Eq. (2.3), Eq. (2.5) may be easily inverted to Eq. (2.4) by using Table 2.1. Therefore, what is required is a method for expanding the common-denominator form of Eq. (2.3) to the separated form of Eq. (2.5). This method is provided by the technique of partial fractions, which is developed in Chap. 3.

SUMMARY

To summarize, the basis for solving *linear, ordinary, differential equations with constant coefficients* with Laplace transforms has been established.
    The procedure is:

1. Take the Laplace transform of both sides of the equation. The initial conditions are incorporated at this step in the transforms of the derivatives.
2. Solve the resulting equation for the Laplace transform of the unknown function algebraically.
3. Find the function of t that has the Laplace transform obtained in step 2. This function satisfies the differential equation and initial conditions and hence is the desired solution. This third step is frequently the most difficult or tedious step and will be developed further in the next chapter. It is called inversion of the transform. Although there are other techniques available for inversion, the one that we shall develop and make consistent use of is that of partial-fraction expansion.

    A simple example will serve to illustrate steps 1 and 2, and a trivial case of step 3.

    Example 2.3. Solve

$$\frac{dx}{dt} + 3x = 0$$

$$x(0) = 2$$

We number our steps according to the discussion in the preceding paragraphs:

    1. $sx(s) - 2 + 3x(s) = 0$

    2. $x(s) = \dfrac{2}{s + 3} = 2\dfrac{1}{s + 3}$

    3. $x(t) = 2e^{-3t}$

CHAPTER

3

INVERSION
BY PARTIAL
FRACTIONS

Our study of the application of Laplace transforms to linear differential equations with constant coefficients has enabled us to rapidly establish the Laplace transform of the solution. We now wish to develop methods for inverting the transforms to obtain the solution in the time domain. The first part of this chapter will be a series of examples that illustrate the partial-fraction technique. After a generalization of these techniques, we proceed to a discussion of the qualitative information that can be obtained from the transform of the solution without inverting it.

The equations to be solved are all of the general form

$$a_n \frac{d^n x}{dt^n} + a_{n-1} \frac{d^{n-1} x}{dt^{n-1}} + \cdots + a_1 \frac{dx}{dt} + a_0 x = f(t)$$

The unknown function of time is $x(t)$, and a_n, $a_{n-1}, \ldots, a_1$, a_0, are constants. The given function $f(t)$ is called the *forcing function*. In addition, for all problems of interest in control system analysis, the initial conditions are given. In other words, values of x, $dx/dt, \ldots, d^{n-1}x/dt^{n-1}$ are specified at time zero. The problem is to determine $x(t)$ for all $t \geq 0$.

Partial Fractions

In the series of examples that follow, the technique of partial-fraction inversion for solution of this class of differential equations is presented.

Example 3.1. Solve

$$\frac{dx}{dt} + x = 1$$

$$x(0) = 0$$

Application of the Laplace transform yields

$$sx(s) + x(s) = \frac{1}{s}$$

or

$$x(s) = \frac{1}{s(s + 1)}$$

The theory of partial fractions enables us to write this as

$$x(s) = \frac{1}{s(s + 1)} = \frac{A}{s} + \frac{B}{s + 1} \tag{3.1}$$

where A and B are constants. Hence, using Table 2.1, it follows that

$$x(t) = A + Be^{-t} \tag{3.2}$$

Therefore, if A and B were known, we would have the solution. The conditions on A and B are that they must be chosen to make Eq. (3.1) an identity in s.

To determine A, multiply both sides of Eq. (3.1) by s.

$$\frac{1}{s + 1} = A + \frac{Bs}{s + 1} \tag{3.3}$$

Since this must hold for all s, it must hold for $s = 0$. Putting $s = 0$ in Eq. (3.3) yields

$$A = 1$$

To find B, multiply both sides of Eq. (3.1) by $(s + 1)$.

$$\frac{1}{s} = \frac{A}{s}(s + 1) + B \tag{3.4}$$

Since this must hold for all s, it must hold for $s = -1$. This yields

$$B = -1$$

Hence,

$$\frac{1}{s(s + 1)} = \frac{1}{s} - \frac{1}{s + 1} \tag{3.5}$$

and therefore,

$$x(t) = 1 - e^{-t} \tag{3.6}$$

Equation (3.5) may be checked by putting the right side over a common denominator, and Eq. (3.6) by substitution into the original differential equation and initial condition.

Example 3.2. Solve

$$\frac{d^3x}{dt^3} + 2\frac{d^2x}{dt^2} - \frac{dx}{dt} - 2x = 4 + e^{2t}$$

$$x(0) = 1 \qquad x'(0) = 0 \qquad x''(0) = -1$$

Taking the Laplace transform of both sides,

$$[s^3x(s) - s^2 + 1] + 2[s^2x(s) - s] - [sx(s) - 1] - 2x(s) = \frac{4}{s} + \frac{1}{s-2}$$

Solving algebraically for $x(s)$,

$$x(s) = \frac{s^4 - 6s^2 + 9s - 8}{s(s-2)(s^3 + 2s^2 - s - 2)}$$

The cubic in the denominator may be factored, and $x(s)$ expanded in partial fractions

$$x(s) = \frac{s^4 - 6s^2 + 9s - 8}{s(s-2)(s+1)(s+2)(s-1)} = \frac{A}{s} + \frac{B}{s-2} + \frac{C}{s+1} + \frac{D}{s+2} + \frac{E}{s-1}$$

$$(3.7)$$

To find A, multiply both sides of Eq. (3.7) by s and then set $s = 0$; the result is

$$A = \frac{-8}{(-2)(1)(2)(-1)} = -2$$

The other constants are determined in the same way. The procedure and results are summarized in the following table.

To determine	multiply (3.7) by	and set s to	Result
B	$s - 2$	2	$B = \frac{1}{12}$
C	$s + 1$	-1	$C = 1\frac{1}{3}$
D	$s + 2$	-2	$D = -\frac{17}{12}$
E	$s - 1$	1	$E = \frac{2}{3}$

Accordingly, the solution to the problem is

$$x(t) = -2 + \tfrac{1}{12}e^{2t} + \tfrac{11}{3}e^{-t} - \tfrac{17}{12}e^{-2t} + \tfrac{2}{3}e^t$$

A comparison between this method and the classical method, as applied to Example 3.2, may be profitable. In the classical method for solution of differential equations we first write down the characteristic function of the homogeneous equation:

$$s^3 + 2s^2 - s - 2 = 0$$

This must be factored, as was also required in the Laplace transform method, to obtain the roots -1, -2, and $+1$. Thus, the complementary solution is

$$x_c(t) = C_1 e^{-t} + C_2 e^{-2t} + C_3 e^t$$

Furthermore, by inspection of the forcing function, we know that the particular solution has the form

$$x_p(t) = A + B e^{2t}$$

The constants A and B are determined by substitution into the differential equation and, as expected, are found to be -2 and $\frac{1}{12}$, respectively. Then

$$x(t) = -2 + \frac{1}{12} e^{2t} + C_1 e^{-t} + C_2 e^{-2t} + C_3 e^t$$

and the constants C_1, C_2, and C_3 are determined by the three initial conditions. The Laplace transform method has systematized the evaluation of these constants, avoiding the solution of three simultaneous equations. Four points are worth noting:

1. In both methods, one must find the roots of the characteristic equation. The roots give rise to terms in the solution *whose form is independent of the forcing function*. These terms make up the *complementary solution*.
2. The forcing function gives rise to terms in the solution *whose form depends on the form of the forcing function and is independent of the left side of the equation*. These terms comprise the *particular solution*.
3. The only interaction between these sets of terms, i.e., between the right side and left side of the differential equation, occurs in the evaluation of the constants involved.
4. The only effect of the initial conditions is in the evaluation of the constants. This is because the initial conditions affect only the numerator of $x(s)$, as may be seen from the solution of this example.

In the two examples we have discussed, the denominator of $x(s)$ factored into real factors only. In the next example, we consider the complications that arise when the denominator of $x(s)$ has complex factors.

Example 3.3. Solve

$$\frac{d^2 x}{dt^2} + 2\frac{dx}{dt} + 2x = 2$$

$$x(0) = x'(0) = 0$$

Application of the Laplace transform yields

$$x(s) = \frac{2}{s(s^2 + 2s + 2)}$$

The quadratic term in the denominator may be factored by use of the quadratic formula. The roots are found to be $(-1 - j)$ and $(-1 + j)$. This gives the partial-fraction expansion

$$x(s) = \frac{2}{s(s + 1 + j)(s + 1 - j)} = \frac{A}{s} + \frac{B}{(s + 1 + j)} + \frac{C}{(s + 1 - j)} \qquad (3.8)$$

where A, B, and C are constants to be evaluated, so that this relation is an identity in s. The presence of complex factors does not alter the procedure at all. However, the computations may be slightly more tedious.

To obtain A, multiply Eq. (3.8) by s and set $s = 0$:

$$A = \frac{2}{(1 + j)(1 - j)} = 1$$

To obtain B, multiply Eq. (3.8) by $(s + 1 + j)$ and set $s = (-1 - j)$:

$$B = \frac{2}{(-1 - j)(-2j)} = \frac{-1 - j}{2}$$

To obtain C, multiply Eq. (3.8) by $(s + 1 - j)$ and set $s = (-1 + j)$:

$$C = \frac{2}{(-1 + j)(2j)} = \frac{-1 + j}{2}$$

Therefore,

$$x(s) = \frac{1}{s} + \frac{-1 - j}{2}\frac{1}{s + 1 + j} + \frac{-1 + j}{2}\frac{1}{s + 1 - j}$$

This is the desired result. To invert $x(s)$, we may now use the fact that $1/(s + a)$ is the transform of e^{-t}. The fact that a is complex does not invalidate this result, as can be seen by returning to the derivation of the transform of e^{-at}. The result is

$$x(t) = 1 + \frac{-1 - j}{2}e^{-(1+j)t} + \frac{-1 + j}{2}e^{-(1-j)t}$$

Using the identity

$$e^{(a+jb)t} = e^{at}(\cos bt + j \sin bt)$$

this can be converted to

$$x(t) = 1 - e^{-t}(\cos t + \sin t)$$

The details of this conversion are recommended as an exercise for the reader.

A more general discussion of this case will promote understanding. It was seen in Example 3.3 that the complex conjugate roots of the denominator of $x(s)$ gave rise to a pair of complex terms in the partial-fraction expansion. The constants in these terms, B and C, proved to be complex conjugates $(-1 - j)/2$ and $(-1 + j)/2$. When these terms were combined through a trigonometric identity, it was found that the complex terms canceled, leaving a real result for $x(t)$. Of course, it is necessary that $x(t)$ be real, since the original differential equation and initial conditions are real.

This information may be utilized as follows: the general case of complex conjugate roots arises in the form

$$x(s) = \frac{F(s)}{(s + k_1 + jk_2)(s + k_1 - jk_2)} \tag{3.9}$$

where $F(s)$ is some real function of s.

For instance, in Example 3.3 we had

$$F(s) = \frac{2}{s} \qquad k_1 = 1 \qquad k_2 = 1$$

Expanding (3.9) in partial fractions,

$$\frac{F(s)}{(s + k_1 + jk_2)(s + k_1 - jk_2)} = F_1(s)$$

$$+ \left(\frac{a_1 + jb_1}{s + k_1 + jk_2} + \frac{a_2 + jb_2}{s + k_1 - jk_2} \right) \qquad (3.10)$$

where a_1, a_2, b_1, b_2 are the constants to be evaluated in the partial-fraction expansion and $F_1(s)$ is a series of fractions arising from $F(s)$.

Again, in Example 3.3,

$$a_1 = -\frac{1}{2} \qquad a_2 = -\frac{1}{2} \qquad b_1 = -\frac{1}{2} \qquad b_2 = \frac{1}{2} \qquad F_1(s) = \frac{1}{s}$$

Now, since the left side of Eq. (3.10) is real for all real s, the right side must also be real for all real s. Since two complex numbers will add to form a real number if they are complex conjugates, it is seen that the right side will be real *for all real s* if and only if the two terms are complex conjugates. Since the denominators of the terms are conjugates, this means that the numerators must also be conjugates, or

$$a_2 = a_1$$
$$b_2 = -b_1$$

This is exactly the result obtained in the specific case of Example 3.3. With this information, Eq. (3.10) becomes

$$\frac{F(s)}{(s + k_1 + jk_2)(s + k_1 - jk_2)} = F_1(s)$$

$$+ \left(\frac{a_1 + jb_1}{s + k_1 + jk_2} + \frac{a_1 - jb_1}{s + k_1 - jk_2} \right) \qquad (3.11)$$

Hence, it has been established that terms in the inverse transform arising from the complex conjugate roots may be written in the form

$$(a_1 + jb_1)e^{(-k_1 - jk_2)t} + (a_1 - jb_1)e^{(-k_1 + jk_2)t}$$

Again, using the identity

$$e^{(C_1 + jC_2)t} = e^{C_1 t}(\cos C_2 t + j \sin C_2 t)$$

this reduces to

$$2e^{-k_1 t}(a_1 \cos k_2 t + b_1 \sin k_2 t) \qquad (3.12)$$

Let us now rework Example 3.3 using Eq. (3.12). We return to the point at which we arrived, by our usual techniques, with the conclusion that

$$B = \frac{-1 - j}{2}$$

Comparison of Eqs. (3.8) and (3.11) and the result for B show that we have two possible ways to assign a_1, b_1, k_1, and k_2 so that we match the form of Eq. (3.11). They are

$$a_1 = -\tfrac{1}{2} \qquad\qquad a_1 = -\tfrac{1}{2}$$

$$b_1 = -\tfrac{1}{2} \qquad\qquad b_1 = \tfrac{1}{2}$$

or

$$k_1 = 1 \qquad\qquad k_1 = 1$$

$$k_2 = 1 \qquad\qquad k_2 = -1$$

The first way corresponds to matching the term involving B with the first term of the conjugates of Eq. (3.11), and the second to matching it with the second term. *In either case,* substitution of these constants into Eq. (3.12) yields

$$-e^{-t}(\cos t + \sin t)$$

which is, as we have discovered, the correct term in $x(t)$.

What this means is that one can proceed directly from the evaluation of one of the partial-fraction constants, in this case B, to the complete term in the inverse transform, in this case $-e^{-t}(\cos t + \sin t)$. It is not necessary to perform all the algebra, since it has been done in the general case to arrive at Eq. (3.12).

Another example will serve to emphasize the application of this technique.

Example 3.4. Solve

$$\frac{d^2x}{dt^2} + 4x = 2e^{-t}$$

$$x(0) = x'(0) = 0$$

The Laplace transform method yields

$$x(s) = \frac{2}{(s^2 + 4)(s + 1)}$$

Factoring and expanding into partial fractions,

$$\frac{2}{(s + 1)(s + 2j)(s - 2j)} = \frac{A}{s + 1} + \frac{B}{s + 2j} + \frac{C}{s - 2j} \qquad (3.13)$$

Multiplying Eq. (3.13) by $(s + 1)$ and setting $s = -1$ yield

$$A = \frac{2}{(-1 + 2j)(-1 - 2j)} = \frac{2}{5}$$

Multiplying Eq. (3.13) by $(s + 2j)$ and setting $s = -2j$ yield

$$B = \frac{2}{(-2j + 1)(-4j)} = \frac{-2 + j}{10}$$

Matching the term

$$\frac{(-2 + j)/10}{s + 2j}$$

with the first term of the conjugates of Eq. (3.11) requires that

$$a_1 = -\tfrac{2}{10} = -\tfrac{1}{5}$$

$$b_1 = \tfrac{1}{10}$$

$$k_1 = 0$$

$$k_2 = 2$$

Substituting in (3.12) results in

$$-\tfrac{2}{5}\cos 2t + \tfrac{1}{5}\sin 2t$$

Hence the complete answer is

$$x(t) = \tfrac{2}{5}e^{-t} - \tfrac{2}{5}\cos 2t + \tfrac{1}{5}\sin 2t$$

Readers should verify that this answer satisfies the differential equation and boundary conditions. In addition, they should show that it can also be obtained by matching the term with the second term of the conjugates of Eq. (3.11) or by determining C instead of B.

ALTERNATE METHOD USING QUADRATIC TERM. Another method for solving Example 3.3, which avoids some of the manipulation of complex numbers, is as follows. Expand the expression for $x(s)$:

$$x(s) = \frac{2}{s(s^2 + 2s + 2)} = \frac{A}{s} + \frac{Bs + C}{s^2 + 2s + 2} \tag{3.14}$$

In this expression, the quadratic term is retained and the second term on the right side is the most general expression for the expansion. The reader will find this form of expansion for a quadratic term in books on advanced algebra.

Solve for A by multiplying both sides of Eq. (3.14) by s and let $s = 0$. The result is $A = 1$. Determine B and C algebraically by placing the two terms on the right side over a common denominator; thus

$$x(s) = \frac{2}{s(s^2 + 2s + 2)} = \frac{(s^2 + 2s + 2)A + Bs^2 + Cs}{s(s^2 + 2s + 2)}$$

Equating the numerators on each side gives

$$2 = (A + B)s^2 + (2A + C)s + 2A$$

We now equate the coefficients of like powers of s to obtain

$$A + B = 0$$
$$2A + C = 0$$
$$2A = 2$$

Solving these equations gives $A = 1$, $B = -1$, and $C = -2$. Equation (3.14) now becomes

$$x(s) = \frac{1}{s} - \frac{s}{s^2 + 2s + 2} - \frac{2}{s^2 + 2s + 2}$$

We now rearrange the second and third terms to match the following transform pairs from Table 2.1:

$$e^{-at}\sin kt \qquad k/[(s + a)^2 + k^2] \tag{3.15a}$$

$$e^{-at}\cos kt \qquad (s + a)/[(s + a)^2 + k^2] \tag{3.15b}$$

The result of the rearrangement gives

$$x(s) = \frac{1}{s} - \frac{s + 1}{(s + 1)^2 + 1^2} - \frac{1}{(s + 1)^2 + 1^2}$$

We see from the quadratic terms that $a = 1$ and $k = 1$, and using the table of transforms, one can easily invert each term to give

$$x(t) = 1 - e^{-t}\cos t - e^{-t}\sin t$$

which is the same result obtained before.

A general discussion of this case follows. Consider the general expression involving a quadratic term

$$x(s) = \frac{F(s)}{s^2 + \alpha s + \beta} \tag{3.16}$$

where $F(s)$ is some function of s (e.g. $1/s$). Expanding the terms on the right side gives

$$x(s) = F_1(s) + \frac{Bs + C}{s^2 + \alpha s + \beta} \tag{3.17}$$

where $F_1(s)$ represents other terms in the partial-fraction expansion. First solve for B and C algebraically by placing the right side over a common denominator and equating the coefficients of like powers of s. The next step is to express the quadratic term in the form

$$s^2 + \alpha s + \beta = (s + a)^2 + k^2$$

The terms a and k can be found by solving for the roots of $s^2 + \alpha s + \beta = 0$ by the quadratic formula to give $s_1 = -a + jk$, $s_2 = -a - jk$. The quadratic term can now be written

$$s^2 + \alpha s + \beta = (s - s_1)(s - s_2) = (s + a - jk)(s + a + jk) = (s + a)^2 + k^2$$

Equation (3.17) now becomes

$$x(s) + F_1(s) + \frac{Bs + C}{(s + a)^2 + k^2} \tag{3.18}$$

The numerator of the quadratic term is now written to correspond to the transform pairs given by Eqs. (3.15a and b)

$$Bs + C = B\left[s + a + \frac{(C/B) - a}{k}k\right] = B(s + a) + \frac{C - aB}{k}k$$

Equation (3.18) becomes

$$x(s) = F_1(s) + B\frac{(s + a)}{(s + a)^2 + k^2} + \left(\frac{C - aB}{k}\right)\frac{k}{(s + a)^2 + k^2}$$

Applying the transform pairs of Eqs. (3.15a and b) to the quadratic terms on the right gives

$$x(t) = F_1(t) + Be^{-at}\cos kt + \left(\frac{C - aB}{k}\right)e^{-at}\sin kt \tag{3.19}$$

where $F_1(t)$ is the result of inverting $F_1(s)$. We now apply this method to the following example.

Example 3.5. Solve

$$x(s) = \frac{1}{s(s^2 - 2s + 5)} = \frac{A}{s} + \frac{Bs + C}{s^2 - 2s + 5}$$

Applying the quadratic equation to the quadratic term gives:

$$s_{1,2} = \frac{2 \pm \sqrt{4 - 20}}{2} = 1 \pm 2j$$

Using the method just presented, we find that $a = -1, k = 2$. Solving for A, B, and C gives $A = 1/5, B = -1/5, C = 2/5$. Introducing these values into the expression for $x(s)$ and applying Eq. (3.19) gives

$$x(t) = \frac{1}{5} - \frac{1}{5}e^t\cos 2t + \frac{1}{10}e^t\sin 2t$$

The reader should solve Example 3.4 with this alternate method, which uses Eq. (3.19).

In the next example, an exceptional case is considered; the denominator of $x(s)$ has repeated roots. The procedure in this case will vary slightly from that of the previous cases.

Example 3.6. Solve

$$\frac{d^3x}{dt^3} + \frac{3d^2x}{dt^2} + \frac{3dx}{dt} + x = 1$$

$$x(0) = x'(0) = x''(0) = 0$$

Application of the Laplace transform yields

$$x(s) = \frac{1}{s(s^3 + 3s^2 + 3s + 1)}$$

Factoring and expanding in partial fractions,

$$x(s) = \frac{1}{s(s + 1)^3} = \frac{A}{s} + \frac{B}{(s + 1)^3} + \frac{C}{(s + 1)^2} + \frac{D}{s + 1} \tag{3.20}$$

As in the previous cases, to determine A, multiply both sides by s and then set s to zero. This yields

$$A = 1$$

Multiplication of both sides of Eq. (3.20) by $(s + 1)^3$ results in

$$\frac{1}{s} = \frac{A(s + 1)^3}{s} + B + C(s + 1) + D(s + 1)^2 \tag{3.21}$$

Setting $s = -1$ in Eq. (3.15) gives

$$B = -1$$

Having found A and B, introduce these values into Eq. (3.20) and place the right side of the equation over a common denominator; the result is:

$$\frac{1}{s(s + 1)^3} = \frac{(s + 1)^3 - s + Cs(s + 1) + Ds(s + 1)^2}{s(s + 1)^3} \tag{3.22}$$

Expanding the numerator of the right side gives

$$\frac{1}{s(s + 1)^3} = \frac{(1 + D)s^3 + (3 + C + 2D)s^2 + (2 + C + D)s + 1}{s(s + 1)^3} \tag{3.23}$$

We now equate the numerators on each side to get

$$1 = (1 + D)s^3 + (3 + C + 2D)s^2 + (2 + C + D)s + 1$$

Equating the coefficients of like powers of s gives

$$1 + D = 0$$
$$3 + C + 2D = 0$$
$$2 + C + D = 0$$

Solving these equations gives $C = -1$ and $D = -1$.
    The final result is then

$$x(s) = \frac{1}{s} - \frac{1}{(s + 1)^3} - \frac{1}{(s + 1)^2} - \frac{1}{s + 1} \tag{3.24}$$

Referring to Table 2.1, this can be inverted to

$$x(t) = 1 - e^{-t}\left(\frac{t^2}{2} + t + 1\right) \tag{3.25}$$

The reader should verify that Eq. (3.24) placed over a common denominator results in the original form

$$x(s) = \frac{1}{s(s + 1)^3}$$

and that Eq. (3.25) satisfies the differential equation and initial conditions.

The result of Example 3.6 may be generalized. The appearance of the factor $(s + a)^n$ in the denominator of $x(s)$ leads to n terms in the partial-fraction expansion:

$$\frac{C_1}{(s + a)^n}, \frac{C_2}{(s + a)^{n-1}}, \cdots, \frac{C_n}{s + a}$$

The constant C_1 can be determined as usual by multiplying the expansion by $(s + a)^n$ and setting $s = -a$. The other constants are determined by the method shown in Example 3.6. These terms, according to Table 2.1, lead to the following expression as the inverse transform:

$$\left[\frac{C_1}{(n - 1)!}t^{n-1} + \frac{C_2}{(n - 2)!}t^{n-2} + \cdots + C_{n-1}t + C_n\right]e^{-at} \qquad (3.26)$$

It is interesting to recall that in the classical method for solving these equations, one treats repeated roots of the characteristic equation by postulating the form of Eq. (3.26) and selecting the constants to fit the initial conditions.

Qualitative Nature of Solutions

If we are interested *only in the form* of the solution $x(t)$, which is often the case in our work, *this information may be obtained directly from the roots of the denominator of $x(s)$*. As an illustration of this "qualitative" approach to differential equations consider Example 3.3 in which

$$x(s) = \frac{2}{s(s^2 + 2s + 2)} = \frac{A}{s} + \frac{B}{s + 1 + j} + \frac{C}{s + 1 - j}$$

is the transformed solution of

$$\frac{d^2x}{dt^2} + \frac{2dx}{dt} + 2x = 2$$

It is evident by inspection of the partial-fraction expansion, *without* evaluation of the constants, that the s in the denominator of $x(s)$ will give rise to a constant in $x(t)$. Also, since the roots of the quadratic term are $-1 \pm j$, it is known that $x(t)$ must contain terms of the form $e^{-t}(C_1\cos t + C_2\sin t)$. This may be sufficient information for our purposes. Alternatively, we may be interested in the behavior of $x(t)$ as $t \to \infty$. It is clear that the terms involving sin and cos vanish because of the factor e^{-t}. Therefore, $x(t)$ ultimately approaches the constant, which by inspection must be unity.

The qualitative nature of the solution $x(t)$ can be related to the location of the roots of the denominator of $x(s)$ in the complex plane. These roots are the roots

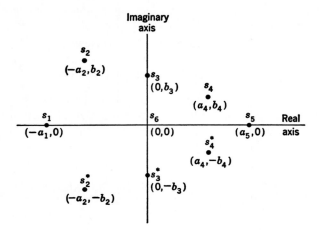

FIGURE 3-1
Location of typical roots of characteristic equation.

of the characteristic equation and the roots of the denominator of the transformed forcing function. Consider Fig. 3.1, a drawing of the complex plane, in which several typical roots are located and labeled with their coordinates. Table 3.1 gives the form of the terms in the equation for $x(t)$, corresponding to these roots. Note that all constants, $a_1, a_2, \ldots, b_1, b_2, \ldots$, are taken as positive. The constants C_1 and C_2 are arbitrary and can be determined by the partial-fraction expansion techniques. As discussed above, this determination is often not necessary for our work.

If any of these roots are repeated, the term given in Table 3.1 is multiplied by a power series in t,

$$K_1 + K_2 t + K_3 t^2 + \cdots + K_r t^{r-1}$$

where r is the number of repetitions of the root and the constants $K_1, K_2, \ldots, K_r$ can be evaluated by partial-fraction expansion.

It is thus evident that the imaginary axis divides the root locations into distinct areas, with regard to the behavior of the corresponding terms in $x(t)$ as t becomes large. Terms corresponding to roots to the left of the imaginary axis vanish exponentially in time, while those corresponding to roots to the right of

TABLE 3.1

Roots	Terms in $x(t)$ for $t > 0$
s_1	$C_1 e^{-a_1 t}$
s_2, s_2^*	$e^{-a_2 t}(C_1 \cos b_2 t + C_2 \sin b_2 t)$
s_3, s_3^*	$C_1 \cos b_3 t + C_2 \sin b_3 t$
s_4, s_4^*	$e^{a_4 t}(C_1 \cos b_4 t + C_2 \sin b_4 t)$
s_5	$C_1 e^{a_5 t}$
s_6	C_1

the imaginary axis increase exponentially in time. Terms corresponding to roots at the origin behave as power series in time, a constant being considered as a degenerate power series. Terms corresponding to roots located elsewhere on the imaginary axis oscillate with constant amplitude in time unless they are multiple roots in which case the amplitude of oscillation increases as a power series in time. Much use will be made of this information in later sections of the text.

SUMMARY

The reader now has available the basic tools for the use of Laplace transforms to solve differential equations. In addition, it is now possible to obtain considerable information about the qualitative nature of the solution with a minimum of labor. It should be pointed out that it is always necessary to factor the denominator of $x(s)$ in order to obtain any information about $x(t)$. If this denominator is a polynomial of order three or more, this may be far from a trivial problem. Chapter 15 is largely devoted to a solution of this problem within the context of control applications.

The next chapter is a grouping of several Laplace transform theorems that will find later application. In addition, a discussion of the impulse function $\delta(t)$ is presented there. Unavoidably, this chapter is rather dry. It may be desirable for the reader to skip directly to Chap. 5, where our control studies begin. At each point where a theorem of Chap. 4 is applied, reference to the appropriate section of Chap. 4 can be made.

PROBLEMS

3.1. Solve the following using Laplace transforms:

(a) $\dfrac{d^2x}{dt^2} + \dfrac{dx}{dt} + x = 1 \qquad x(0) = x'(0) = 0$

(b) $\dfrac{d^2x}{dt^2} + \dfrac{2dx}{dt} + x = 1 \qquad x(0) = x'(0) = 0$

(c) $\dfrac{d^2x}{dt^2} + \dfrac{3dx}{dt} + x = 1 \qquad x(0) = x'(0) = 0$

Sketch the behavior of these solutions on a single graph. What is the effect of the coefficient of dx/dt?

3.2. Solve the following differential equations by Laplace transforms:

(a) $\dfrac{d^4x}{dt^4} + \dfrac{d^3x}{dt^3} = \cos t \qquad x(0) = x'(0) = x'''(0) = 0 \qquad x''(0) = 1$

(b) $\dfrac{d^2q}{dt^2} + \dfrac{dq}{dt} = t^2 + 2t \qquad q(0) = 4 \qquad q'(0) = -2$

3.3. Invert the following transforms:

(a) $\dfrac{3s}{(s^2 + 1)(s^2 + 4)}$

(b) $\dfrac{1}{s(s^2 - 2s + 5)}$

(c) $\dfrac{3s^3 - s^2 - 3s + 2}{s^2(s - 1)^2}$

3.4. Expand the following functions by partial-fraction expansion. Do *not* evaluate coefficients or invert expressions.

(a) $X(s) = \dfrac{2}{(s + 1)(s^2 + 1)^2(s + 3)}$

(b) $X(s) = \dfrac{1}{s^3(s + 1)(s + 2)(s + 3)^3}$

(c) $X(s) = \dfrac{1}{(s + 1)(s + 2)(s + 3)(s + 4)}$

3.5. (a) Invert: $x(s) = 1/[s(s + 1)(0.5s + 1)]$
   (b) Solve: $dx/dt + 2x = 2, x(0) = 0$

3.6. Obtain $y(t)$ for

(a) $y(s) = \dfrac{s + 1}{s^2 + 2s + 5}$

(b) $y(s) = \dfrac{s^2 + 2s}{s^4}$

(c) $y(s) = \dfrac{2s}{(s - 1)^3}$

3.7. (a) Invert the following function

$$y(s) = 1/(s^2 + 1)^2$$

   (b) Plot y versus t from 0 to 3π.
3.8. Determine $f(t)$ for $f(s) = 1/[s^2(s + 1)]$.

FURTHER
PROPERTIES
OF TRANSFORMS

This chapter is a collection of theorems and results relative to the Laplace transformation. The theorems are selected because of their applicability to problems in control theory. Other theorems and properties of the Laplace transformation are available in standard texts [see Churchill (1972)]. In later chapters, the theorems presented here will be used as needed.

Final-Value Theorem

If $f(s)$ is the Laplace transform of $f(t)$, then

$$\lim_{t \uparrow \infty}[f(t)] = \lim_{s \downarrow 0}[sf(s)]$$

provided that $sf(s)$ does not become infinite for any value of s satisfying $\mathrm{Re}(s) \geq 0$. If this condition does not hold, $f(t)$ does not approach a limit as $t \uparrow \infty$. In the practical application of this theorem, the limit of $f(t)$ that is found by use of the theorem is correct only if $f(t)$ is bounded as t approaches infinity.

Proof. From the Laplace transform of a derivative, we have

$$\int_0^\infty \frac{df}{dt} e^{-st} dt = sf(s) - f(0)$$

Hence,

$$\lim_{s \downarrow 0} \int_0^\infty \frac{df}{dt} e^{-st} \, dt = \lim_{s \downarrow 0} [sf(s)] - f(0)$$

It can be shown that the order of the integration and limit operation on the left side of this equation can be interchanged if the conditions of the theorem hold. Doing this gives

$$\int_0^\infty \frac{df}{dt} \, dt = \lim_{s \downarrow 0} [sf(s)] - f(0)$$

Evaluating the integral,

$$\lim_{t \to \infty} [f(t)] - f(0) = \lim_{s \downarrow 0} [sf(s)] - f(0)$$

which immediately yields the desired result.

Example 4.1. Find the final value of the function $x(t)$ for which the Laplace transform is

$$x(s) = \frac{1}{s(s^3 + 3s^2 + 3s + 1)}$$

Direct application of the final-value theorem yields

$$\lim_{t \to \infty} [x(t)] = \lim_{s \downarrow 0} \frac{1}{s^3 + 3s^2 + 3s + 1} = 1$$

As a check, note that this transform was inverted in Example 3.6 to give

$$x(t) = 1 - e^{-t} \left(\frac{t^2}{2} + t + 1 \right)$$

which approaches unity as t approaches infinity. Note that since the denominator of $sx(s)$ can be factored to $(s + 1)^3$, the conditions of the theorem are satisfied; that is, $(s + 1)^3 \neq 0$ unless $s = -1$.

Example 4.2. Find the final value of the function $x(t)$ for which the Laplace transform is

$$x(s) = \frac{s^4 - 6s^2 + 9s - 8}{s(s - 2)(s^3 + 2s^2 - s - 2)}$$

In this case, the function $sx(s)$ can be written

$$sx(s) = \frac{s^4 - 6s^2 + 9s - 8}{(s + 1)(s + 2)(s - 1)(s - 2)}$$

Since this becomes infinite for $s = 1$ and $s = 2$, the conditions of the theorem are not satisfied. Note that we inverted this transform in Example 3.2, where it was found that

$$x(t) = -2 + \frac{1}{12} e^{2t} + \frac{11}{3} e^{-t} - \frac{17}{12} e^{-2t} + \frac{2}{3} e^t$$

This function continues to grow exponentially with t and, as expected, does not approach a limit.

The proof of the next theorem closely parallels the proof of the last one and is left as an exercise for the reader.

Initial-Value Theorem

$$\lim_{t \downarrow 0}[f(t)] = \lim_{s \uparrow \infty}[sf(s)]$$

The conditions on this theorem are not so stringent as those for the previous one because for functions of interest to us the order of integration and limiting process need not be interchanged to establish the result.

Example 4.3. Find the initial value $x(0)$ of the function that has the following transform

$$x(s) = \frac{s^4 - 6s^2 + 9s - 8}{s(s-2)(s^3 + 2s^2 - s - 2)}$$

The function $sx(s)$ is written in the form

$$sx(s) = \frac{s^4 - 6s^2 + 9s - 8}{s^4 - 5s^2 + 4}$$

After performing the indicated long division, this becomes

$$sx(s) = 1 - \frac{s^2 - 9s + 12}{s^4 - 5s^2 + 4}$$

which clearly goes to unity as s becomes infinite. Hence

$$x(0) = 1$$

which again checks Example 3.2.

Translation of Transform

If $L\{f(t)\} = f(s)$, then

$$L\{e^{-at} f(t)\} = f(s + a)$$

In other words, the variable in the transform s is translated by a.

Proof.

$$L\{e^{-at} f(t)\} = \int_0^\infty f(t)e^{-(s+a)t}\, dt = f(s + a)$$

Example 4.4. Find $L\{e^{-at}\cos kt\}$. Since

$$L\{\cos kt\} = \frac{s}{s^2 + k^2}$$

then by the previous theorem,

$$L\{e^{-at}\cos kt\} = \frac{s + a}{(s + a)^2 + k^2}$$

which checks Table 2.1.

A primary use for this theorem is in the inversion of transforms. For example, using this theorem the transform

$$x(s) = \frac{1}{(s + a)^2}$$

can be immediately inverted to

$$x(t) = te^{-at}$$

In obtaining this result, we made use of the following transform pair from Table 2.1:

$$L\{t\} = \frac{1}{s^2}$$

Translation of Function

If $L\{f(t)\} = f(s)$, then

$$L\{f(t - t_0)\} = e^{-st_0}f(s)$$

provided that

$$f(t) = 0 \qquad \text{for } t < 0$$

(which will always be true for functions we use).

Before proving this theorem, it may be desirable to clarify the relationship between $f(t - t_0)$ and $f(t)$. This is done for an arbitrary function $f(t)$ in Fig. 4.1, where it can be seen that $f(t - t_0)$ is simply translated horizontally from $f(t)$ through a distance t_0.

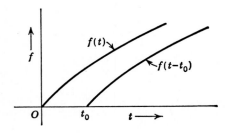

FIGURE 4-1
Illustration of $f(t - t_0)$ as related to $f(t)$.

Proof.

$$L\{f(t - t_0)\} = \int_0^\infty f(t - t_0)e^{-st}dt$$

$$= e^{-st_0} \int_{-t_0}^\infty f(t - t_0)e^{-s(t-t_0)}d(t - t_0)$$

But since $f(t) = 0$ for $t < 0$, the lower limit of this integral may be replaced by zero. Since $(t - t_0)$ is now the dummy variable of integration, the integral may be recognized as the Laplace transform of $f(t)$; thus, the theorem is proved.

This result is also useful in inverting transforms. It follows that, if $f(t)$ is the inverse transform of $f(s)$, then the inverse transform of

$$e^{-st_0}f(s)$$

is the function

$$g(t) = \begin{cases} 0 & t < t_0 \\ f(t - t_0) & t > t_0 \end{cases}$$

Example 4.5. Find the Laplace transform of

$$f(t) = \begin{cases} 0 & t < 0 \\ \frac{1}{h} & 0 < t < h \\ 0 & t > h \end{cases}$$

This function is pictured in Fig. 4.2. It is clear that $f(t)$ may be represented by the difference of two functions,

$$f(t) = \frac{1}{h}[u(t) - u(t - h)]$$

where $u(t - h)$ is the unit-step function translated h units to the right. We may now use the linearity of the transform and the previous theorem to write immediately

$$f(s) = \frac{1}{h}\frac{1 - e^{-hs}}{s}$$

This result is of considerable value in establishing the transform of the unit-impulse function, as will be described in the next section.

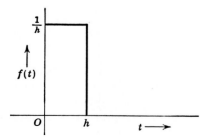

FIGURE 4-2
Pulse function of Example 4.5.

Transform of the Unit-impulse Function

Consider again the function of Example 4.5. If we allow h to shrink to zero, we obtain a new function which is zero everywhere except at the origin, where it is infinite. However, it is important to note that the area under this function always remains equal to unity. We call this new function $\delta(t)$, and the fact that its area is unity means that

$$\int_{-\infty}^{\infty} \delta(t)\,dt = 1$$

The graph of $\delta(t)$ appears as a line of infinite height at the origin, as indicated in Table 2.1. The function $\delta(t)$ is called the unit-impulse function or, alternatively, the delta function.

It is mentioned here that, in the strict mathematical sense of a limit, the function $f(t)$ does not possess a limit as h goes to zero. Hence, the function $\delta(t)$ does not fit the strict mathematical definition of a function. To assign a mathematically precise meaning to the unit-impulse function requires use of the theory of distributions, which is clearly beyond the scope of this text. However, for our work in automatic control, we shall be able to obtain useful results by formal manipulation of the delta function, and hence we ignore these mathematical difficulties.

We have derived in Example 4.5 the Laplace transform of $f(t)$ as

$$L\{f(t)\} = \frac{1 - e^{-hs}}{hs}$$

Formally, then, the Laplace transform of $\delta(t)$ can be obtained by letting h go to zero in $L\{f(t)\}$. Applying L'Hôpital's rule,

$$L\{\delta(t)\} = \lim_{h \to 0} \frac{1 - e^{-hs}}{hs} = \lim_{h \to 0} \frac{se^{-hs}}{s} = 1 \qquad (4.1)$$

This "verifies" the entry in Table 2.1.

It is interesting to note that, since we rewrote $f(t)$ in Example 4.5 as

$$f(t) = \frac{1}{h}[u(t) - u(t - h)]$$

then $\delta(t)$ can be written as

$$\delta(t) = \lim_{h \to 0} \frac{u(t) - u(t - h)}{h}$$

In this form, the delta function appears as the derivative of the unit-step function. The reader may find it interesting to ponder this statement in relation to the graphs of $\delta(t)$ and $u(t)$ and in relation to the integral of $\delta(t)$ discussed previously.

The unit-impulse function finds use as an idealized disturbance in control systems analysis and design.

Transform of an Integral

If $L\{f(t)\} = f(s)$, then

$$L\left\{\int_0^t f(t)dt\right\} = \frac{f(s)}{s}$$

This important theorem is closely related to the theorem on differentiation. Since the operations of differentiation and integration are inverses of each other when applied to the time functions, i.e.,

$$\frac{d}{dt}\int_0^t f(t)dt = \int_0^t \frac{df}{dt}dt = f(t) \tag{4.2}$$

it is to be expected that these operations when applied to the transforms will also be inverses. Thus assuming the theorem to be valid, Eq. (4.2) in the transformed variable s becomes

$$s\frac{f(s)}{s} = \frac{1}{s}sf(s) = f(s)$$

In other words, multiplication of $f(s)$ by s corresponds to differentiation of $f(t)$ with respect to t, and division of $f(s)$ by s corresponds to integration of $f(t)$ with respect to t.

The proof follows from a straightforward integration by parts.

$$f(s) = \int_0^\infty f(t)e^{-st}dt$$

Let

$$u = e^{-st} \qquad dv = f(t)dt$$

Then

$$du = -se^{-st}dt \qquad v = \int_0^t f(t)dt$$

Hence,

$$f(s) = e^{-st}\int_0^t f(t)dt\bigg|_0^\infty + s\int_0^\infty \left[\int_0^t f(t)dt\right]e^{-st}dt$$

Since $f(t)$ must satisfy the requirements for possession of a transform, it can be shown that the first term on the right, when evaluated at the upper limit of ∞, vanishes because of the factor e^{-st}. Furthermore, the lower limit clearly vanishes, and hence, there is no contribution from the first term. The second term may be recognized as $sL\{\int_0^t f(t)dt\}$, and the theorem follows immediately.

Example 4.6. Solve the following equation for $x(t)$:

$$\frac{dx}{dt} = \int_0^t x(t)dt - t$$

$$x(0) = 3$$

Taking the Laplace transform of both sides, and making use of the previous theorem

$$sx(s) - 3 = \frac{x(s)}{s} - \frac{1}{s^2}$$

Solving for $x(s)$,

$$x(s) = \frac{3s^2 - 1}{s(s^2 - 1)} = \frac{3s^2 - 1}{s(s + 1)(s - 1)}$$

This may be expanded into partial fractions according to the usual procedure to give

$$x(s) = \frac{1}{s} + \frac{1}{s + 1} + \frac{1}{s - 1}$$

Hence,

$$x(t) = 1 + e^{-t} + e^t$$

The reader should verify that this function satisfies the original equation.

PROBLEMS

4.1. If a forcing function $f(t)$ has the Laplace transform

$$f(s) = \frac{1}{s} + \frac{e^{-s} - e^{-2s}}{s^2} - \frac{e^{-3s}}{s}$$

graph the function $f(t)$.

4.2. Solve the following equation for $y(t)$:

$$\int_0^t y(\tau)d\tau = \frac{dy(t)}{dt} \qquad y(0) = 1$$

4.3. Express the function given in Fig. P4.3 in the t-domain and the s-domain.

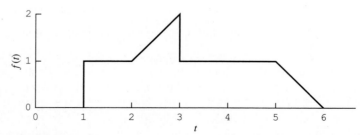

FIGURE P4-3

4.4. Sketch the following functions:

$$f(t) = u(t) - 2u(t - 1) + u(t - 3)$$
$$f(t) = 3tu(t) - 3u(t - 1) - u(t - 2)$$

4.5. The function $f(t)$ has the Laplace transform

$$f(s) = \left(1 - 2e^{-s} + e^{-2s}\right)/s^2$$

Obtain the function $f(t)$ and graph $f(t)$.

4.6. Determine $f(t)$ at $t = 1.5$ and at $t = 3$ for the following function:

$$f(t) = 0.5u(t) - 0.5u(t - 1) + (t - 3)u(t - 2)$$

PART
II

LINEAR
OPEN-LOOP
SYSTEMS

CHAPTER
5

RESPONSE OF FIRST-ORDER SYSTEMS

Before discussing a complete control system, it is necessary to become familiar with the responses of some of the simple, basic systems that often are the building blocks of a control system. This chapter and the three that follow describe in detail the behavior of several basic systems and show that a great variety of physical systems can be represented by a combination of these basic systems. Some of the terms and conventions that have become well established in the field of automatic control will also be introduced.

By the end of this part of the book, systems for which a transient must be calculated will be of high-order and require calculations that are time-consuming if done by hand. The reader should start now using Chap. 34 to see how the digital computer can be used to simulate the dynamics of control systems.

TRANSFER FUNCTION

MERCURY THERMOMETER. We shall develop the *transfer function* for a *first-order system* by considering the unsteady-state behavior of an ordinary mercury-in-glass thermometer. A cross-sectional view of the bulb is shown in Fig. 5.1.

Consider the thermometer to be located in a flowing stream of fluid for which the temperature x varies with time. Our problem is to calculate the *response* or the time variation of the thermometer reading y for a particular change in x.*

---

*In order that the result of the analysis of the thermometer be general and therefore applicable to other first-order systems, the symbols x and y have been selected to represent surrounding temperature and thermometer reading, respectively.

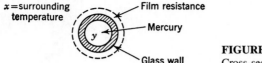

x =surrounding temperature
Film resistance
Mercury
y
Glass wall

FIGURE 5-1
Cross-sectional view of thermometer.

The following assumptions* will be used in this analysis:

1. All the resistance to heat transfer resides in the film surrounding the bulb (i.e., the resistance offered by the glass and mercury is neglected).
2. All the thermal capacity is in the mercury. Furthermore, at any instant the mercury assumes a uniform temperature throughout.
3. The glass wall containing the mercury does not expand or contract during the transient response. (In an actual thermometer, the expansion of the wall has an additional effect on the response of the thermometer reading. (See Iinoya and Altpeter (1962).)

It is assumed that the thermometer is initially at steady state. This means that, before time zero, there is no change in temperature with time. At time zero the thermometer will be subjected to some change in the surrounding temperature $x(t)$.

By applying the unsteady-state energy balance

Input rate − output rate = rate of accumulation

we get the result

$$hA(x - y) - 0 = mC\frac{dy}{dt} \tag{5.1}$$

where A = surface area of bulb for heat transfer, ft^2
  C = heat capacity of mercury, Btu/(lb$_m$)(°F)
  m = mass of mercury in bulb, lb$_m$
  t = time, hr
  h = film coefficient of heat transfer, Btu/(hr)(ft^2)(°F)

For illustrative purposes, typical engineering units have been used.

---

*Making the first two assumptions is often referred to as the *lumping of parameters* because all the resistance is "lumped" into one location and all the capacitance into another. As shown in the analysis, these assumptions make it possible to represent the dynamics of the system by an ordinary differential equation. If such assumptions were not made, the analysis would lead to a partial differential equation, and the representation would be referred to as a *distributed-parameter system*. In Chap. 21, distributed-parameter systems will be considered in detail.

Equation (5.1) states that the rate of flow of heat through the film resistance surrounding the bulb causes the internal energy of the mercury to increase at the same rate. The increase in internal energy is manifested by a change in temperature and a corresponding expansion of mercury, which causes the mercury column, or "reading" of the thermometer, to rise.

The coefficient h will depend on the flow rate and properties of the surrounding fluid and the dimensions of the bulb. We shall assume that h is constant for a particular installation of the thermometer.

Our analysis has resulted in Eq. (5.1), which is a first-order differential equation. Before solving this equation by means of the Laplace transform, *deviation variables* will be introduced into Eq. (5.1). The reason for these new variables will soon become apparent. Prior to the change in x, the thermometer is at steady state and the derivative dy/dt is zero. For the steady-state condition, Eq. (5.1) may be written

$$hA(x_s - y_s) = 0 \qquad t < 0 \tag{5.2}$$

The subscript s is used to indicate that the variable is the steady-state value. Equation (5.2) simply states that $y_s = x_s$, or the thermometer reads the true, bath temperature. Subtracting Eq. (5.2) from Eq. (5.1) gives

$$hA[(x - x_s) - (y - y_s)] = mC\frac{d(y - y_s)}{dt} \tag{5.3}$$

Notice that $d(y - y_s)/dt = dy/dt$ because y_s is a constant.

If we define the deviation variables to be the differences between the variables and their steady-state values

$$X = x - x_s$$
$$Y = y - y_s$$

Eq. (5.3) becomes

$$hA(X - Y) = mC\frac{dY}{dt} \tag{5.4}$$

If we let $mC/hA = \tau$, Eq. (5.4) becomes

$$X - Y = \tau\frac{dY}{dt} \tag{5.5}$$

Taking the Laplace transform of Eq. (5.5) gives

$$X(s) - Y(s) = \tau s Y(s) \tag{5.6}$$

Rearranging Eq. (5.6) as a ratio of $Y(s)$ to $X(s)$ gives

$$\frac{Y(s)}{X(s)} = \frac{1}{\tau s + 1} \tag{5.7}$$

The parameter τ is called the *time constant* of the system and has the units of time.

The expression on the right side of Eq. (5.7) is called the *transfer function* of the system. It is the ratio of the Laplace transform of the deviation in thermometer reading to the Laplace transform of the deviation in the surrounding temperature. In examining other physical systems, we shall usually attempt to obtain a transfer function.

Any physical system for which the relation between Laplace transforms of input and output deviation variables is of the form given by Eq. (5.7) is called a *first-order system*. Synonyms for first-order system are first-order lag and single exponential stage. The naming of all these terms is motivated by the fact that Eq. (5.7) results from a first-order, linear differential equation, Eq. (5.5). In Chap. 6 is a discussion of a number of other physical systems which are first-order.

By reviewing the steps leading to Eq. (5.7), one can discover that the introduction of deviation variables prior to taking the Laplace transform of the differential equation results in a transfer function that is free of initial conditions because the initial values of X and Y are zero. In control system engineering, we are primarily concerned with the deviations of system variables from their steady-state values. The use of deviation variables is, therefore, natural as well as convenient.

PROPERTIES OF TRANSFER FUNCTIONS. In general, a transfer function relates two variables in a physical process; one of these is the cause (forcing function or input variable) and the other is the effect (response or output variable). In terms of the example of the mercury thermometer, the surrounding temperature is the cause or input, whereas the thermometer reading is the effect or output. We may write

$$\text{Transfer function} = G(s) = \frac{Y(s)}{X(s)}$$

where $G(s)$ = symbol for transfer function

$\quad\quad X(s)$ = transform of forcing function or input, in deviation form

$\quad\quad Y(s)$ = transform of response or output, in deviation form

The transfer function completely describes the dynamic characteristics of the system. If we select a particular input variation $X(t)$ for which the transform is $X(s)$, the response of the system is simply

$$Y(s) = G(s)X(s) \tag{5.8}$$

By taking the inverse of $Y(s)$, we get $Y(t)$, the response of the system.

The transfer function results from a linear differential equation; therefore, the principle of superposition is applicable. This means that the transformed response of a system with transfer function $G(s)$ to a forcing function

$$X(s) = a_1 X_1(s) + a_2 X_2(s)$$

where X_1 and X_2 are particular forcing functions and a_1 and a_2 are constants, is

$$\begin{aligned} Y(s) &= G(s)X(s) \\ &= a_1 G(s)X_1(s) + a_2 G(s)X_2(s) \\ &= a_1 Y_1(s) + a_2 Y_2(s) \end{aligned}$$

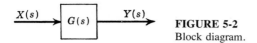

FIGURE 5-2
Block diagram.

$Y_1(s)$ and $Y_2(s)$ are the responses to X_1 and X_2 alone, respectively. For example, the response of the mercury thermometer to a sudden change in surrounding temperature of 10°F is simply twice the response to a sudden change of 5°F in surrounding temperature.

The functional relationship contained in a transfer function is often expressed by a *block-diagram* representation, as shown in Fig. 5.2. The arrow entering the box is the forcing function or input variable, and the arrow leaving the box is the response or output variable. Inside the box is placed the transfer function. We state that the transfer function $G(s)$ in the box "operates" on the input function $X(s)$ to produce an output function $Y(s)$. The usefulness of the block diagram will be appreciated in Chap. 9, when a complete control system containing several blocks is analyzed.

TRANSIENT RESPONSE

Now that the transfer function of a first-order system has been established, we can easily obtain its transient response to *any* forcing function. Since this type of system occurs so frequently in practice, it is worthwhile to study its response to several common forcing functions: step, impulse, and sinusoidal. These forcing functions have been found to be very useful in theoretical and experimental aspects of process control. They will be used extensively in our studies and hence, each one is explored before studying the transient response of the first-order system to these forcing functions.

Forcing Functions

STEP FUNCTION. Mathematically, the step function of magnitude A can be expressed as

$$X(t) = Au(t)$$

where $u(t)$ is the unit-step function defined in Chap. 2. A graphical representation is shown in Fig. 5.3.

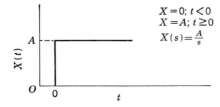

$X = 0; t < 0$
$X = A; t \geq 0$

$X(s) = \frac{A}{s}$

FIGURE 5-3
Step input.

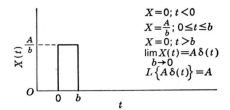

$$X=0;\ t<0$$
$$X=\frac{A}{b};\ 0\le t\le b$$
$$X=0;\ t>b$$
$$\lim_{b\to 0} X(t)=A\delta(t)$$
$$L\{A\delta(t)\}=A$$

FIGURE 5-4
Impulse function.

The transform of this function is $X(s) = A/s$. A step function can be approximated very closely in practice. For example, a step change in flow rate can be obtained by the sudden opening of a valve.

IMPULSE FUNCTION. Mathematically, the impulse function of magnitude A is defined as

$$X(t) = A\delta(t)$$

where $\delta(t)$ is the unit-impulse function defined and discussed in Chap. 4. A graphical representation of this function, before the limit is taken, is shown in Fig. 5.4.

The true impulse function, obtained by letting $b\ l\ 0$ in Fig. 5.4, has Laplace transform A. It is used more frequently as a mathematical aid than as an actual input to a physical system. For some systems it is difficult even to approximate an impulse forcing function. For this reason the representation of Fig. 5.4 is valuable, since this form can usually be approximated physically by application and removal of a step function. If the time duration b is sufficiently small, we shall see in Chap. 6 that the forcing function of Fig. 5.4 gives a response that closely resembles the response to a true impulse. In this sense, we often justify the use of A as the Laplace transform of the physically realizable forcing function of Fig. 5.4.

SINUSOIDAL INPUT. This function is represented mathematically by the equations

$$X = 0 \qquad\qquad t < 0$$
$$X = A \sin \omega t \qquad t \ge 0$$

where A is the amplitude and ω is the radian frequency. The radian frequency ω is related to the frequency f in cycles per unit time by $\omega = 2\pi f$. Figure 5.5 shows the graphical representation of this function. The transform is $X(s) = A\omega/(s^2 + \omega^2)$. This forcing function forms the basis of an important branch of control theory known as *frequency response*. Historically, a large segment of the development of control theory was based on frequency-response methods, which will be presented in Chaps. 16 and 17. Physically, it is more difficult to obtain a sinusoidal forcing function in most process variables than to obtain a step function.

This completes the discussion of some of the common forcing functions. We shall now devote our attention to the transient response of the first-order system to each of the forcing functions just discussed.

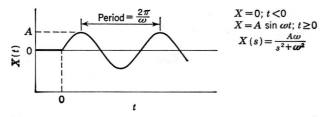

$X=0;\ t<0$

$X=A\sin\omega t;\ t\geq0$

$X(s)=\dfrac{A\omega}{s^2+\omega^2}$

FIGURE 5-5
Sinusoidal input.

Step Response

If a step change of magnitude A is introduced into a first-order system, the transform of $X(t)$ is

$$X(s) = \frac{A}{s} \tag{5.9}$$

The transfer function, which is given by Eq. (5.7), is

$$\frac{Y(s)}{X(s)} = \frac{1}{\tau s + 1} \tag{5.7}$$

Combining Eqs. (5.7) and (5.9) gives

$$Y(s) = \frac{A}{s}\frac{1}{\tau s + 1} \tag{5.10}$$

This can be expanded by partial fractions to give

$$Y(s) = \frac{A/\tau}{(s)(s + 1/\tau)} = \frac{C_1}{s} + \frac{C_2}{s + 1/\tau} \tag{5.11}$$

Solving for the constants C_1 and C_2 by the techniques covered in Chap. 3 gives $C_1 = A$ and $C_2 = -A$. Inserting these constants into Eq. (5.11) and taking the inverse transform give the time response for Y:

$$\begin{aligned} Y(t) &= 0 & t < 0 \\ Y(t) &= A(1 - e^{-t/\tau}) & t \geq 0 \end{aligned} \tag{5.12}$$

Hereafter, for the sake of brevity, it will be understood that, as in Eq. (5.12), the response is zero before $t = 0$. Equation (5.12) is plotted in Fig. 5.6 in terms of the dimensionless quantities $Y(t)/A$ and t/τ.

Having obtained the step response, Eq. (5.12), from a purely mathematical approach, we should consider whether or not the result seems to be correct from physical principles. Immediately after the thermometer is placed in the new environment, the temperature difference between the mercury in the bulb and the bath temperature is at its maximum value. With our simple lumped-parameter model,

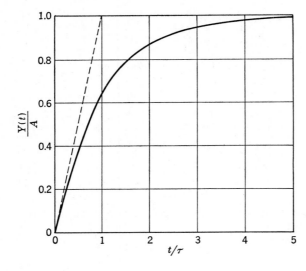

FIGURE 5-6
Response of a first-order system to a step input.

we should expect the flow of heat to commence immediately, with the result that the mercury temperature rises, causing a corresponding rise in the column of mercury. As the mercury temperature rises, the driving force causing heat to flow into the mercury will diminish, with the result that the mercury temperature changes at a slower rate as time proceeds. We see that this description of the response based on physical grounds does agree with the response given by Eq. (5.12) and shown graphically in Fig. 5.6.

Several features of this response, worth remembering, are

1. The value of $Y(t)$ reaches 63.2 percent of its ultimate value when the time elapsed is equal to one time constant τ. When the time elapsed is $2\tau, 3\tau$, and 4τ, the percent response is 86.5, 95, and 98, respectively. From these facts, one can consider the response essentially completed in three to four time constants.

2. One can show from Eq. (5.12) that the slope of the response curve at the origin in Fig. 5.6 is 1. This means that, if the initial rate of change of $Y(t)$ were maintained, the response would be complete in one time constant. (See the dotted line in Fig. 5.6.)

3. A consequence of the principle of superposition is that the response to a step input of any magnitude A may be obtained directly from Fig. 5.6 by multiplying the ordinate by A. Figure 5.6 actually gives the response to a unit-step function input, from which all other step responses are derived by superposition.

These results for the step response of a first-order system will now be applied to the following example.

Example 5.1. A thermometer having a time constant* of 0.1 min is at a steady-state temperature of 90°F. At time $t = 0$, the thermometer is placed in a temperature bath maintained at 100°F. Determine the time needed for the thermometer to read 98°.

In terms of symbols used in this chapter, we have

$$\tau = 0.1 \text{ min} \qquad x_s = 90° \qquad A = 10°$$

The ultimate thermometer reading will, of course, be 100°, and the ultimate value of the deviation variable $Y(\infty)$ is 10°. When the thermometer reads 98°, $Y(t) = 8°$. Substituting into Eq. (5.12) the appropriate values of Y, A, and τ gives

$$8 = 10(1 - e^{-t/0.1})$$

Solving this equation for t yields

$$t = 0.161 \text{ min}$$

The same result can also be obtained by referring to Fig. 5.6, where it is seen that $Y/A = 0.8$ at $t/\tau = 1.6$.

Impulse Response

The impulse response of a first-order system will now be developed. Anticipating the use of superposition, we consider a unit impulse for which the Laplace transform is

$$X(s) = 1 \tag{5.13}$$

Combining this with the transfer function for a first-order system, which is given by Eq. (5.7), results in

$$Y(s) = \frac{1}{\tau s + 1} \tag{5.14}$$

This may be rearranged to

$$Y(s) = \frac{1/\tau}{s + 1/\tau} \tag{5.15}$$

The inverse of $Y(s)$ can be found directly from the table of transforms and can be written in the form

$$\tau Y(t) = e^{-t/\tau} \tag{5.16}$$

A plot of this response is shown in Fig. 5.7 in terms of the variables t/τ and $\tau Y(t)$. The response to an impulse of magnitude A is obtained, as usual, by multiplying $\tau Y(t)$ from Fig. 5.7 by A/τ.

---

*The time constant given in this problem applies to the thermometer when it is located in the temperature bath. The time constant for the thermometer in air will be considerably different from that given because of the lower heat-transfer coefficient in air.

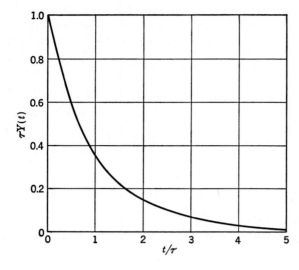

FIGURE 5-7
Unit-impulse response of a first-order system.

Notice that the response rises immediately to 1.0 and then decays exponentially. Such an abrupt rise is, of course, physically impossible, but as we shall see in Chap. 6, it is approached by the response to a finite pulse of narrow width, such as that of Fig. 5.4.

Sinusoidal Response

To investigate the response of a first-order system to a sinusoidal forcing function, the example of the mercury thermometer will be considered again. Consider a thermometer to be in equilibrium with a temperature bath at temperature x_s. At some time $t = 0$, the bath temperature begins to vary according to the relationship

$$x = x_s + A \sin \omega t \qquad t > 0 \qquad (5.17)$$

where x = temperature of bath
x_s = temperature of bath before sinusoidal disturbance is applied
A = amplitude of variation in temperature
ω = radian frequency, rad/time

In anticipation of a simple result we shall introduce a deviation variable X which is defined as

$$X = x - x_s \qquad (5.18)$$

Using this new variable in Eq. (5.17) gives

$$X = A \sin \omega t \qquad (5.19)$$

By referring to a table of transforms, the transform of Eq. (5.19) is

$$X(s) = \frac{A\omega}{s^2 + \omega^2} \qquad (5.20)$$

Combining Eqs. (5.7) and (5.20) to eliminate $X(s)$ yields

$$Y(s) = \frac{A\omega}{s^2 + \omega^2} \frac{1/\tau}{s + 1/\tau} \tag{5.21}$$

This equation can be solved for $Y(t)$ by means of a partial-fraction expansion, as described in Chap. 3. The result is

$$Y(t) = \frac{A\omega\tau e^{-t/\tau}}{\tau^2\omega^2 + 1} - \frac{A\omega\tau}{\tau^2\omega^2 + 1} \cos \omega t + \frac{A}{\tau^2\omega^2 + 1} \sin \omega t \tag{5.22}$$

Equation (5.22) can be written in another form by using the trigonometric identity

$$p \cos A + q \sin A = r \sin (A + \theta) \tag{5.23}$$

where

$$r = \sqrt{p^2 + q^2} \qquad \tan \theta = \frac{p}{q}$$

Applying the identity of Eq. (5.23) to (5.22) gives

$$Y(t) = \frac{A\omega\tau}{\tau^2\omega^2 + 1} e^{-t/\tau} + \frac{A}{\sqrt{\tau^2\omega^2 + 1}} \sin (\omega t + \phi) \tag{5.24}$$

where

$$\phi = \tan^{-1}(-\omega\tau)$$

As $t \to \infty$, the first term on the right side of Eq. (5.24) vanishes and leaves only the ultimate periodic solution, which is sometimes called the steady-state solution

$$Y(t)\Big|_s = \frac{A}{\sqrt{\tau^2\omega^2 + 1}} \sin (\omega t + \phi) \tag{5.25}$$

By comparing Eq. (5.19) for the input forcing function with Eq. (5.25) for the ultimate periodic response, we see that

1. The output is a sine wave with a frequency ω equal to that of the input signal.
2. The ratio of output amplitude to input amplitude is $1/\sqrt{\tau^2\omega^2 + 1}$. This is always smaller than 1. We often state this by saying that the signal is *attenuated*.
3. The output lags behind the input by an angle $|\phi|$. It is clear that lag occurs, for the sign of ϕ is always negative.*

---

*By convention, the output sinusoid lags the input sinusoid if ϕ in Eq. (5.25) is negative. In terms of a recording of input and output, this means that the input peak occurs before the output peak. If ϕ is positive in Eq. (5.25), the system exhibits phase *lead*, or the output leads the input. In this book we shall always use the term phase angle (ϕ) and interpret whether there is lag or lead by the convention

$$\phi < 0 \qquad \text{phase lag}$$
$$\phi > 0 \qquad \text{phase lead}$$

For a particular system for which the time constant τ is a fixed quantity, it is seen from Eq. (5.25) that the attenuation of amplitude and the phase angle ϕ depend only on the frequency ω. The attenuation and phase lag increase with frequency, but the phase lag can never exceed 90° and approaches this value asymptotically.

The sinusoidal response is interpreted in terms of the mercury thermometer by the following example.

Example 5.2. A mercury thermometer having a time constant of 0.1 min is placed in a temperature bath at 100°F and allowed to come to equilibrium with the bath. At time $t = 0$, the temperature of the bath begins to vary sinusoidally about its average temperature of 100°F with an amplitude of 2°F. If the frequency of oscillation is $10/\pi$ cycles/min, plot the ultimate response of the thermometer reading as a function of time. What is the phase lag?

In terms of the symbols used in this chapter

$$\tau = 0.1$$

$$x_s = 100°F$$

$$A = 2°F$$

$$f = \frac{10}{\pi} \text{ cycles/min}$$

$$\omega = 2\pi f = 2\pi \frac{10}{\pi} = 20 \text{ rad/min}$$

From Eq. (5.25), the amplitude of the response and the phase angle are calculated; thus

$$\frac{A}{\sqrt{\tau^2\omega^2 + 1}} = \frac{2}{\sqrt{4 + 1}} = 0.896°F$$

$$\phi = -\tan^{-1} 2 = -63.5°$$

or

$$\text{Phase lag} = 63.5°$$

The response of the thermometer is therefore

$$Y(t) = 0.896 \sin (20t - 63.5°)$$

or

$$y(t) = 100 + 0.896 \sin (20t - 63.5°)$$

To obtain the lag in terms of time rather than angle, we proceed as follows: A frequency of $10/\pi$ cycles/min means that a complete cycle (peak to peak) occurs in $(10/\pi)^{-1}$ min. Since one cycle is equivalent to 360° and the lag is 63.5°, the time corresponding to this lag is

$$\frac{63.5}{360} \times (\text{time for 1 cycle})$$

or

$$\text{Lag} = \frac{63.5}{360}\frac{\pi}{10} = 0.0555 \text{ min}$$

In general, the lag in units of time is given by

$$\text{Lag} = \frac{|\phi|}{360f}$$

when ϕ is expressed in degrees.

The response of the thermometer reading and the variation in bath temperature are shown in Fig. 5.8. It should be noted that the response shown in this figure holds only after sufficient time has elapsed for the nonperiodic term of Eq. (5.24) to become negligible. For all practical purposes this term becomes negligible after a time equal to about 3τ. If the response were desired beginning from the time the bath temperature begins to oscillate, it would be necessary to plot the complete response as given by Eq. (5.24).

SUMMARY

In this chapter several basic concepts and definitions of control theory have been introduced. These include input variable, output variable, deviation variable, transfer function, response, time constant, first-order system, block diagram, attenuation, and phase lag. Each of these ideas arose naturally in the study of the dynamics of the first-order system, which was the basic subject matter of the chapter. As might be expected, the concepts will find frequent use in succeeding chapters.

In addition to introducing new concepts, we have listed the response of the first-order system to forcing functions of major interest. This information on

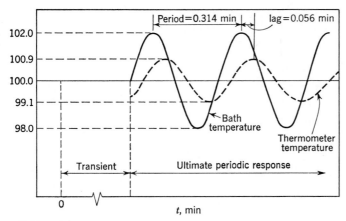

FIGURE 5-8
Response of thermometer in Example 5.2.

the dynamic behavior of the first-order system will be of significant value in the remainder of our studies.

PROBLEMS

5.1. A thermometer having a time constant of 0.2 min is placed in a temperature bath, and after the thermometer comes to equilibrium with the bath, the temperature of the bath is increased linearly with time at a rate of 1°/min. What is the difference between the indicated temperature and the bath temperature (a) 0.1 min, (b) 1.0 min after the change in temperature begins?

(c) What is the maximum deviation between indicated temperature and bath temperature, and when does it occur?

(d) Plot the forcing function and response on the same graph. After a long enough time, by how many minutes does the response lag the input?

5.2. A mercury thermometer bulb is $\frac{1}{2}$ in. long by $\frac{1}{8}$ in. diameter. The glass envelope is very thin. Calculate the time constant in water flowing at 10 ft/sec at a temperature of 100°F. In your solution, give a summary which includes

(a) Assumptions used

(b) Source of data

(c) Results

5.3. Given a system with the transfer function $Y(s)/X(s) = (T_1s + 1)/(T_2s + 1)$. Find $Y(t)$ if $X(t)$ is a unit-step function. If $T_1/T_2 = 5$, sketch $Y(t)$ versus t/T_2. Show the numerical values of minimum, maximum, and ultimate values that may occur during the transient. Check these using the initial-value and final-value theorems of Chap. 4.

5.4. A thermometer having first-order dynamics with a time constant of 1 min is placed in a temperature bath at 100°F. After the thermometer reaches steady state, it is suddenly placed in a bath at 110°F at $t = 0$ and left there for 1 min, after which it is immediately returned to the bath at 100°F.

(a) Draw a sketch showing the variation of the thermometer reading with time.

(b) Calculate the thermometer reading at $t = 0.5$ min and at $t = 2.0$ min.

5.5. Repeat Prob. 5.4 if the thermometer is in the 110°F bath for only 10 sec.

5.6. A mercury thermometer, which has been on a table for some time, is registering the room temperature, 75°F. Suddenly, it is placed in a 400°F oil bath. The following data are obtained for the response of the thermometer.

Time, sec	Thermometer reading, °F
0	75
1	107
2.5	140
5	205
8	244
10	282
15	328
30	385

Give two independent estimates of the thermometer time constant.

5.7. Rewrite the sinusoidal response of a first-order system [Eq. (5.24)] in terms of a cosine wave. Reexpress the forcing function [Eq. (5.19)] as a cosine wave, and compute the phase difference between input and output cosine waves.

5.8. The mercury thermometer of Prob. 5.6 is again allowed to come to equilibrium in the room air at 75°F. Then it is placed in the 400°F oil bath for a length of time less than 1 sec, and quickly removed from the bath and reexposed to the 75°F ambient conditions. It may be estimated that the heat-transfer coefficient to the thermometer in air is one-fifth that in the oil bath. If 10 sec after the thermometer is removed from the bath it reads 98°F, estimate the length of time that the thermometer was in the bath.

5.9. A thermometer having a time constant of 1 min is initially at 50°C. It is immersed in a bath maintained at 100°C at $t = 0$. Determine the temperature reading at $t = 1.2$ min.

5.10. In problem 5.9, if at $t = 1.5$ min, the thermometer is removed from the bath and put in a bath at 75°C, determine the maximum temperature indicated by the thermometer. What will be the indicated temperature at $t = 20$ min?

5.11. A process of unknown transfer function is subjected to a unit-impulse input. The output of the process is measured accurately and is found to be represented by the function $y(t) = te^{-t}$. Determine the unit-step response of this process.

CHAPTER
6

PHYSICAL EXAMPLES OF FIRST-ORDER SYSTEMS

In the first part of this chapter, we shall consider several physical systems that can be represented by a first-order transfer function. In the second part, a method for approximating the dynamic response of a nonlinear system by a linear response will be presented. This approximation is called linearization.

EXAMPLES OF FIRST-ORDER SYSTEMS

Liquid Level

Consider the system shown in Fig. 6.1, which consists of a tank of uniform cross-sectional area A to which is attached a flow resistance R such as a valve, a pipe, or a weir. Assume that q_o, the volumetric flow rate (volume/time) through the resistance, is related to the head h by the linear relationship

$$q_o = \frac{h}{R} \tag{6.1}$$

A resistance that has this linear relationship between flow and head is referred to as a *linear resistance*.* A time-varying volumetric flow q of liquid of constant density ρ enters the tank. Determine the transfer function that relates head to flow.

---

*A pipe is a linear resistance if the flow is in the laminar range. A specially contoured weir, called a Sutro weir, produces a linear head-flow relationship. Formulas used to prepare the shape of

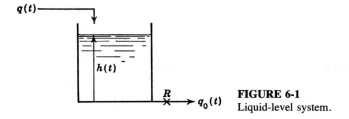

FIGURE 6-1
Liquid-level system.

We can analyze this system by writing a transient mass balance around the tank:

Mass flow in − mass flow out = rate of accumulation of mass in the tank

In terms of the variables used in this analysis, the mass balance becomes

$$\rho q(t) - \rho q_o(t) = \frac{d(\rho Ah)}{dt}$$

$$q(t) - q_o(t) = A\frac{dh}{dt} \qquad (6.2)$$

Combining Eqs. (6.1) and (6.2) to eliminate $q_o(t)$ gives the following linear differential equation:

$$q - \frac{h}{R} = A\frac{dh}{dt} \qquad (6.3)$$

We shall introduce deviation variables into the analysis before proceeding to the transfer function. Initially, the process is operating at steady state, which means that $dh/dt = 0$ and we can write Eq. (6.3) as

$$q_s - \frac{h_s}{R} = 0 \qquad (6.4)$$

where the subscript s has been used to indicate the steady-state value of the variable.

Subtracting Eq. (6.4) from Eq. (6.3) gives

$$(q - q_s) = \frac{1}{R}(h - h_s) + A\frac{d(h - h_s)}{dt} \qquad (6.5)$$

---

such a weir have been reported in the literature; see Soucek, Howe, and Mavis (1936). Turbulent flow through pipes and valves is generally proportional to $\sqrt{h}$. Flow through weirs having simple geometric shapes can be expressed as Kh^n, where K and n are positive constants. For example, the flow through a rectangular-shaped weir is proportional to $h^{3/2}$.

If we define the deviation variables as

$$Q = q - q_s$$
$$H = h - h_s$$

Eq. (6.5) can be written

$$Q = \frac{1}{R}H + A\frac{dH}{dt} \tag{6.6}$$

Taking the transform of Eq. (6.6) gives

$$Q(s) = \frac{1}{R}H(s) + AsH(s) \tag{6.7}$$

Notice that $H(0)$ is zero and therefore the transform of dH/dt is simply $sH(s)$.

Equation (6.7) can be rearranged into the standard form of the first-order lag to give

$$\frac{H(s)}{Q(s)} = \frac{R}{\tau s + 1} \tag{6.8}$$

where $\tau = AR$.

In comparing the transfer function of the tank given by Eq. (6.8) with the transfer function for the thermometer given by Eq. (5.7), we see that Eq. (6.8) contains the factor R. The term R is simply the conversion factor that relates $h(t)$ to $q(t)$ when the system is at steady state. For this reason, a factor K in the transfer function $K/(\tau s + 1)$ is often called the steady-state gain. We can readily show this name to be appropriate by applying the final-value theorem of Chap. 4 to the determination of the steady-state value of H when the flow rate $Q(t)$ changes according to a unit-step change; thus

$$Q(t) = u(t)$$

where $u(t)$ is the symbol for the unit-step change. The transform of $Q(t)$ is

$$Q(s) = \frac{1}{s}$$

Combining this forcing function with Eq. (6.8) gives

$$H(s) = \frac{1}{s}\frac{R}{\tau s + 1}$$

Applying the final-value theorem, proved in Chap. 4, to $H(s)$ gives

$$H(t)\bigg|_{t\to\infty} = \lim_{s\to 0}[sH(s)] = \lim_{s\to 0}\frac{R}{\tau s + 1} = R$$

This shows that the ultimate change in $H(t)$ for a unit change in $Q(t)$ is simply R.

If the transfer function relating the inlet flow $q(t)$ to the outlet flow is desired, note that we have from Eq. (6.1)

$$q_{os} = \frac{h_s}{R} \qquad (6.9)$$

Subtracting Eq. (6.9) from Eq. (6.1) and using the deviation variable $Q_o = q_o - q_{os}$ gives

$$Q_o = \frac{H}{R} \qquad (6.10)$$

Taking the transform of Eq. (6.10) gives

$$Q_o(s) = \frac{H(s)}{R} \qquad (6.11)$$

Combining Eqs. (6.11) and (6.8) to eliminate $H(s)$ gives

$$\frac{Q_o(s)}{Q(s)} = \frac{1}{\tau s + 1} \qquad (6.12)$$

Notice that the steady-state gain for this transfer function is dimensionless, which is to be expected because the input variable $q(t)$ and the output variable $q_o(t)$ have the same units (volume/time).

The possibility of approximating an impulse forcing function in the flow rate to the liquid-level system is quite real. Recall that the unit-impulse function is defined as a pulse of unit area as the duration of the pulse approaches zero, the impulse function can be approximated by suddenly increasing the flow to a large value for a very short time; i.e. we may pour very quickly a volume of liquid into the tank. The nature of the impulse response for a liquid-level system will be described by the following example.

Example 6.1. A tank having a time constant of 1 min and a resistance of $\frac{1}{9}$ ft/cfm is operating at steady state with an inlet flow of 10 ft^3/min. At time $t = 0$, the flow is suddenly increased to 100 ft^3/min for 0.1 min by adding an additional 9 ft^3 of water to the tank uniformly over a period of 0.1 min. (See Fig. 6.2 for this input disturbance.) Plot the response in tank level and compare with the impulse response.

Before proceeding with the details of the computation, we should observe that, as the time interval over which the 9 ft^3 of water is added to the tank is shortened, the input approaches an impulse function having a magnitude of 9.

From the data given in this example, the transfer function of the process is

$$\frac{H(s)}{Q(s)} = \frac{1}{9} \frac{1}{s + 1}$$

The input may be expressed as the difference in step functions, as was done in Example 4.5.

$$Q(t) = 90[u(t) - u(t - 0.1)]$$

The transform of this is

$$Q(s) = \frac{90}{s}(1 - e^{-0.1s})$$

Combining this and the transfer function of the process, we obtain

$$H(s) = 10\left(\frac{1}{s(s+1)} - \frac{e^{-0.1s}}{s(s+1)}\right) \tag{6.13}$$

The first term in Eq. (6.13) can be inverted as shown in Eq. (5.12) to give $10(1 - e^{-t})$. The second term, which includes $e^{-0.1s}$, must be inverted by use of the theorem on translation of functions given in Chap. 4. According to this theorem, the inverse of $e^{-st_0}f(s)$ is $f(t - t_0)$ with $f(t) = 0$ for $t - t_0 < 0$ or $t < t_0$. The inverse of the second term in Eq. (6.13) is

$$L^{-1}\left\{\frac{e^{-0.1s}}{s(s+1)}\right\} = 0 \qquad \text{for } t < 0.1$$
$$= 10[1 - e^{-(t-0.1)}] \qquad \text{for } t > 0.1$$

The complete solution to this problem, which is the inverse of Eq. (6.13), is

$$H(t) = 10(1 - e^{-t}) \qquad\qquad\qquad t < 0.1$$
$$H(t) = 10\{(1 - e^{-t}) - [1 - e^{-(t-0.1)}]\} \qquad t > 0.1 \tag{6.14}$$

Simplifying the expression for $H(t)$ for $t > 0.1$ gives

$$H(t) = 1.052e^{-t} \qquad t > 0.1$$

From Eq. (5.16), the response of the system to an impulse of magnitude 9 is given by

$$H(t)|_{\text{impulse}} = (9)\tfrac{1}{9}e^{-t} = e^{-t}$$

In Fig. 6.2, the pulse response of the liquid-level system and the ideal impulse response are shown for comparison. Notice that the level rises very rapidly during the 0.1 min that additional flow is entering the tank; the level then decays exponentially and follows very closely the ideal impulse response.

The responses to step and sinusoidal forcing functions are the same for the liquid-level system as for the mercury thermometer of Chap. 5. Hence, they need

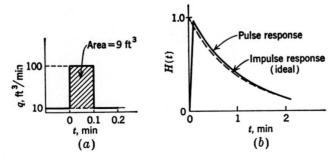

FIGURE 6-2
Approximation of an impulse function in a liquid-level system. (Example 6-1) (a) Pulse input; (b) response of tank level.

not be rederived. This is the advantage of characterizing all first-order systems by the same transfer function.

Liquid-Level Process with Constant-flow Outlet

An example of a transfer function that often arises in control systems may be developed by considering the liquid-level system shown in Fig. 6.3. The resistance shown in Fig. 6.1 is replaced by a constant-flow pump. The same assumptions of constant cross-sectional area and constant density that were used before also apply here. For this system, Eq. (6.2) still applies, but $q(t)$ is now a constant; thus

$$q(t) - q_o = A\frac{dh}{dt} \tag{6.15}$$

At steady state, Eq. (6.15) becomes

$$q_s - q_o = 0 \tag{6.16}$$

Subtracting Eq. (6.16) from Eq. (6.15) and introducing the deviation variables $Q = q - q_s$ and $H = h - h_s$ gives

$$Q = A\frac{dH}{dt} \tag{6.17}$$

Taking the Laplace transform of each side of Eq. (6.17) and solving for H/Q gives

$$\frac{H(s)}{Q(s)} = \frac{1}{As} \tag{6.18}$$

Notice that the transfer function, $1/As$, in Eq. (6.18) is equivalent to integration. One realizes this from the discussion on the transform of an integral presented in Chap. 4. Therefore, the solution of Eq. (6.18) is

$$h(t) = h_s + \frac{1}{A}\int_0^t Q(t)\,dt \tag{6.19}$$

If a step change $Q(t) = u(t)$ were applied to the system shown in Fig. 6.3 the result is

$$h(t) = h_s + t/A \tag{6.20}$$

The step response given by Eq. (6.20) is a ramp function that grows without limit. Such a system that grows without limit for a sustained change in input is

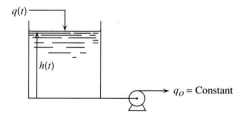

$q(t)$

$h(t)$

$q_o = \text{Constant}$

FIGURE 6-3
Liquid-level system with constant flow outlet.

said to have *nonregulation*. Systems that have a limited change in output for a sustained change in input are said to have *regulation*. An example of a system having regulation is the step response of a first-order system, which is shown in Fig. 5.6.

The transfer function for the liquid-level system with constant outlet flow given by Eq. (6.18) can be considered as a special case of Eq. (6.8) as $R \to \infty$. The next example of a first-order system is a mixing process.

Mixing Process

Consider the mixing process shown in Fig. 6.4 in which a stream of solution containing dissolved salt flows at a constant volumetric flow rate q into a tank of constant holdup volume V. The concentration of the salt in the entering stream, x (mass of salt/volume), varies with time. It is desired to determine the transfer function relating the outlet concentration y to the inlet concentration x.

Assuming the density of the solution to be constant, the flow rate in must equal the flow rate out, since the holdup volume is fixed. We may analyze this system by writing a transient mass balance for the salt; thus

Flow rate of salt in $-$ flow rate of salt out
= rate of accumulation of salt in the tank

Expressing this mass balance in terms of symbols gives

$$qx - qy = \frac{d(Vy)}{dt} \tag{6.21}$$

We shall again introduce deviation variables as we have in the previous examples. At steady state, Eq. (6.21) may be written

$$qx_s - qy_s = 0 \tag{6.22}$$

Subtracting Eq. (6.22) from Eq. (6.21) and introducing the deviation variables

$$X = x - x_s$$
$$Y = y - y_s$$

give

$$qX - qY = V\frac{dY}{dt}$$

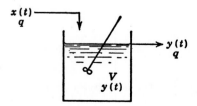

FIGURE 6-4
Mixing process.

Taking the Laplace transform of this expression and rearranging the result give

$$\frac{Y(s)}{X(s)} = \frac{1}{\tau s + 1} \tag{6.23}$$

where $\tau = V/q$.

This mixing process is, therefore, another first-order process, for which the dynamics are now well known. We next bring in an example from DC circuit theory.

RC Circuit

Consider the simple *RC* circuit shown in Fig. 6.5 in which a voltage source $v(t)$ is applied to a series combination of a resistance R and a capacitance C. For $t < 0$, $v(t) = v_s$. Determine the transfer function relating $e_c(t)$ to $v(t)$, where $e_c(t)$ is the voltage across the capacitor.

Applying Kirchhoff's law, which states that in any loop the sum of voltage rises [$v(t)$ in this example] must equal the sum of the voltage drops, gives

$$v(t) = Ri(t) + \frac{1}{C}\int i \, dt \tag{6.24}$$

Recalling that the current is the rate of change of charge with respect to time (coulombs per second), we may replace i by dq/dt in Eq. (6.24) to obtain

$$v(t) = R\frac{dq(t)}{dt} + \frac{1}{C}q(t) \tag{6.25}$$

Since the voltage across the capacitance is given by the relationship

$$e_c = \frac{q}{C} \tag{6.26}$$

the initial charge on the capacitor is simply

$$q_s = Ce_{c_s}$$

Initially, when the circuit is at steady state and the capacitor is fully charged, the voltage across the capacitor is equal to the source voltage v_s; therefore, Eq. (6.25) can be written for these steady-state conditions as

$$v_s = \frac{1}{C}q_s = e_{c_s} \tag{6.27}$$

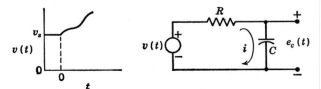

FIGURE 6-5
RC circuit.

Subtracting Eq. (6.27) from Eq. (6.25) and introducing the deviation variables

$$V = v - v_s$$
$$Q = q - q_s \tag{6.28}$$
$$E_c = e_c - e_{c_s} = \frac{Q}{C}$$

we obtain the result

$$V = R\frac{dQ}{dt} + \frac{Q}{C} \tag{6.29}$$

or

$$V = RC\frac{dE_c}{dt} + E_c \tag{6.30}$$

Taking the transform of Eq. (6.30) and rearranging the result give

$$\frac{E_c(s)}{V(s)} = \frac{1}{\tau s + 1} \tag{6.31}$$

where $\tau = RC$. Again we obtain a first-order transfer function.

The three examples that have been presented in this section are intended to show that the dynamic characteristics of many physical systems can be represented by a first-order transfer function. In the remainder of the book, more examples of first-order systems will appear as we discuss a variety of control systems.

Summary

In each example of a first-order system, the time constant has been expressed in terms of system parameters; thus

$$\tau = \frac{mC}{hA} \qquad \text{for thermometer, Eq. (5.5)}$$

$$\tau = AR \qquad \text{for liquid-level process, Eq. (6.8)}$$

$$\tau = \frac{V}{q} \qquad \text{for mixing process, Eq. (6.23)}$$

$$\tau = RC \qquad \text{for RC circuit, Eq. (6.31)}$$

LINEARIZATION

Thus far, all the examples of physical systems, including the liquid-level system of Fig. 6.1, have been linear. Actually, most physical systems of practical importance are nonlinear.

Characterization of a dynamic system by a transfer function can be done only for linear systems (those described by linear differential equations). The convenience of using transfer functions for dynamic analysis, which we have already seen in applications, provides significant motivation for approximating nonlinear systems by linear ones. A very important technique for such approximation is illustrated by the following discussion of the liquid-level system of Fig. 6.1.

We now assume that the resistance follows the square-root relationship

$$q_o = Ch^{1/2} \tag{6.32}$$

where C is a constant.

For a liquid of constant density and a tank of uniform cross-sectional area A, a material balance around the tank gives

$$q(t) - q_o(t) = A\frac{dh}{dt} \tag{6.33}$$

Combining Eqs. (6.32) and (6.33) gives the nonlinear differential equation

$$q - Ch^{1/2} = A\frac{dh}{dt} \tag{6.34}$$

At this point, we cannot proceed as before and take the Laplace transform. This is owing to the presence of the nonlinear term $h^{1/2}$, for which there is no simple transform. This difficulty can be circumvented as follows.

By means of a Taylor-series expansion, the function $q_o(h)$ may be expanded around the steady-state value h_s; thus

$$q_o = q_o(h_s) + q_o'(h_s)(h - h_s) + \frac{q_o''(h_s)(h - h_s)^2}{2!} + \cdots$$

where $q_o'(h_s)$ is the first derivative of q_o evaluated at h_s, $q_o''(h_s)$ the second derivative, etc. If we keep only the linear term, the result is

$$q_o \cong q_o(h_s) + q_o'(h_s)(h - h_s) \tag{6.35}$$

Taking the derivative of q_o with respect to h in Eq. (6.32) and evaluating the derivative at $h = h_s$ gives

$$q_o'(h_s) = (1/2)Ch_s^{-1/2}$$

Introducing this into Eq. (6.35) gives

$$q_o = q_{os} + \frac{1}{R_1}(h - h_s) \tag{6.36}$$

where $q_{os} = q_o(h_s)$

$(R_1)^{-1} = \frac{1}{2}Ch_s^{-1/2}$

Substituting Eq. (6.36) into (6.33) gives

$$q - q_{os} - \frac{h - h_s}{R_1} = A\frac{dh}{dt} \tag{6.37}$$

At steady state the flow entering the tank equals the flow leaving the tank; thus

$$q_o = q_{os} \tag{6.38}$$

Introducing this last equation into Eq. (6.37) gives

$$A\frac{dh}{dt} + \frac{h - h_s}{R_1} = q - q_s \tag{6.39}$$

Introducing deviation variables $Q = q - q_s$ and $H = h - h_s$ into Eq. (6.39) and transforming give

$$\frac{H(s)}{Q(s)} = \frac{R_1}{\tau s + 1} \tag{6.40}$$

where $R_1 = 2h_s^{1/2}/C$
$\qquad \tau = R_1 A$

We see that a transfer function is obtained that is identical in form with that of the linear system, Eq. (6.8). However, in this case, the resistance R_1 depends on the steady-state conditions around which the process operates. Graphically, the resistance R_1 is the reciprocal of the slope of the tangent line passing through the point (q_{os}, h_s) as shown in Fig. 6.6. Furthermore, the linear approximation given by Eq. (6.35) is the equation of the tangent line itself. From the graphical representation, it should be clear that the linear approximation improves as the deviation in h becomes smaller. If one does not have an analytic expression such as $h^{1/2}$ for the nonlinear function, but only a graph of the function, the technique can still be applied by representing the function by the tangent line passing through the point of operation.

Whether or not the linearized result is a valid representation depends on the operation of the system. If the level is being maintained by a controller at or close to a fixed level h_s, then by the very nature of the control imposed on the system, deviations in level should be small (for good control) and the linearized equation is adequate. On the other hand, if the level should change over a wide range, the linear approximation may be very poor and the system may deviate significantly from the prediction of the linear transfer function. In such cases, it may be necessary to use the more difficult methods of nonlinear analysis, some of which are discussed in Chaps. 31 through 33. We shall extend the discussion of linearization to more complex systems in Chap. 21.

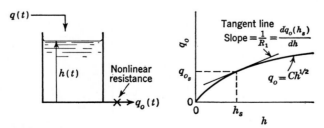

FIGURE 6-6
Liquid-level system with nonlinear resistance.

In summary, we have characterized, in an approximate sense, a nonlinear system by a linear transfer function. In general, this technique may be applied to any nonlinearity that can be expressed in a Taylor series (or, equivalently, has a unique slope at the operating point). Since this includes most nonlinearities arising in process control, we have ample justification for studying linear systems in considerable detail.

PROBLEMS

6.1. Derive the transfer function $H(s)/Q(s)$ for the liquid-level system of Fig. P6.1 when
   (a) The tank level operates about the steady-state value of $h_s = 1$ ft.
   (b) The tank level operates about the steady-state value of $h_s = 3$ ft.
   The pump removes water at a constant rate of 10 cfm (cubic feet per minute); this rate is independent of head. The cross-sectional area of the tank is 1.0 ft^2 and the resistance R is 0.5 ft/cfm.

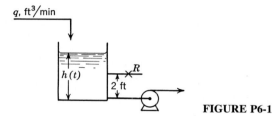

FIGURE P6-1

6.2. A liquid-level system, such as the one shown in Fig. 6.1, has a cross-sectional area of 3.0 ft^2. The valve characteristics are

$$q = 8\sqrt{h}$$

where q = flow rate cfm
         h = level above the valve, ft

Calculate the time constant for this system if the average operating level is
   (a) 3 ft
   (b) 9 ft

6.3. A tank having a cross-sectional area of 2 ft^2 is operating at steady state with an inlet flow rate of 2.0 cfm. The flow-head characteristics are shown in Fig. P6.3.
   (a) Find the transfer function $H(s)/Q(s)$.
   (b) If the flow to the tank increases from 2.0 to 2.2 cfm according to a step change, calculate the level h two minutes after the change occurs.

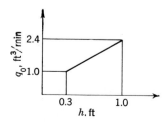

FIGURE P6-3

6.4. Develop a formula for finding the time constant of the liquid-level system shown in Fig. P6.4 when the average operating level is h_o. The resistance R is linear. The tank has three vertical walls and one which slopes at an angle α from the vertical as shown. The distance separating the parallel walls is 1.

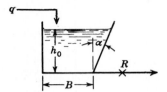

FIGURE P6-4

6.5. Consider the stirred-tank reactor shown in Fig. P6.5. The reaction occurring is

$$A \rightarrow B$$

and it proceeds at a rate

$$r = kC_o$$

where

$\quad r = $ moles A reacting/(volume)(time)
$\quad k = $ reaction velocity constant
$\quad C_o(t) = $ concentration of A in reactor, moles/volume
$\quad V = $ volume of mixture in reactor
Further let $F = $ constant feed rate, volume/time
$\quad C_i(t) = $ concentration of A in feed stream

Assuming constant density and constant V, derive the transfer function relating the concentration in the reactor to the feed-stream concentration. Prepare a block diagram for the reactor. Sketch the response of the reactor to a unit-step change in C_i.

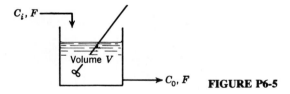

Volume V

C_o, F **FIGURE P6-5**

6.6. A thermocouple junction of area A, mass m, heat capacity C, and emissivity e is located in a furnace that normally is at T_{i_s} °C. At these temperatures convective and conductive heat transfer to the junction are negligible compared with radiative heat transfer. Determine the linearized transfer function between the furnace temperature T_i and the junction temperature T_o. For the case

$$m = 0.1\text{g}$$
$$C = 0.12\text{cal/(g)(°C)}$$
$$e = 0.7$$
$$A = 0.1\text{cm}^2$$
$$T_{i_s} = 1100\text{°C}$$

plot the response of the thermocouple to a 10°C step change in furnace temperature. Compare this with the true response obtained by integration of the differential equation.

6.7. A liquid-level system has the following properties:
Tank dimensions: 10 ft high by 5 ft diameter
Steady-state operating characteristics:

Inflow, gal/hr	Steady-state level, ft
0	0
5,000	.7
10,000	1.1
15,000	2.3
20,000	3.9
25,000	6.3
30,000	8.8

(a) Plot the level response of the tank under the following circumstances: The inlet flow rate is held at 300 gal/min for 1 hr and then suddenly raised to 400 gal/min.

(b) How accurate is the steady-state level calculated from the dynamic response in part (a) when compared with the value given by the table above?

(c) The tank is now connected in series with a second tank that has identical operating characteristics, but which has dimensions 8 ft high by 4 ft diameter. Plot the response of the original tank (which is upstream of the new tank) to the change described in part (a) when the connection is such that the tanks are (1) interacting, (2) noninteracting. (See Chap. 7.)

6.8. A mixing process may be described as follows: a stream with solute concentration C_i (pounds/volume) is fed to a perfectly stirred tank at a constant flow rate of q (volume/time). The perfectly mixed product is withdrawn from the tank, also at the flow rate q at the same concentration as the material in the tank, C_o. The total volume of solution in the tank is constant at V. Density may be considered to be independent of concentration.

A trace of the tank concentration versus time appears as shown in Fig. P6.8.

(a) Plot on this same figure your best guess of the *quantitative* behavior of the inlet concentration versus time. Be sure to label the graph with *quantitative* information regarding times and magnitudes and any other data that will demonstrate your understanding of the situation.

(b) Write an equation for C_i as a function of time.

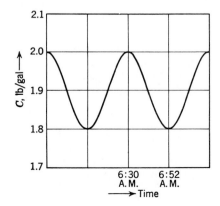

FIGURE P6-8

Data: Tank dimensions: 8 ft high by 5 ft diameter

Tank volume V: 700 gal

Flow rate q: 100 gal/min

Average density: 70 lb/ft^3

6.9. The liquid-level process shown in Fig. P6.9 is operating at steady state when the following disturbance occurs: at time $t = 0$, 1 ft^3 water is added suddenly (unit impulse) to the tank; at $t = 1$, 2 ft^3 of water is added suddenly to the tank. Sketch the response of the level in the tank versus time and determine the level at $t = 0.5$, 1, and 1.5.

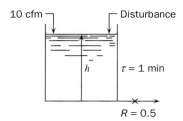

R = 0.5   **FIGURE P6-9**

6.10. A tank having a cross-sectional area of 2 ft^2 and a linear resistance of R $= 1$ ft/cfm is operating at steady state with a flow rate of 1 cfm. At time zero, the flow varies as shown in Fig. P6.10.
   (a) Determine $Q(t)$ and $Q(s)$ by combining simple functions. Note that Q is the deviation in flow rate.
   (b) Obtain an expression for $H(t)$ where H is the deviation in level.
   (c) Determine $H(t)$ at $t = 2$ and $t = \infty$.

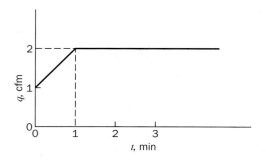

FIGURE P6-10

6.11. Determine $Y(5)$ if $Y(s) = e^{-3s}/[s(7s + 1)]$.
6.12. Derive the transfer function H/Q for the liquid level system shown in Fig. P6.12. The resistances are linear. H and Q are deviation variables. Show clearly how you derived the transfer function. You are expected to give numerical values in the transfer function.

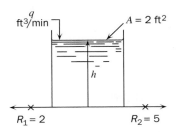

$$\text{ft}^3/\text{min}$$

$A = 2 \text{ ft}^2$

h

$R_1 = 2$ $R_2 = 5$ **FIGURE P6-12**

6.13. The liquid-level system shown in Fig. P6.13 is initially at steady state with the inlet flow rate at 1 cfm. At time zero, one ft^3 of water is suddenly added to the tank; at $t = 1$, one ft^3 is added, etc. In other words, a train of unit impulses is applied to the tank at intervals of one minute. Ultimately the output wave train becomes periodic as shown in the sketch. Determine the maximum and minimum values of this output.

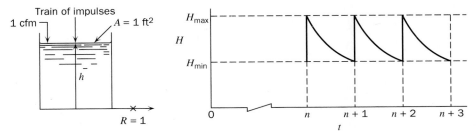

Train of impulses
1 cfm $A = 1 \text{ ft}^2$ H_{max}

H

H_{min}

h

$R = 1$

0 n $n + 1$ $n + 2$ $n + 3$

t

FIGURE P6-13

6.14. The two-tank mixing process shown in Fig. P6.14 contains a recirculation loop that transfers solution from tank 2 to tank 1 at a flow rate of αq_o.

(a) Develop a transfer function that relates the concentration in tank 2, c_2, to the concentration in the feed, x; i.e. $C_2(s)/X(s)$ where C_2 and X are deviation variables. For convenience, assume that the initial concentrations are $x = c_1 = c_2 = 0$.

(b) If a unit-step change in x occurs, determine the time needed for c_2 to reach 60 percent of its ultimate value for the cases where $\alpha = 0, 1$, and ∞.

(c) Sketch the response for $\alpha = \infty$.

Assume that each tank has a constant holdup volume of 1 ft^3. Neglect transportation lag in the line connecting the tanks and the recirculation line. Try to answer parts (b) and (c) by intuition.

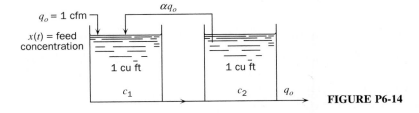

αq_o

$q_o = 1$ cfm

$x(t) = $ feed concentration

1 cu ft   1 cu ft

c_1 c_2 q_o **FIGURE P6-14**

CHAPTER
7

RESPONSE
OF FIRST-ORDER
SYSTEMS
IN SERIES

Introductory Remarks

Very often a physical system can be represented by several first-order processes connected in series. To illustrate this type of system, consider the liquid-level systems shown in Fig. 7.1 in which two tanks are arranged so that the outlet flow from the first tank is the inlet flow to the second tank.

Two possible piping arrangements are shown in Fig. 7.1. In Fig. 7.1a the outlet flow from tank 1 discharges directly into the atmosphere before spilling into tank 2 and the flow through R_1 depends only on h_1. The variation in h_2 in tank 2 does not affect the transient response occurring in tank 1. This type of system is referred to as a *noninteracting* system. In contrast to this, the system shown in Fig. 7.1b is said to be *interacting* because the flow through R_1 now depends on the difference between h_1 and h_2. We shall consider first the noninteracting system of Fig. 7.1a.

Noninteracting System

As in the previous liquid-level example, we shall assume the liquid to be of constant density, the tanks to have uniform cross-sectional area, and the flow resistances to be linear. Our problem is to find a transfer function that relates h_2 to q, that is, $H_2(s)/Q(s)$. The approach will be to obtain a transfer function

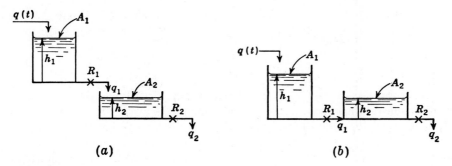

FIGURE 7-1
Two-tank liquid-level system: (*a*) noninteracting; (*b*) interacting.

for each tank, $Q_1(s)/Q(s)$ and $H_2(s)/Q_1(s)$, by writing a transient mass balance around each tank; these transfer functions will then be combined to eliminate the intermediate flow $Q_1(s)$ and produce the desired transfer function.
A balance on tank 1 gives

$$q - q_1 = A_1 \frac{dh_1}{dt} \tag{7.1}$$

A balance on tank 2 gives

$$q_1 - q_2 = A_2 \frac{dh_2}{dt} \tag{7.2}$$

The flow-head relationships for the two linear resistances are given by the expressions

$$q_1 = \frac{h_1}{R_1} \tag{7.3}$$

$$q_2 = \frac{h_2}{R_2} \tag{7.4}$$

Combining Eqs. (7.1) and (7.3) in exactly the same manner as was done in Chap. 6 and introducing deviation variables give the transfer function for tank 1; thus

$$\frac{Q_1(s)}{Q(s)} = \frac{1}{\tau_1 s + 1} \tag{7.5}$$

where $Q_1 = q_1 - q_{1_s}$, $Q = q - q_s$, and $\tau_1 = R_1 A_1$.
In the same manner, we can combine Eqs. (7.2) and (7.4) to obtain the transfer function for tank 2; thus

$$\frac{H_2(s)}{Q_1(s)} = \frac{R_2}{\tau_2 s + 1} \tag{7.6}$$

where $H_2 = h_2 - h_{2_s}$ and $\tau_2 = R_2 A_2$.

Having the transfer function for each tank, we can obtain the overall transfer function $H_2(s)/Q(s)$ by multiplying Eqs. (7.5) and (7.6) to eliminate $Q_1(s)$:

$$\frac{H_2(s)}{Q(s)} = \frac{1}{\tau_1 s + 1}\frac{R_2}{\tau_2 s + 1} \tag{7.7}$$

Notice that the overall transfer function of Eq. (7.7) is the product of two first-order transfer functions, each one of which is the transfer function of a single tank operating independently of the other. In the case of the interacting system of Fig. 7.1b, the overall transfer function *cannot* be found by simply multiplying together the separate transfer functions; this will become apparent when the interacting system is analyzed later.

Example 7.1. Two noninteracting tanks are connected in series as shown in Fig. 7.1a. The time constants are $\tau_2 = 1$ and $\tau_1 = 0.5$; $R_2 = 1$. Sketch the response of the level in tank 2 if a unit-step change is made in the inlet flow rate to tank 1.

The transfer function for this system is found directly from Eq. (7.7); thus

$$\frac{H_2(s)}{Q(s)} = \frac{R_2}{(\tau_1 s + 1)(\tau_2 s + 1)} \tag{7.8}$$

For a unit-step change in Q, we obtain

$$H_2(s) = \frac{1}{s}\frac{R_2}{(\tau_1 s + 1)(\tau_2 s + 1)} \tag{7.9}$$

Inversion by means of partial-fraction expansion gives

$$H_2(t) = R_2\left[1 - \frac{\tau_1 \tau_2}{\tau_1 - \tau_2}\left(\frac{1}{\tau_2}e^{-t/\tau_1} - \frac{1}{\tau_1}e^{-t/\tau_2}\right)\right] \tag{7.10}$$

Substituting in the values of τ_1, τ_2, and R_2 gives

$$H_2(t) = 1 - (2e^{-t} - e^{-2t}) \tag{7.11}$$

A plot of this response is shown in Fig. 7.2. Notice that the response is S-shaped and the slope dH_2/dt at origin is zero. If the change in flow rate were introduced into the second tank, the response would be first-order and is shown for comparison in Fig. 7.2 by the dotted curve.

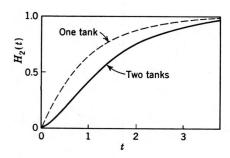

FIGURE 7-2
Transient response of liquid-level system (Example 7.1).

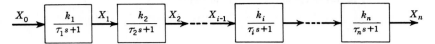

FIGURE 7-3
Noninteracting first-order systems.

Generalization for Several Noninteracting Systems in Series

Having observed that the overall transfer function for two noninteracting first-order systems connected in series is simply the product of the individual transfer functions, we may now generalize by considering n noninteracting first-order systems as represented by the block diagram of Fig. 7.3.

The block diagram is equivalent to the relationships

$$\frac{X_1(s)}{X_0(s)} = \frac{k_1}{\tau_1 s + 1}$$

$$\frac{X_2(s)}{X_1(s)} = \frac{k_2}{\tau_2 s + 1}$$

$$\cdots\cdots\cdots$$

$$\frac{X_n(s)}{X_{n-1}(s)} = \frac{k_n}{\tau_n s + 1}$$

To obtain the overall transfer function, we simply multiply together the individual transfer functions; thus

$$\frac{X_n(s)}{X_0(s)} = \prod_{i=1}^{n} \frac{k_i}{\tau_i s + 1} \tag{7.12}$$

From Example 7.1, notice that the step response of a system consisting of two first-order systems is S-shaped and that the response changes very slowly just after introduction of the step input. This sluggishness or delay is sometimes called *transfer lag* and is always present when two or more first-order systems are connected in series. For a single first-order system, there is no transfer lag; i.e., the response begins immediately after the step change is applied, and the rate of change of the response (slope of response curve) is maximal at $t = 0$.

In order to show how the transfer lag is increased as the number of stages increases, Fig. 7.4 gives the unit-step response curves for several systems containing one or more first-order stages in series.

Interacting System

To illustrate an interacting system, we shall derive the transfer function for the system shown in Fig. 7.1b. The analysis is started by writing mass balances on the tanks as was done for the noninteracting case. The balances on tanks 1 and

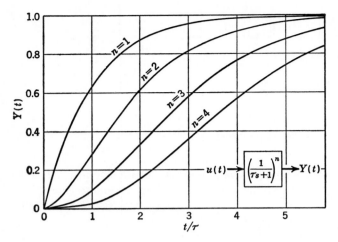

FIGURE 7-4
Step response of noninteracting first-order systems.

2 are the same as before and are given by Eqs. (7.1) and (7.2). However, the flow-head relationship for tank 1 is now

$$q_1 = \frac{1}{R_1}(h_1 - h_2) \tag{7.13}$$

The flow-head relationship for R_2 is the same as before and is expressed by Eq. (7.4). A simple way to combine Eqs. (7.1), (7.2), (7.4), and (7.13) is to first express them in terms of deviation variables, transform the resulting equations, and then combine the transformed equations to eliminate the unwanted variables.

At steady state, Eqs. (7.1) and (7.2) can be written

$$q_s - q_{1_s} = 0 \tag{7.14}$$
$$q_{1_s} - q_{2_s} = 0 \tag{7.15}$$

Subtracting Eq. (7.14) from Eq. (7.1) and Eq. (7.15) from Eq. (7.2) and introducing deviation variables give

$$Q - Q_1 = A_1 \frac{dH_1}{dt} \tag{7.16}$$

$$Q_1 - Q_2 = A_2 \frac{dH_2}{dt} \tag{7.17}$$

Expressing Eqs. (7.13) and (7.4) in terms of deviation variables gives

$$Q_1 = \frac{H_1 - H_2}{R_1} \tag{7.18}$$

$$Q_2 = \frac{H_2}{R_2} \tag{7.19}$$

Transforming Eqs. (7.16) through (7.19) gives

$$Q(s) - Q_1(s) = A_1 s H_1(s) \tag{7.20}$$

$$Q_1(s) - Q_2(s) = A_2 s H_2(s) \tag{7.21}$$

$$R_1 Q_1(s) = H_1(s) - H_2(s) \tag{7.22}$$

$$R_2 Q_2(s) = H_2(s) \tag{7.23}$$

The analysis has produced four algebraic equations containing five unknowns: $(Q, Q_1, Q_2, H_1,$ and $H_2)$. These equations may be combined to eliminate $Q_1, Q_2,$ and H_1 and arrive at the desired transfer function:

$$\frac{H_2(s)}{Q(s)} = \frac{R_2}{\tau_1 \tau_2 s^2 + (\tau_1 + \tau_2 + A_1 R_2)s + 1} \tag{7.24}$$

Notice that the product of the transfer functions for the tanks operating separately, Eqs. (7.5) and (7.6), does not produce the correct result for the interacting system. The difference between the transfer function for the noninteracting system, Eq. (7.7), and the interacting system, Eq. (7.24), is the presence of the term $A_1 R_2$ in the coefficient of s.

The term *interacting* is often referred to as *loading*. The second tank of Fig. 7.1b is said to *load* the first tank.

To understand the effect of interaction on the transient response of a system, consider a two-tank system for which the time constants are equal ($\tau_1 = \tau_2 = \tau$). If the tanks are noninteracting, the transfer function relating inlet flow to outlet flow is

$$\frac{Q_2(s)}{Q(s)} = \left(\frac{1}{\tau s + 1}\right)^2 \tag{7.25}$$

The unit-step response for this transfer function can be obtained by the usual procedure to give

$$Q_2(t) = 1 - e^{-t/\tau} - \frac{t}{\tau} e^{-t/\tau} \tag{7.26}$$

If the tanks are interacting, the overall transfer function, according to Eq. (7.24), is (assuming further that $A_1 = A_2$)

$$\frac{Q_2(s)}{Q(s)} = \frac{1}{\tau^2 s^2 + 3\tau s + 1} \tag{7.27}$$

By application of the quadratic formula, the denominator of this transfer function can be written as

$$\frac{Q_2(s)}{Q(s)} = \frac{1}{(0.38\tau s + 1)(2.62\tau s + 1)} \tag{7.28}$$

For this example, we see that the effect of interaction has been to change the effective time constants of the interacting system. One time constant has become considerably larger and the other smaller than the time constant τ of either tank in the noninteracting system. The response of $Q_2(t)$ to a unit-step change in $Q(t)$ for the interacting case [Eq. (7.28)] is

$$Q_2(t) = 1 + 0.17e^{-t/0.38\tau} - 1.17e^{-t/2.62\tau} \qquad (7.29)$$

In Fig. 7.5, the unit-step responses [Eqs. (7.26) and (7.29)] for the two cases are plotted to show the effect of interaction. From this figure, it can be seen that interaction slows up the response. This result can be understood on physical grounds in the following way: if the same size step change is introduced into the two systems of Fig. 7.1, the flow from tank 1 (q_1) for the noninteracting case will not be reduced by the increase in level in tank 2. However, for the interacting case, the flow q_1 will be reduced by the build-up of level in tank 2. At any time t_1 following the introduction of the step input, q_1 for the interacting case will be less than for the noninteracting case with the result that h_2 (or q_2) will increase at a slower rate.

In general, the effect of interaction on a system containing two first-order lags is to change the ratio of effective time constants in the interacting system. In terms of the transient response, this means that the interacting system is more sluggish than the noninteracting system.

This chapter concludes our specific discussion of first-order systems. We shall make continued use of the material developed here in the succeeding chapters.

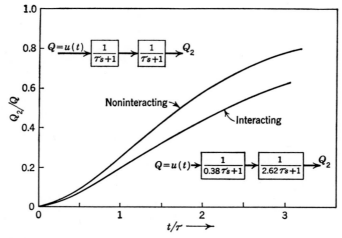

FIGURE 7-5
Effect of interaction on step response of two-tank system.

PROBLEMS

7.1. Determine the transfer function $H(s)/Q(s)$ for the liquid-level system shown in Fig. P7.1. Resistances R_1 and R_2 are linear. The flow rate from tank 3 is maintained constant at b by means of a pump; i.e., the flow rate from tank 3 is independent of head h. The tanks are noninteracting.

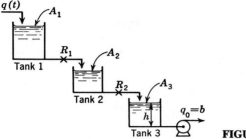

FIGURE P7-1

7.2. The mercury thermometer in Chap. 5 was considered to have all its resistance in the convective film surrounding the bulb and all its capacitance in the mercury. A more detailed analysis would consider both the convective resistance surrounding the bulb and that between the bulb and mercury. In addition, the capacitance of the glass bulb would be included. Let

A_i = inside area of bulb, for heat transfer to mercury

A_o = outside area of bulb, for heat transfer from surrounding fluid

m = mass of mercury in bulb

m_b = mass of glass bulb

C = heat capacity of mercury

C_b = heat capacity of glass bulb

h_i = convective coefficient between bulb and mercury

h_o = convective coefficient between bulb and surrounding fluid

T = temperature of mercury

T_b = temperature of glass bulb

T_f = temperature of surrounding fluid

Determine the transfer function between T_f and T. What is the effect of the bulb resistance and capacitance on the thermometer response? Note that the inclusion of the bulb results in a pair of interacting systems, which give an overall transfer function somewhat different from that of Eq. (7.24).

7.3. There are N storage tanks of volume V arranged so that when water is fed into the first tank, an equal volume of liquid overflows from the first tank into the second tank, and so on. Each tank initially contains component A at some concentration C_o and is equipped with a perfect stirrer. At time zero, a stream of zero concentration is fed into the first tank at a volumetric rate q. Find the resulting concentration in each tank as a function of time.

7.4. (*a*) Find the transfer functions H_2/Q and H_3/Q for the three-tank system shown in Fig. P7.4 where H_2, H_3 and Q are deviation variables. Tank 1 and Tank 2 are interacting.

(*b*) For a unit-step change in q (i.e., $Q = 1/s$), determine $H_3(0), H_3(\infty)$, and sketch $H_3(t)$ versus t.

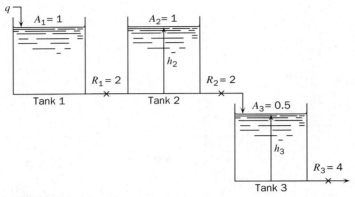

FIGURE P7-4

7.5. Three identical tanks are operated in series in a noninteracting fashion as shown in Fig. P7.5. For each tank, $R = 1, \tau = 1$. If the deviation in flow rate to the first tank is an impulse function of magnitude 2, determine

(*a*) An expression for $H(s)$ where H is the deviation in level in the third tank.

(*b*) Sketch the response $H(t)$.

(*c*) Obtain an expression for $H(t)$.

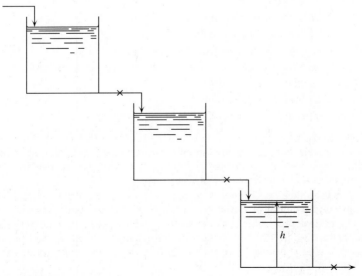

FIGURE P7-5

7.6. In the two-tank mixing process shown in Fig. P7.6, x varies from 0 lb salt/ft^3 to 1 lb salt/ft^3 according to a step function. At what time does the salt concentration in tank 2 reach 0.6 lb salt/ft^3? The holdup volume of each tank is 6 ft^3.

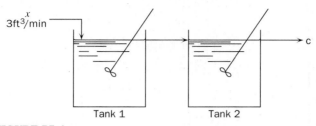

FIGURE P7-6

7.7. Starting from first principles, derive the transfer functions $H_1(s)/Q(s)$ and $H_2(s)/Q(s)$ for the liquid level system shown in Fig. P7.7. The resistances are linear and $R_1 = R_2 = 1$. Note that two streams are flowing from tank 1, one of which flows into tank 2. You are expected to give numerical values of the parameters in the transfer functions and to show clearly how you derived the transfer functions.

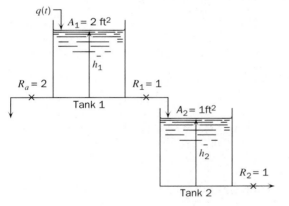

FIGURE P7-7

CHAPTER
8

HIGHER-ORDER SYSTEMS: SECOND-ORDER AND TRANSPORTATION LAG

SECOND-ORDER SYSTEM

Transfer Function

This section introduces a basic system called a *second-order system* or a *quadratic lag*. A second-order transfer function will be developed by considering a classical example from mechanics. This is the damped vibrator, which is shown in Fig. 8.1.

A block of mass W resting on a horizontal, frictionless table is attached to a linear spring. A viscous damper (dashpot) is also attached to the block. Assume that the system is free to oscillate horizontally under the influence of a forcing function $F(t)$. The origin of the coordinate system is taken as the right edge of the block when the spring is in the relaxed or unstretched condition. At time zero, the block is assumed to be at rest at this origin.* Positive directions for force and displacement are indicated by the arrows in Fig. 8.1.

Consider the block at some instant when it is to the right of $Y = 0$ and when it is moving toward the right (positive direction). Under these conditions,

---

*In effect, this assumption makes the displacement variable $Y(t)$ a deviation variable. Also, the assumption that the block is initially at rest permits derivation of the second-order transfer function in its standard form. An initial velocity has the same effect as a forcing function. Hence, this assumption is in no way restrictive.

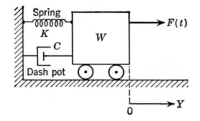

FIGURE 8-1
Damped vibrator.

the position Y and the velocity dY/dt are both positive. At this particular instant, the following forces are acting on the block:

1. The force exerted by the spring (toward the left) of $-KY$ where K is a positive constant, called Hooke's constant.
2. The viscous friction force (acting to the left) of $-C\ dY/dt$, where C is a positive constant called the damping coefficient.
3. The external force $F(t)$ (acting toward the right).

Newton's law of motion, which states that the sum of all forces acting on the mass is equal to the rate of change of momentum (mass × acceleration), takes the form

$$\frac{W}{g_c}\frac{d^2Y}{dt^2} = -KY - C\frac{dY}{dt} + F(t) \tag{8.1}$$

Rearrangement gives

$$\frac{W}{g_c}\frac{d^2Y}{dt^2} + C\frac{dY}{dt} + KY = F(t) \tag{8.2}$$

where W = mass of block, lb_m
   g_c = $32.2(\text{lb}_m)(\text{ft})/(\text{lb}_f)(\text{sec}^2)$
   C = viscous damping coefficient, $\text{lb}_f/(\text{ft/sec})$
   K = Hooke's constant, lb_f/ft
   $F(t)$ = driving force, a function of time, lb_f

Dividing Eq. (8.2) by K gives

$$\frac{W}{g_cK}\frac{d^2Y}{dt^2} + \frac{C}{K}\frac{dY}{dt} + Y = \frac{F(t)}{K} \tag{8.3}$$

For convenience, this is written as

$$\tau^2\frac{d^2Y}{dt^2} + 2\zeta\tau\frac{dY}{dt} + Y = X(t) \tag{8.4}$$

where

$$\tau^2 = \frac{W}{g_cK} \tag{8.5}$$

$$2\zeta\tau = \frac{C}{K} \tag{8.6}$$

$$X(t) = \frac{F(t)}{K} \tag{8.7}$$

Solving for τ and ζ from Eqs. (8.5) and (8.6) gives

$$\tau = \sqrt{\frac{W}{g_c K}} \quad \text{sec} \tag{8.8}$$

$$\zeta = \sqrt{\frac{g_c C^2}{4WK}} \quad \text{dimensionless} \tag{8.9}$$

By definition, both τ and ζ must be positive. The reason for introducing τ and ζ in the particular form shown in Eq. (8.4) will become clear when we discuss the solution of Eq. (8.4) for particular forcing functions $X(t)$.

Equation (8.4) is written in a standard form that is widely used in control theory. Notice that, because of superposition, $X(t)$ can be considered as a forcing function because it is proportional to the force $F(t)$.

If the block is motionless $(dY/dt = 0)$ and located at its rest position $(Y = 0)$ before the forcing function is applied, the Laplace transform of Eq. (8.4) becomes

$$\tau^2 s^2 Y(s) + 2\zeta\tau s Y(s) + Y(s) = X(s) \tag{8.10}$$

From this, the transfer function follows:

$$\frac{Y(s)}{X(s)} = \frac{1}{\tau^2 s^2 + 2\zeta\tau s + 1} \tag{8.11}$$

The transfer function given by Eq. (8.11) is written in standard form, and we shall show later that other physical systems can be represented by a transfer function having the denominator $\tau^2 s^2 + 2\zeta\tau s + 1$. All such systems are defined as second-order. Note that it requires two parameters, τ and ζ, to characterize the dynamics of a second-order system in contrast to only one parameter for a first-order system. For the time being, the variables and parameters of Eq. (8.11) can be interpreted in terms of the damped vibrator. We shall now discuss the response of a second-order system to some of the common forcing functions, namely, step, impulse, and sinusoidal.

Step Response

If the forcing function is a unit-step function, we have

$$X(s) = \frac{1}{s} \tag{8.12}$$

In terms of the damped vibrator shown in Fig. 8.1 this is equivalent to suddenly applying a force of magnitude K directed toward the right at time $t = 0$. This follows from the fact that X is defined by the relationship $X(t) = F(t)/K$. Superposition will enable us to determine easily the response to a step function of any other magnitude.

Combining Eq. (8.12) with the transfer function of Eq. (8.11) gives

$$Y(s) = \frac{1}{s} \frac{1}{\tau^2 s^2 + 2\zeta\tau s + 1} \tag{8.13}$$

The quadratic term in this equation may be factored into two linear terms that contain the roots

$$s_1 = -\frac{\zeta}{\tau} + \frac{\sqrt{\zeta^2 - 1}}{\tau} \tag{8.14}$$

$$s_2 = -\frac{\zeta}{\tau} - \frac{\sqrt{\zeta^2 - 1}}{\tau} \tag{8.15}$$

Equation (8.13) can now be written

$$Y(s) = \frac{1/\tau^2}{(s)(s - s_1)(s - s_2)} \tag{8.16}$$

The response of the system $Y(t)$ can be found by inverting Eq. (8.16). The roots s_1 and s_2 will be real or complex depending on the parameter ζ. The nature of the roots will, in turn, affect the form of $Y(t)$. The problem may be divided into the three cases shown in Table 8.1. Each case will now be discussed.

CASE I   STEP RESPONSE FOR $\zeta < 1$. For this case, the inversion of Eq. (8.16) yields the result

$$Y(t) = 1 - \frac{1}{\sqrt{1 - \zeta^2}} e^{-\zeta t/\tau} \sin\left(\sqrt{1 - \zeta^2}\frac{t}{\tau} + \tan^{-1}\frac{\sqrt{1 - \zeta^2}}{\zeta}\right) \tag{8.17}$$

To derive Eq. (8.17), use is made of the techniques of Chap. 3. Since $\zeta < 1$, Eqs. (8.14) to (8.16) indicate a pair of complex conjugate roots in the left-half plane and a root at the origin. In terms of the symbols of Fig. 3.1, the complex roots correspond to s_2 and s_2^* and the root at the origin to s_6. By referring to Table 3.1, we see that $Y(t)$ has the form

$$Y(t) = C_1 + e^{-\zeta t/\tau}\left(C_2 \cos \sqrt{1 - \zeta^2}\frac{t}{\tau} + C_3 \sin \sqrt{1 - \zeta^2}\frac{t}{\tau}\right) \tag{8.18}$$

The constants C_1, C_2, and C_3 are found by partial fractions. The resulting equation is then put in the form of Eq. (8.17) by applying the trigonometric identity used

TABLE 8.1

Case	ζ	Nature of roots	Description of response
I	< 1	Complex	Underdamped or oscillatory
II	$= 1$	Real and equal	Critically damped
III	> 1	Real	Overdamped or nonoscillatory

in Chap. 5, Eq. (5.23). The details are left as an exercise for the reader. It is evident from Eq. (8.17) that $Y(t) \rightarrow 1$ as $t \rightarrow \infty$.

The nature of the response can be understood most clearly by plotting Eq. (8.17) as shown in Fig. 8.2, where $Y(t)$ is plotted against the dimensionless variable t/τ for several values of ζ, including those above unity, which will be considered in the next section. Note that, for $\zeta < 1$, all the response curves are oscillatory in nature and become less oscillatory as ζ is increased. The slope at the origin in Fig. 8.2 is zero for all values of ζ. The response of a second-order system for $\zeta < 1$ is said to be *underdamped*.

CASE II   STEP RESPONSE FOR $\zeta = 1$. For this case, the response is given by the expression

$$Y(t) = 1 - \left(1 + \frac{t}{\tau}\right)e^{-t/\tau} \tag{8.19}$$

This is derived as follows: Equations (8.14) and (8.15) show that the roots s_1 and s_2 are real and equal. Referring to Fig. 3.1 and Table 3.1, it is seen that

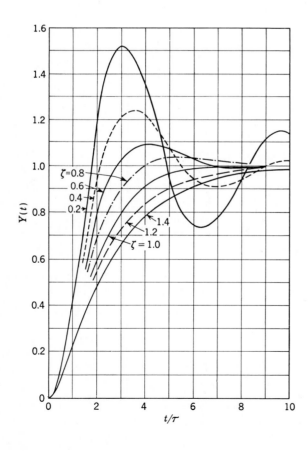

FIGURE 8-2
Response of a second-order system to a unit-step forcing function.

Eq. (8.19) is in the correct form. The constants are obtained, as usual, by partial fractions.

The response, which is plotted in Fig. 8.2, is nonoscillatory. This condition, $\zeta = 1$, is called *critical damping* and allows most rapid approach of the response to $Y = 1$ without oscillation.

CASE III   STEP RESPONSE FOR $\zeta > 1$. For this case, the inversion of Eq. (8.16) gives the result

$$Y(t) = 1 - e^{-\zeta t/\tau}\left(\cosh \sqrt{\zeta^2 - 1}\frac{t}{\tau} + \frac{\zeta}{\sqrt{\zeta^2 - 1}}\sinh \sqrt{\zeta^2 - 1}\frac{t}{\tau}\right) \qquad (8.20)$$

where the hyperbolic functions are defined as

$$\sinh a = \frac{e^a - e^{-a}}{2}$$

$$\cosh a = \frac{e^a + e^{-a}}{2}$$

The procedure for obtaining Eq. (8.20) is parallel to that used in the previous cases.

The response has been plotted in Fig. 8.2 for several values of ζ. Notice that the response is nonoscillatory and becomes more "sluggish" as ζ increases. This is known as an *overdamped* response. As in previous cases, all curves eventually approach the line $Y = 1$.

Actually, the response for $\zeta > 1$ is not new. We met it previously in the discussion of the step response of a system containing two first-order systems in series, for which the transfer function is

$$\frac{Y(s)}{X(s)} = \frac{1}{(\tau_1 s + 1)(\tau_2 s + 1)} \qquad (8.21)$$

This is true for $\zeta > 1$ because the roots s_1 and s_2 are real, and the denominator of Eq. (8.11) may be factored into two real linear factors. Therefore, Eq. (8.11) is equivalent to Eq. (8.21) in this case. By comparing the linear factors of the denominator of Eq. (8.11) with those of Eq. (8.21), it follows that

$$\tau_1 = (\zeta + \sqrt{\zeta^2 - 1})\tau \qquad (8.22)$$

$$\tau_2 = (\zeta - \sqrt{\zeta^2 - 1})\tau \qquad (8.23)$$

Note that, if $\tau_1 = \tau_2$, then $\tau = \tau_1 = \tau_2$ and $\zeta = 1$. The reader should verify these results.

Terms Used to Describe an Underdamped System

Of these three cases, the underdamped response occurs most frequently in control systems. Hence a number of terms are used to describe the underdamped response

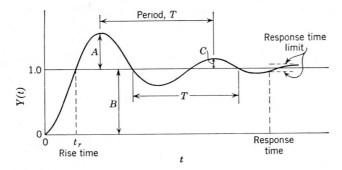

FIGURE 8-3
Terms used to describe an underdamped second-order response.

quantitatively. Equations for some of these terms are listed below for future reference. In general, the terms depend on ζ and/or τ. All these equations can be derived from the time response as given by Eq. (8.17); however, the mathematical derivations are left to the reader as exercises.

1. **Overshoot.** Overshoot is a measure of how much the response exceeds the ultimate value following a step change and is expressed as the ratio A/B in Fig. 8.3.

　　The overshoot for a unit step is related to ζ by the expression

$$\text{Overshoot} = \exp(-\pi\zeta/\sqrt{1-\zeta^2}) \qquad (8.24)$$

This relation is plotted in Fig. 8.4. The overshoot increases for decreasing ζ.

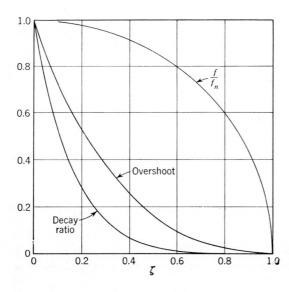

FIGURE 8-4
Characteristics of a step response of an underdamped second-order system.

2. **Decay ratio.** The decay ratio is defined as the ratio of the sizes of successive peaks and is given by C/A in Fig. 8.3. The decay ratio is related to ζ by the expression

$$\text{Decay ratio} = \exp(-2\pi\zeta/\sqrt{1-\zeta^2}) = (\text{overshoot})^2 \qquad (8.25)$$

which is plotted in Fig. 8.4. Notice that larger ζ means greater damping, hence greater decay.

3. **Rise time.** This is the time required for the response to first reach its ultimate value and is labeled t_r in Fig. 8.3. The reader can verify from Fig. 8.2 that t_r increases with increasing ζ.

4. **Response time.** This is the time required for the response to come within ± 5 percent of its ultimate value and remain there. The response time is indicated in Fig. 8.3. The limits ± 5 percent are arbitrary, and other limits have been used in other texts for defining a response time.

5. **Period of oscillation.** From Eq. (8.17), the radian frequency (radians/time) is the coefficient of t in the sine term; thus,

$$\omega, \text{ radian frequency} = \frac{\sqrt{1-\zeta^2}}{\tau} \qquad (8.26)$$

Since the radian frequency ω is related to the cyclical frequency f by $\omega = 2\pi f$, it follows that

$$f = \frac{1}{T} = \frac{1}{2\pi}\frac{\sqrt{1-\zeta^2}}{\tau} \qquad (8.27)$$

where T is the period of oscillation (time/cycle). In terms of Fig. 8.3, T is the time elapsed between peaks. It is also the time elapsed between alternate crossings of the line $Y = 1$.

6. **Natural period of oscillation.** If the damping is eliminated [$C = 0$ in Eq. (8.1), or $\zeta = 0$], the system oscillates continuously without attenuation in amplitude. Under these "natural" or undamped conditions, the radian frequency is $1/\tau$, as shown by Eq. (8.26) when $\zeta = 0$. This frequency is referred to as the natural frequency ω_n:

$$\omega_n = \frac{1}{\tau} \qquad (8.28)$$

The corresponding natural cyclical frequency f_n and period T_n are related by the expression

$$f_n = \frac{1}{T_n} = \frac{1}{2\pi\tau} \qquad (8.29)$$

Thus, τ has the significance of the undamped period.

From Eqs. (8.27) and (8.29), the natural frequency is related to the actual frequency by the expression

$$\frac{f}{f_n} = \sqrt{1-\zeta^2}$$

which is plotted in Fig. 8.4. Notice that, for $\zeta < 0.5$, the natural frequency is nearly the same as the actual frequency.

In summary, it is evident that ζ is a measure of the degree of damping, or the oscillatory character, and τ is a measure of the period, or speed, of the response of a second-order system.

Impulse Response

If a unit impulse $\delta(t)$ is applied to the second-order system, then from Eqs. (8.11) and (4.1) the transform of the response is

$$Y(s) = \frac{1}{\tau^2 s^2 + 2\zeta\tau s + 1} \tag{8.30}$$

As in the case of the step input, the nature of the response to a unit impulse will depend on whether the roots of the denominator of Eq. (8.30) are real or complex. The problem is again divided into the three cases shown in Table 8.1, and each is discussed below.

CASE I   IMPULSE RESPONSE FOR $\zeta < 1$. The inversion of Eq. (8.30) for $\zeta < 1$ yields the result

$$Y(t) = \frac{1}{\tau}\frac{1}{\sqrt{1-\zeta^2}}e^{-\zeta t/\tau}\sin\sqrt{1-\zeta^2}\frac{t}{\tau} \tag{8.31}$$

which is plotted in Fig. 8.5. The slope at the origin in Fig. 8.5 is 1.0 for all values of ζ.

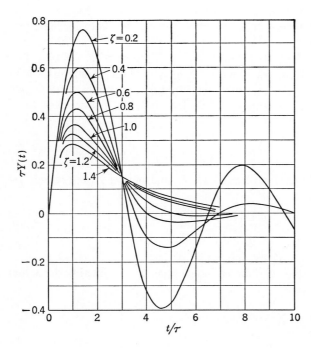

FIGURE 8-5
Response of a second-order system to a unit-impulse forcing function.

A simple way to obtain Eq. (8.31) from the step response of Eq. (8.17) is to take the derivative of Eq. (8.17). Comparison of Eqs. (8.13) and (8.30) shows that

$$Y(s)\big|_{\text{impulse}} = sY(s)\big|_{\text{step}} \qquad (8.32)$$

The presence of s on the right side of Eq. (8.32) implies differentiation with respect to t in the time response. In other words, the inverse transform of Eq. (8.32) is

$$Y(t)\big|_{\text{impulse}} = \frac{d}{dt}\left(Y(t)\big|_{\text{step}}\right) \qquad (8.33)$$

Application of Eq. (8.33) to Eq. (8.17) yields Eq. (8.31). This principle also yields the results for the next two cases.

CASE II   IMPULSE RESPONSE FOR $\zeta = 1$. For the critically damped case, the response is given by

$$Y(t) = \frac{1}{\tau^2}te^{-t/\tau} \qquad (8.34)$$

which is plotted in Fig. 8.5.

CASE III   IMPULSE RESPONSE FOR $\zeta > 1$. For the overdamped case, the response is given by

$$Y(t) = \frac{1}{\tau}\frac{1}{\sqrt{\zeta^2 - 1}}e^{-\zeta t/\tau}\sinh\sqrt{\zeta^2 - 1}\frac{t}{\tau} \qquad (8.35)$$

which is plotted in Fig. 8.5.

To summarize, the impulse-response curves of Fig. 8.5 show the same general behavior as the step-response curves of Fig. 8.2. However, the impulse response always returns to zero. Terms such as decay ratio, period of oscillation, etc., may also be used to describe the impulse response. Many control systems exhibit transient responses such as those of Fig. 8.5. This is illustrated by Fig. 1.7 for the stirred-tank heat exchanger.

Sinusoidal Response

If the forcing function applied to the second-order system is sinusoidal,

$$X(t) = A\sin \omega t$$

then it follows from Eqs. (8.11) and (5.20) that

$$Y(s) = \frac{A\omega}{(s^2 + \omega^2)(\tau^2 s^2 + 2\zeta\tau s + 1)} \qquad (8.36)$$

The inversion of Eq. (8.36) may be accomplished by first factoring the two quadratic terms to give

$$Y(s) = \frac{A\omega/\tau^2}{(s - j\omega)(s + j\omega)(s - s_1)(s - s_2)} \qquad (8.37)$$

Here s_1 and s_2 are the roots of the denominator of the transfer function and are given by Eqs. (8.14) and (8.15). For the case of an underdamped system ($\zeta < 1$), the roots of the denominator of Eq. (8.37) are a pair of pure imaginary roots ($+j\omega, -j\omega$) contributed by the forcing function and a pair of complex roots ($-\zeta/\tau + j\sqrt{1 - \zeta^2}/\tau, -\zeta/\tau - j\sqrt{1 - \zeta^2}/\tau$). We may write the form of the response $Y(t)$ by referring to Fig. 3.1 and Table 3.1; thus

$$Y(t) = C_1 \cos \omega t + C_2 \sin \omega t + e^{-\zeta t/\tau}\left(C_3 \cos \sqrt{1 - \zeta^2}\frac{t}{\tau} + C_4 \sin \sqrt{1 - \zeta^2}\frac{t}{\tau}\right)$$

(8.38)

The constants are evaluated by partial fractions. Notice in Eq. (8.38) that, as $t \to \infty$, only the first two terms do not become zero. These remaining terms are the ultimate periodic solution; thus

$$Y(t)|_{t \to \infty} = C_1 \cos \omega t + C_2 \sin \omega t \qquad (8.39)$$

The reader should verify that Eq. (8.39) is also true for $\zeta \geq 1$. From this little effort, we see already that the response of the second-order system to a sinusoidal driving function is ultimately sinusoidal and has the same frequency as the driving function. If the constants C_1 and C_2 are evaluated, we get from Eqs. (5.23) and (8.39)

$$Y(t) = \frac{A}{\sqrt{[1 - (\omega\tau)^2]^2 + (2\zeta\omega\tau)^2}} \sin(\omega t + \phi) \qquad (8.40)$$

where

$$\phi = -\tan^{-1}\frac{2\zeta\omega\tau}{1 - (\omega\tau)^2}$$

By comparing Eq. (8.40) with the forcing function

$$X(t) = A\sin \omega t$$

it is seen that:

1. The ratio of the output amplitude to the input amplitude is

$$\frac{1}{\sqrt{[1 - (\omega\tau)^2]^2 + (2\zeta\omega\tau)^2}}$$

It will be shown in Chap. 16 that this may be greater or less than 1, depending on ζ and $\omega\tau$. This is in direct contrast to the sinusoidal response of the first-order system, where the ratio of the output amplitude to the input amplitude is always *less than* 1.

2. The output lags the input by phase angle $|\phi|$. It can be seen from Eq. (8.40), and will be shown in Chap. 16, that $|\phi|$ approaches 180° asymptotically as ω increases. The phase lag of the first-order system, on the other hand, can never exceed 90°. Discussion of other characteristics of the sinusoidal response will be deferred until Chap. 16.

We now have at our disposal considerable information about the dynamic behavior of the second-order system. It happens that many control systems that are not truly second-order exhibit step responses very similar to those of Fig. 8.2. Such systems are often characterized by second-order equations for approximate mathematical analysis. Hence, the second-order system is quite important in control theory, and frequent use will be made of the material in this chapter.

TRANSPORTATION LAG

A phenomenon that is often present in flow systems is the *transportation lag*. Synonyms for this term are *dead time* and *distance velocity lag*. As an example, consider the system shown in Fig. 8.6, in which a liquid flows through an insulated tube of uniform cross-sectional area A and length L at a constant volumetric flow rate q. The density ρ and the heat capacity C are constant. The tube wall has negligible heat capacity, and the velocity profile is flat (plug flow).

The temperature x of the entering fluid varies with time, and it is desired to find the response of the outlet temperature $y(t)$ in terms of a transfer function.

As usual, it is assumed that the system is initially at steady state; for this system, it is obvious that the inlet temperature equals the outlet temperature; i.e.,

$$x_s = y_s \tag{8.41}$$

If a step change were made in $x(t)$ at $t = 0$, the change would not be detected at the end of the tube until τ sec later, where τ is the time required for the entering fluid to pass through the tube. This simple step response is shown in Fig. 8.7a.

If the variation in $x(t)$ were some arbitrary function, as shown in Fig. 8.7b, the response $y(t)$ at the end of the pipe would be identical with $x(t)$ but again delayed by τ units of time. The transportation lag parameter τ is simply the time needed for a particle of fluid to flow from the entrance of the pipe to the exit, and it can be calculated from the expression

$$\tau = \frac{\text{volume of pipe}}{\text{volumetric flow rate}}$$

or

$$\tau = \frac{AL}{q} \tag{8.42}$$

It can be seen from Fig. 8.7 that the relationship between $y(t)$ and $x(t)$ is

$$y(t) = x(t - \tau) \tag{8.43}$$

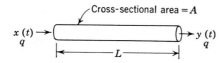

FIGURE 8-6
System with transportation lag.

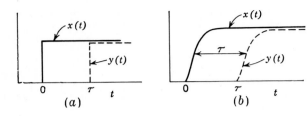

FIGURE 8-7
Response of transportation lag to various inputs.

Subtracting Eq. (8.41) from (8.43) and introducing the deviation variables $X = x - x_s$ and $Y = y - y_s$ give

$$Y(t) = X(t - \tau) \tag{8.44}$$

If the Laplace transform of $X(t)$ is $X(s)$, the Laplace transform of $X(t - \tau)$ is $e^{-s\tau}X(s)$. This result follows from the theorem on translation of a function, which was discussed in Chap. 4. Equation (8.44) becomes

$$Y(s) = e^{-s\tau}X(s)$$

or

$$\frac{Y(s)}{X(s)} = e^{-s\tau} \tag{8.45}$$

Therefore, the transfer function of a transportation lag is $e^{-s\tau}$.

The transportation lag is quite common in the chemical process industries where a fluid is transported through a pipe. We shall see in a later chapter that the presence of a transportation lag in a control system can make it much more difficult to control. In general, such lags should be avoided if possible by placing equipment close together. They can seldom be entirely eliminated.

APPROXIMATION OF TRANSPORT LAG. The transport lag is quite different from the other transfer functions (first-order, second-order, etc.) that we have discussed in that it is not a rational function (i.e., a ratio of polynomials.) As shown in Chap. 14, a system containing a transport lag cannot be analyzed for stability by the Routh test. The transport lag is also difficult to simulate by computer as explained in Chap. 34. For these reasons, several approximations of transport lag that are useful in control calculations are presented here.

One approach to approximating the transport lag is to write $e^{-\tau s}$ as $1/e^{\tau s}$ and to express the denominator as a Taylor series; the result is

$$e^{-\tau s} = \frac{1}{e^{\tau s}} = \frac{1}{1 + \tau s + \tau^2 s^2/2 + \tau^3 s^3/3! + \cdots}$$

Keeping only the first two terms in the denominator gives

$$e^{-\tau s} \cong \frac{1}{1 + \tau s} \tag{8.46}$$

This approximation, which is simply a first-order lag, is a crude approximation of a transport lag. An improvement can be made by expressing the transport lag as

$$e^{-\tau s} = \frac{e^{-\tau s/2}}{e^{\tau s/2}}$$

Expanding numerator and denominator in a Taylor series and keeping only terms of first-order give

$$e^{-\tau s} \cong \frac{1 - \tau s/2}{1 + \tau s/2} \qquad \text{1st-order Padé} \qquad (8.47)$$

This expression is also known as a *first-order Padé* approximation.

Another well known approximation for a transport lag is the second-order Padé approximation:

$$e^{-\tau s} \cong \frac{1 - \tau s/2 + \tau^2 s^2/12}{1 + \tau s/2 + \tau^2 s^2/12} \qquad \text{2nd-order Padé} \qquad (8.48)$$

The step responses of the three approximations of transport lag presented here are shown in Fig. 8.8. The step response of $e^{-\tau s}$ is also shown for comparison. Notice that the response for the first-order Padé approximation drops to -1 before rising exponentially toward $+1$. The response for the second-order Padé approximation jumps to $+1$ and then descends to below zero before returning gradually back to $+1$.

Although none of the approximations for $e^{-\tau s}$ is very accurate, the approximation for $e^{-\tau s}$ is more useful when it is multiplied by several first-order or second-order transfer functions. In this case, the other transfer functions filter out the high frequency content of the signals passing through the transport lag with the result that the transport lag approximation, when combined with other transfer functions, provides a satisfactory result in many cases. The accuracy of a transport lag can be evaluated most clearly in terms of frequency response, a topic to be covered later in this book.

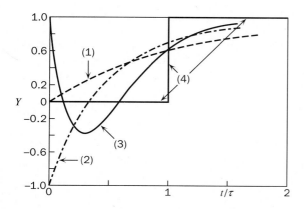

FIGURE 8-8
Step response to approximations
of the transport lag $e^{-\tau s}$.
(1) $1/(\tau s + 1)$, (2) 1st-order Padé,
(3) 2nd-order Padé, (4) $e^{-\tau s}$.

PROBLEMS

8.1. A step change of magnitude 4 is introduced into a system having the transfer function

$$\frac{Y(s)}{X(s)} = \frac{10}{s^2 + 1.6s + 4}$$

Determine
(a) Percent overshoot
(b) Rise time
(c) Maximum value of $Y(t)$
(d) Ultimate value of $Y(t)$
(e) Period of oscillation

8.2. The two-tank system shown in Fig. P8.2 is operating at steady state. At time $t = 0$, 10 ft^3 of water are quickly added to the first tank. Using appropriate figures and equations in the text, determine the maximum deviation in level (feet) in both tanks from the ultimate steady-state values and the time at which each maximum occurs. Data:

$$A_1 = A_2 = 10\text{ft}^2$$
$$R_1 = 0.1 \text{ ft/cfm}$$
$$R_2 = 0.35 \text{ ft/cfm}$$

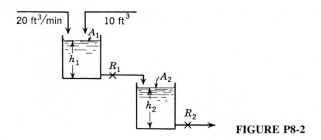

FIGURE P8-2

8.3. The two-tank liquid-level system shown in Fig. P8.3 is operating at steady state when a step change is made in the flow rate to tank 1. The transient response is critically damped, and it takes 1.0 min for the change in level of the second tank to reach 50 percent of the total change.

If the ratio of the cross-sectional areas of the tanks is $A_1/A_2 = 2$, calculate the ratio R_1/R_2. Calculate the time constant for each tank. How long does it take for the change in level of the first tank to reach 90 percent of the total change?

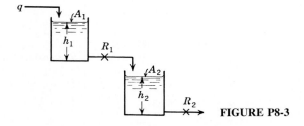

FIGURE P8-3

8.4. A mercury manometer is depicted in Fig. P8.4. Assuming the flow in the manometer to be laminar and the steady-state friction law for drag force in laminar flow to apply at each instant, determine a transfer function between the applied pressure p_1 and the manometer reading h. It will simplify the calculations if, for inertial terms, the velocity profile is assumed to be flat. From your transfer function, written in standard second-order form, list (a) the steady-state gain, (b) τ, and (c) ζ. Comment on these parameters as they are related to the physical nature of the problem.

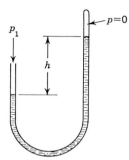

FIGURE P8-4

8.5. Design a mercury manometer that will measure pressures up to 2 atm absolute and will give responses that are slightly underdamped (that is, $\zeta \approx 0.7$).

8.6. Verify Eqs. (8.17), (8.19), and (8.20).

8.7. Verify Eqs. (8.24) and (8.25).

8.8. Verify Eq. (8.40).

8.9. If a second-order system is overdamped, it is more difficult to determine the parameters ζ and τ experimentally. One method for determining the parameters from a step response has been suggested by R. C. Oldenbourg and H. Sartorius (*The Dynamics of Automatic Controls.* ASME, p. 78, 1948), as described below.

(a) Show that the unit-step response for the overdamped case may be written in the form

$$S(t) = 1 - \frac{r_1 e^{r_2 t} - r_2 e^{r_1 t}}{r_1 - r_2}$$

where r_1 and r_2 are the (real and negative) roots of

$$\tau^2 s^2 + 2\zeta\tau s + 1 = 0$$

(b) Show that $S(t)$ has an inflection point at

$$t_i = \frac{\ln(r_2/r_1)}{r_1 - r_2}$$

(c) Show that the slope of the step response at the inflection point

$$\left.\frac{dS(t)}{dt}\right|_{t=t_i} = S'(t_i)$$

has the value

$$S'(t_i) = -r_1 e^{r_1 t_i} = -r_2 e^{r_2 t_i}$$

$$= -r_1\left(\frac{r_2}{r_1}\right)^{r_1/(r_1-r_2)}$$

(d) Show that the value of the step response at the inflection point is

$$S(t_i) = 1 + \frac{r_1 + r_2}{r_1 r_2} S'(t_i)$$

and that hence

$$\frac{1 - S(t_i)}{S'(t_i)} = -\frac{1}{r_1} - \frac{1}{r_2}$$

(e) On a typical sketch of a unit-step response, show distances equal to

$$\frac{1}{S'(t_i)} \quad \text{and} \quad \frac{1 - S(t_i)}{S'(t_i)}$$

and hence present two simultaneous equations resulting from a graphical method for determination of r_1 and r_2.

(f) Relate ζ and τ to r_1 and r_2.

8.10. Determine $Y(0)$, $Y(0.6)$, and $Y(\infty)$ if

$$Y(s) = \frac{1}{s} \frac{25(s + 1)}{s^2 + 2s + 25}$$

8.11. In the liquid-level system shown in Fig. P8.11, the deviation in flow rate to the first tank is an impulse function of magnitude 5. The following data apply: $A_1 = 1 \text{ ft}^2$, $A_2 = A_3 = 2 \text{ ft}^2$, $R_1 = 1 \text{ ft/cfm}$, $R_2 = 1.5 \text{ ft/cfm}$.

(a) Determine expressions for $H_1(s)$, $H_2(s)$, and $H_3(s)$ where H_1, H_2, and H_3 are deviations in tank level for tanks, 1, 2, and 3.

(b) Sketch the responses of $H_1(t)$, $H_2(t)$, and $H_3(t)$. (You need show only the shape of the responses; do not plot.)

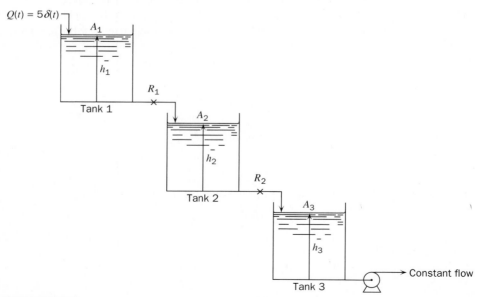

FIGURE P8-11

(c) Determine $H_1(3.46)$, $H_2(3.46)$, and $H_3(3.46)$. For H_2 and H_3, use graphs in Chap. 8 of this text after first finding values of τ and ζ for an equivalent second-order system.

8.12. Sketch the response $Y(t)$ if $Y(s) = e^{-2s}/[s^2 + 1.2s + 1]$. Determine $Y(t)$ for $t = 0, 1, 5$, and ∞.

8.13. The two tanks shown in Fig. P8.13 are connected in an interacting fashion. The system is initially at steady state with $q = 10$ cfm. The following data apply to the tanks: $A_1 = 1$ ft^2, $A_2 = 1.25$ ft^2, $R_1 = 1$ ft/cfm, $R_2 = 0.8$ ft/cfm.

(a) If the flow changes from 10 to 11 cfm according to a step change, determine $H_2(s)$, i.e., the Laplace transform of $H_2(t)$, where H_2 is the deviation in h_2.

(b) Determine $H_2(1)$, $H_2(4)$, and $H_2(\infty)$.

(c) Determine the initial levels (actual levels) $h_1(0)$ and $h_2(0)$ in the tanks.

(d) Obtain an expression for $H_1(s)$ for the unit-step change described above.

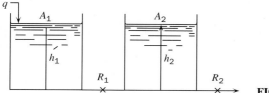

FIGURE P8-13

8.14. From figures in your text, determine $Y(4)$ for the system response expressed by

$$y(s) = \frac{2}{s} \frac{2s + 4}{4s^2 + 0.8s + 1}$$

8.15. A step change of magnitude 3 is introduced into the transfer function

$$Y(s)/X(s) = 10/[2s^2 + 0.3s + 0.5]$$

Determine the overshoot and the frequency of oscillation.

PART
III

LINEAR CLOSED-LOOP SYSTEMS

CHAPTER
9

THE CONTROL SYSTEM

INTRODUCTION

In the previous chapters, the dynamic behavior of several basic systems was examined. With this background, we can extend the discussion to a complete control system and introduce the fundamental concept of feedback. In order to work with a familiar system, the treatment will be based on the illustrative example of Chap. 1, which is concerned with a stirred-tank heater.

Figure 9.1 is a sketch of the apparatus. To orient the reader, the physical description of this control system will be reviewed. A liquid stream at a temperature T_i enters an insulated, well-stirred tank at a constant flow rate w (mass/time). It is desired to maintain (or control) the temperature in the tank at T_R by means of the controller. If the measured tank temperature T_m differs from the desired temperature T_R, the controller senses the difference or *error*, $\epsilon = T_R - T_m$, and changes the heat input in such a way as to reduce the magnitude of ϵ. If the controller changes the heat input to the tank by an amount that is proportional to ϵ, we have *proportional* control.

In Fig. 9.1, it is indicated that the source of heat input q may be electricity or steam. If an electrical source were used, the final control element might be a variable transformer that is used to adjust current to a resistance heating element; if steam were used, the final control element would be a control valve that adjusts the flow of steam. In either case, the output signal from the controller should adjust q in such a way as to maintain control of the temperature in the tank.

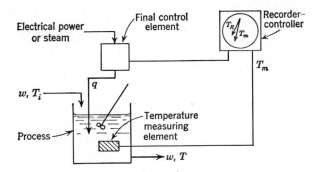

FIGURE 9-1
Control system for a stirred-tank heater.

Components of a Control System

The system shown in Fig. 9.1 may be divided into the following components:

1. Process (stirred-tank heater).
2. Measuring element (thermometer).
3. Controller.
4. Final control element (variable transformer or control valve).

Each of these components can be readily identified as a separate physical item in the process. In general, these four components will constitute most of the control systems that we shall consider in this text; however, the reader should realize that more complex control systems exist in which more components are used. For example, there are some processes which require a cascade control system in which two controllers and two measuring elements are used. A cascade system is discussed in Chap. 18.

Block Diagram

For computational purposes, it is convenient to represent the control system of Fig. 9.1 by means of the block diagram shown in Fig. 9.2. Such a diagram makes it much easier to visualize the relationships among the various signals. New terms, which appear in Fig. 9.2, are *set point* and *load*. The set point is a synonym for the desired value of the controlled variable. The load refers to a change in any variable that may cause the controlled variable of the process to change. In this example, the inlet temperature T_i is a load variable. Other possible loads for this system are changes in flow rate and heat loss from the tank. (These loads are not shown on the diagram.)

The control system shown in Fig. 9.2 is called a *closed-loop* system or a feedback system because the measured value of the controlled variable is returned or "fed back" to a device called the *comparator*. In the comparator, the controlled variable is compared with the desired value or *set point*. If there is any difference

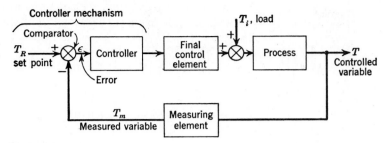

FIGURE 9-2
Block diagram of a simple control system.

between the measured variable and the set point, an error is generated. This error enters a *controller,* which in turn adjusts the *final control element* in order to return the controlled variable to the set point.

Negative Feedback versus Positive Feedback

Several terms have been used that may need further clarification. The feedback principle, which is illustrated by Fig. 9.2, involves the use of the controlled variable T to maintain itself at a desired value T_R. The arrangement of the apparatus of Fig. 9.2 is often described as *negative feedback* to contrast with another arrangement called positive feedback. Negative feedback ensures that the difference between T_R and T_m is used to adjust the control element so that the tendency is to reduce the error. For example, assume that the system is at steady state and that $T = T_m = T_R$. If the load T_i should increase, T and T_m would start to increase, which would cause the error ϵ to become negative. With proportional control, the decrease in error would cause the controller and final control element to *decrease* the flow of heat to the system with the result that the flow of heat would eventually be reduced to a value such that T approaches T_R. A verbal description of the operation of a feedback control system, such as the one just given, is admittedly inadequate, for this description necessarily is given as a sequence of events. Actually all the components operate simultaneously, and the only adequate description of what is occurring is a set of simultaneous differential equations. This more accurate description is the primary subject matter of the present and succeeding chapters.

If the signal to the comparator were obtained by adding T_R and T_m, we would have a *positive feedback* system, which is inherently unstable. To see that this is true, again assume that the system is at steady state and that $T = T_m = T_R$. If T_i were to increase, T and T_m would increase, which would cause the signal the comparator (ϵ in Fig. 9.2) to increase, with the result that the heat to the system would increase. However, this action, which is just the opposite of that needed, would cause T to increase further. It should be clear that this situation would cause T to "run away" and control would not be achieved. For this reason, positive feedback would never be used intentionally in the system of Fig. 9.2. However, in more complex systems it may arise naturally. An example of this is discussed in Chap. 21.

Servo Problem versus Regulator Problem

The control system of Fig. 9.2 can be considered from the point of view of its ability to handle either of two types of situations. In the first situation, which is called the servomechanism-type (or servo) problem, we assume that there is no change in load T_i and that we are interested in changing the bath temperature according to some prescribed function of time. For this problem, the set point T_R would be changed in accordance with the desired variation in bath temperature. If the variation is sufficiently slow, the bath temperature may be expected to follow the variation in T_R very closely. There are occasions when a control system in the chemical industry will be operated in this manner. For example, one may be interested in varying the temperature of a reactor according to a prescribed time-temperature pattern. However, the majority of problems that may be described as the servo type come from fields other than the chemical industry. The tracking of missiles and aircraft and the automatic machining of intricate parts from a master pattern are well-known examples of the servo-type problem. The other situation will be referred to as the regulator problem. In this case, the desired value T_R is to remain fixed and the purpose of the control system is to maintain the controlled variable at T_R in spite of changes in load T_i. This problem is very common in the chemical industry, and a complicated industrial process will often have many self-contained control systems, each of which maintains a particular process variable at a desired value. These control systems are of the regulator type.

In considering control systems in the following chapters, we shall frequently discuss the response of a linear control system to a change in set point (servo problem) separately from the response to a change in load (regulator problem). However, it should be realized that this is done only for convenience. The basic approach to obtaining the response of either type is essentially the same, and the two responses may be superimposed to obtain the response to any linear combination of set-point and load changes.

DEVELOPMENT OF BLOCK DIAGRAM

Each block in Fig. 9.2 represents the functional relationship existing between the input and output of a particular component. In the previous chapters, such input-output relations were developed in the form of transfer functions. In block-diagram representations of control systems, the variables selected are *deviation variables,* and inside each block is placed the transfer function relating the input-output pair of variables. Finally, the blocks are combined to give the overall block diagram. This is the procedure to be followed in developing Fig. 9.2.

Process

Consider first the block for the process. This block will be seen to differ somewhat from those presented in previous chapters in that two input variables are present; however, the procedure for developing the transfer function remains the same.

An unsteady-state energy balance* around the tank gives

$$q + wC(T_i - T_o) - wC(T - T_o) = \rho C V \frac{dT}{dt} \tag{9.1}$$

where T_o is the reference temperature.

At steady state, dT/dt is zero, and Eq. (9.1) can be written

$$q_s + wC(T_{i_s} - T_o) - wC(T_s - T_o) = 0 \tag{9.2}$$

where the subscript s has been used to indicate steady state.

Subtracting Eq. (9.2) from Eq. (9.1) gives

$$q - q_s + wC[(T_i - T_{i_s}) - (T - T_s)] = \rho C V \frac{d(T - T_s)}{dt} \tag{9.3}$$

Notice that the reference temperature T_o cancels in the subtraction. If we introduce the deviation variables

$$T_i' = T_i - T_{i_s} \tag{9.4}$$
$$Q = q - q_s \tag{9.5}$$
$$T' = T - T_s \tag{9.6}$$

Eq. (9.3) becomes

$$Q + wC(T_i' - T') = \rho C V \frac{dT'}{dt} \tag{9.7}$$

Taking the Laplace transform of Eq. (9.7) gives

$$Q(s) + wC[T_i'(s) - T'(s)] = \rho C V s T'(s) \tag{9.8}$$

or

$$T'(s)\left(\frac{\rho V}{w} s + 1\right) = \frac{Q(s)}{wC} + T_i'(s) \tag{9.9}$$

---

*In this analysis, it is assumed that the flow rate of heat q is instantaneously available and independent of the temperature in the tank. In some stirred-tank heaters, such as a jacketed kettle, q depends on both the temperature of the fluid in the jacket and the temperature of the fluid in the kettle. In this introductory chapter, systems (electrically heated tank or direct steam-heated tank) are selected for which this complication can be ignored. In Chap. 21, the analysis of a steam-jacketed kettle is given in which the effect of kettle temperature on q is taken into account.

This last expression can be written

$$T'(s) = \frac{1/wC}{\tau s + 1}Q(s) + \frac{1}{\tau s + 1}T_i'(s) \qquad (9.10)$$

where

$$\tau = \frac{\rho V}{w}$$

If there is a change in $Q(t)$ only, then $T_i'(t) = 0$ and the transfer function relating T' to Q is

$$\frac{T'(s)}{Q(s)} = \frac{1/wC}{\tau s + 1} \qquad (9.11)$$

If there is a change in $T_i'(t)$ only, then $Q(t) = 0$ and the transfer function relating T' to T_i' is

$$\frac{T'(s)}{T_i'(s)} = \frac{1}{\tau s + 1} \qquad (9.12)$$

Equation (9.10) is represented by the block diagram shown in Fig. 9.3a. This diagram is simply an alternate way to express Eq. (9.10) in terms of the transfer functions of Eqs. (9.11) and (9.12). Superposition makes this representation possible. Notice that, in Fig. 9.3, we have indicated summation of signals by the symbol shown in Fig. 9.4, which is called a *summing junction*. Subtraction can also be indicated with this symbol by placing a minus sign at the appropriate input. The summing junction was used previously as the symbol for the comparator of the controller (see Fig. 9.2). This symbol, which is standard in the control literature, may have several inputs but only one output.

A block diagram that is equivalent to Fig. 9.3a is shown in Fig. 9.3b. That this diagram is correct can be seen by rearranging Eq. (9.10); thus

$$T'(s) = [Q(s) + wCT_i'(s)]\frac{1/wC}{\tau s + 1} \qquad (9.13)$$

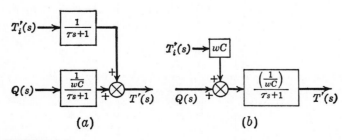

(a)   (b)

FIGURE 9-3
Block diagram for process.

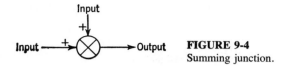

FIGURE 9-4
Summing junction.

In Fig. 9.3b, the input variables $Q(s)$ and $wCT_i'(s)$ are summed before being operated on by the transfer function $1/wC/(\tau s + 1)$.

The physical situation that exists for the control system (Fig. 9.1) if steam heating is used requires more careful analysis to show that Fig. 9.3 is an equivalent block diagram. Assume that a supply of steam at constant conditions is available for heating the tank. One method for introducing heat to the system is to let the steam flow through a control valve and discharge directly into the water in the tank, where it will condense completely and become part of the stream leaving the tank (see Fig. 9.5).

If the flow of steam, f (pounds/time), is small compared with the inlet flow w, the total outlet flow is approximately equal to w. When the system is at steady state, the heat balance may be written

$$wC(T_{i_s} - T_o) - wC(T_s - T_o) + f_s(H_g - H_{l_s}) = 0 \qquad (9.14)$$

where T_o = reference temperature used to evaluate enthalpy of all streams entering and leaving tank

H_g = specific enthalpy of the steam supplied, a constant

H_{l_s} = specific enthalpy of the condensed steam flowing out at T_s, as part of the total stream

The term H_{l_s} may be written in terms of heat capacity and temperature; thus

$$H_{l_s} = C(T_s - T_o) \qquad (9.15)$$

From this, we see that, if the steady-state temperature changes, H_{l_s} changes. In Eq. (9.14), $f_s(H_g - H_{l_s})$ is equivalent to the steady-state input q_s used previously, as can be seen by comparing Eq. (9.2) with (9.14).

Now consider an *unsteady-state* operation in which f is much less than w and the temperature T of the bath does not deviate significantly from the steady-state

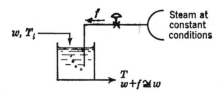

FIGURE 9-5
Supplying heat by steam.

temperature T_s. For these conditions, we may write the unsteady-state balance approximately; thus

$$wC(T_i - T_o) - wC(T - T_o) + f(H_g - H_{l_s}) = \rho C V \frac{dT}{dt} \tag{9.16}$$

In a practical situation for steam, H_g will be about 1000 Btu/lb$_m$. If the temperature of the bath, T, never deviates from T_s by more than $10°$, the error in using the term $f(H_g - H_{l_s})$ instead of $f(H_g - H_l)$ will be no more than 1 percent. Under these conditions, Eq. (9.16) represents the system closely, and by comparing Eq. (9.16) with Eq. (9.1), it is clear that

$$q = f(H_g - H_{l_s}) \tag{9.17}$$

Therefore, q is proportional to the flow of steam f, which may be varied by means of a control valve. It should be emphasized that the analysis presented here is only approximate. Both f and the deviation in T must be small. The smaller they become, the more closely Eq. (9.16) represents the actual physical system. An exact analysis of the problem leads to a differential equation with time-varying coefficients, and the transfer-function approach does not apply. The problem becomes considerably more difficult. A better approximation will be discussed in Chap. 21, where linearization techniques are used.

Measuring Element

The temperature-measuring element, which senses the bath temperature T and transmits a signal T_m to the controller, may exhibit some dynamic lag. From the discussion of the mercury thermometer in Chap. 5, we observed this lag to be first-order. In this example, we shall assume that the temperature-measuring element is a first-order system, for which the transfer function is

$$\frac{T'_m(s)}{T'(s)} = \frac{1}{\tau_m s + 1} \tag{9.18}$$

where the input-output variables T' and T'_m are deviation variables, defined as

$$T' = T - T_s$$
$$T'_m = T_m - T_{m_s}$$

Note that, when the control system is at steady state, $T_s = T_{m_s}$, which means that the temperature-measuring element reads the true bath temperature. The transfer function for the measuring element may be represented by the block diagram shown in Fig. 9.6.

FIGURE 9-6
Block diagram of measuring element.

Controller and Final Control Element

For convenience, the blocks representing the controller and the final control element are combined into one block. In this way, we need be concerned only with the overall response between the error and the heat input to the tank. Also, it is assumed that the controller is a proportional controller. (In the next chapter, the response of other controllers, which are commonly used in control systems, will be described.) The relationship for a proportional controller is

$$q = K_c \epsilon + A \tag{9.19}$$

where $\epsilon = T_R - T_m$

T_R = set-point temperature

K_c = proportional sensitivity or controller gain

A = heat input when $\epsilon = 0$

At steady state, it is assumed* that the set point, the process temperature, and the measured temperature are all equal to each other; thus

$$T_{R_s} = T_s = T_{m_s} \tag{9.20}$$

Let ϵ' be the deviation variable for error; thus

$$\epsilon' = \epsilon - \epsilon_s \tag{9.21}$$

where $\epsilon_s = T_{R_s} - T_{m_s}$

Since $T_{R_s} = T_{m_s}$, $\epsilon_s = 0$ and Eq. (9.21) becomes

$$\epsilon' = \epsilon - 0 = \epsilon \tag{9.22}$$

This result shows that ϵ is itself a deviation variable.

Since $\epsilon_s = 0$, Eq. (9.19) becomes at steady state

$$q_s = K_c\epsilon_s + A = 0 + A = A$$

Equation (9.19) may now be written in terms of q_s; thus

$$q = K_c\epsilon + q_s$$

or

$$Q = K_c\epsilon \tag{9.23}$$

where $Q = q - q_s$

The transform of Eq. (9.23) is simply

$$Q(s) = K_c\epsilon(s) \tag{9.24}$$

---

*In a practical situation, the equality among the three variables, T, T_m, and T_R, at steady state as given by Eq. (9.20) can always be established by adjustment of the instruments. The equality between T and T_m can be achieved by calibration of the measuring element. The equality between T_m and T_R can be achieved by adjustment of the proportional controller.

$T'_R(s) \xrightarrow{\;+\;} \bigotimes \xrightarrow{\;\epsilon\,(s)\;} \boxed{K_c} \longrightarrow Q(s)$

$\uparrow -$

$T'_m(s)$

FIGURE 9-7

Block diagram of proportional controller.

Note that ϵ, which is also equal to ϵ', may be expressed as

$$\epsilon = T_R - T_{R_s} - (T_m - T_{m_s}) \tag{9.25}$$

or

$$\epsilon = T'_R - T'_m \tag{9.26}$$

Equation (9.25) follows from the definition of ϵ and the fact that $T_{R_s} = T_{m_s}$. Taking the transform of Eq. (9.26) gives

$$\epsilon(s) = T'_R(s) - T'_m(s) \tag{9.27}$$

The transfer function for the proportional controller given by Eq. (9.24) and the generation of error given by Eq. (9.27) may be expressed by the block diagram shown in Fig. 9.7.

We have now completed the development of the separate blocks. If these are combined according to Fig. 9.2, there is obtained the block diagram for the complete control system shown in Fig. 9.8. The reader should verify this figure.

SUMMARY

It has been shown that a control system can be translated into a block diagram that includes the transfer functions of the various components. It should be emphasized that a block diagram is simply a systematic way of writing the simultaneous differential and algebraic equations that describe the dynamic behavior of the components. In the present case, these were Eqs. (9.10), (9.18), and (9.24) and the definition of ϵ. The block diagram clarifies the relationships among the variables of these simultaneous equations. Another advantage of the block-diagram representation is that it clearly shows the feedback relationship between measured variable and desired variable and how the difference in these two signals (the

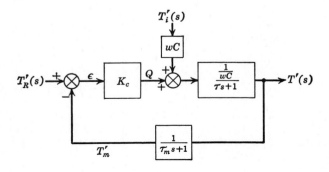

FIGURE 9-8

Block diagram of control system.

error ϵ) is used to maintain control. A set of equations generally does not clearly indicate the relationships shown by the block diagram.

In the next several chapters, tools will be developed that will enable us to reduce a block diagram such as the one in Fig. 9.8 to a single block that relates $T'(s)$ to T_i' or T_R'. We shall then obtain the transient response of the control system shown in Fig. 9.8 to some specific changes in T_R' and T_i'. However, we shall first pause in Chap.10 to look more carefully at the controller and control element blocks, which have been skimmed over in the present chapter.

PROBLEMS

9.1. The two-tank heating process shown in Fig. P9.1 consists of two identical, well-stirred tanks in series. A flow of heat can enter tank 2. At time $t = 0$, the flow rate of heat to tank 2 suddenly increases according to a step function to 1000 Btu/min, and the temperature of the inlet water T_i drops from 60°F to 52°F according to a step function. These changes in heat flow and inlet water temperature occur simultaneously.

(a) Develop a block diagram that relates the outlet temperature of tank 2 to the inlet temperature to tank 1 and the flow of heat to tank 2.

(b) Obtain an expression for $T_2'(s)$ where T_2' is the deviation in the temperature of tank 2. This expression should contain numerical values of the parameters.

(c) Determine $T_2(2)$ and $T_2(\infty)$.

(d) Sketch the response $T_2'(t)$ versus t.

Initially, $T_i = T_1 = T_2 = 60°F$ and $q = 0$. The following data apply:

$w = 250$ lb/min

holdup volume of each tank $= 5$ ft^3

density of fluid $= 50$ lb/ft^3

heat capacity of fluid $= 1$ Btu/(lb) (°F)

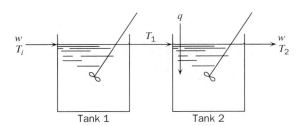

Tank 1   Tank 2   **FIGURE P9-1**

9.2. The two-tank heating process shown in Fig. P9.2 consists of two identical, well-stirred tanks in series. At steady state, $T_a = T_b = 60°F$. At time $t = 0$, the temperature of each stream, entering the tanks changes according to a step function, i.e., $T_a' = 10u(t)$, $T_b' = 20u(t)$ where T_a' and T_b' are deviation variables.

(a) Develop the block diagram that relates T_2', the deviation in temperature in tank 2, to T_a' and T_b'.

(b) Obtain an expression for $T_2'(s)$.

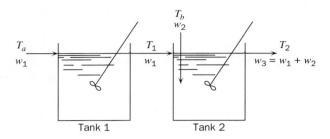

Tank 1   Tank 2   **FIGURE P9-2**

(c) Determine $T_2(2)$.
The following data apply:

$w_1 = w_2 = 250$ lb/min
Holdup volume of each tank $= 10$ ft^3
Density of fluid $= 50$ lb/ft^3
Heat capacity of fluid $= 1$ Btu/(lb)(°F)

9.3. The heat transfer equipment shown in Fig. P9.3 consists of two tanks, one nested inside the other. Heat is transferred by convection through the wall of the inner tank. The contents of each tank are well mixed. The following data and information apply:

1. The holdup volume of the inner tank is 1 ft^3. The holdup of the outer tank is 1 ft^3.
2. The cross-sectional area for heat transfer between the tanks is 1 ft^2.
3. The overall heat transfer coefficient for the flow of heat between the tanks is 10 Btu/(hr)(ft^2)(°F).
4. The heat capacity of fluid in each tank is 1 Btu/(lb)(°F). The density of each fluid is 50 lb/ft^3.

   Initially the temperatures of the feed stream to the outer tank and the contents of the outer tank are equal to 100°F. The contents of the inner tank are initially at 100°F. At time zero, the flow of heat to the inner tank (Q) is changed according to a step change from 0 to 500 Btu/hr.

(a) Obtain an expression for the Laplace transform of the temperature of the inner tank, $T(s)$.
(b) Invert $T(s)$ and obtain T for time $= 0$, 5 hr, 10 hr, and ∞.

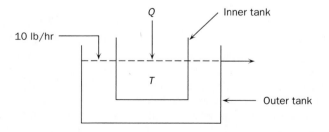

FIGURE P9-3

CHAPTER
10

CONTROLLERS AND FINAL CONTROL ELEMENTS

In the previous chapter, the block-diagram representation of a simple control system (Fig. 9.2) was developed. This chapter will focus attention on the controller and final control element and discuss the dynamic characteristics of some of these components that are in common use. As shown in Fig. 9.2, the input signal to the controller is the error and the output signal of the controller is fed to the final control element. In many process control systems, this output signal is an air pressure and the final control element is a pneumatic valve that opens and closes as air pressure on the diaphragm changes.

For the mathematical analysis of control systems, it is sufficient to regard the controller as a simple computer. For example, a proportional controller may be thought of as a device that receives the error signal and puts out a signal proportional to it. Similarly, the final control element may be regarded as a device that produces corrective action on the process. The corrective action is regarded as mathematically related to the output signal from the controller. However, it is desirable to have some appreciation of the actual physical mechanisms used to accomplish this. For this reason, we begin this chapter with a physical description of a pneumatic control valve and a simplified description of a proportional controller.

Up to about 1960, most controllers were pneumatic. Although pneumatic controllers are still in use and function quite well in many installations, the controllers being installed today are electronic or computer-based instruments. For this reason, the proportional controller to be discussed in this chapter will be electronic or computer-based. The transfer functions that are presented in this chapter apply to either type of controller, and the discussion is in no way restrictive.

After the introductory discussion, transfer functions will be presented for simplified or idealized versions of the control valve and the conventional controllers. These transfer functions, for practical purposes, will adequately represent the dynamic behavior of control valves and controllers. Hence, they will be used in subsequent chapters for mathematical analysis and design of control systems.

MECHANISMS

Control Valve

The control valve shown in Fig. 10.1 contains a pneumatic device (valve motor) that moves the valve stem as the pressure on a spring-loaded diaphragm changes. The stem positions a plug in the orifice of the valve body. As the pressure increases, the plug moves downward and restricts the flow of fluid through the valve. This action is referred to as air-to-close. The valve may also be constructed to have air-to-open action. Valve motors are often constructed so that the valve stem position is proportional to the valve-top pressure. Most commercial valves move from fully open to fully closed as the valve-top pressure changes from 3 to 15 psig.

In general, the flow rate of fluid through the valve depends upon the upstream and downstream fluid pressures and the size of the opening through the valve. The plug and seat (or orifice) can be shaped so that various relationships between stem position and size of opening (hence, flow rate) are obtained. In our example, we shall assume for simplicity that at *steady state* the flow (for fixed upstream and downstream fluid pressures) is proportional to the valve-top pneumatic pressure. A valve having this relation is called a *linear valve*. An extensive discussion of control valves is presented in Chap. 20.

Controller

The control hardware required to control the temperature of a stream leaving a heat exchanger is shown in Fig. 10.2. This hardware, available from manufacturers of

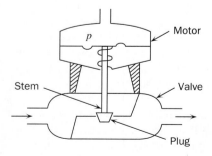

FIGURE 10-1
Pneumatic control valve (air-to-close).

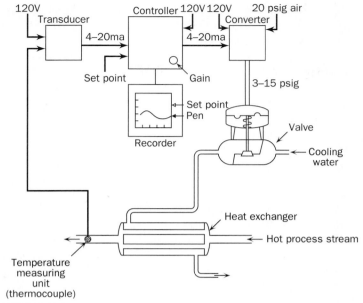

FIGURE 10-2
Schematic diagram of control system.

such equipment, consists of the following components listed here along with their respective conversions:

Transducer (temperature to current).
Controller-recorder (current to current).
Converter (current to pressure).
Control valve (pressure to flow rate).

Figure 10.2 shows that a thermocouple is used to measure the temperature; the signal from the thermocouple is sent to the transducer, which produces an output in the range of 4–20 ma, which is a linear function of the input. The output of the transducer enters the controller where it is compared to the set point to produce an error signal. The controller converts the error to an output in the range of 4–20 ma according to the control law stored in the memory of the computer. The only control law we have considered so far has been proportional. Later in this chapter other control laws will be described. The output of the controller enters the converter, which produces an output in the range of 3–15 psig, which is a linear function of the input. Finally, the output of the converter is sent to the top of the control valve, which adjusts the flow of cooling water to the heat exchanger. We shall assume that the valve is linear and is the pressure-to-open type. The external power (120 V) needed for each component is also shown in Fig. 10.2. Electricity is needed for the transducer, controller, and converter. A source of 20 psig air is needed for the converter.

To see how the components interact with each other, consider the process to be operating at steady state with the outlet temperature equal to the set point. If the temperature of the hot process stream increases, the following events occur: After some delay the thermocouple detects an increase in the outlet temperature and produces a proportional change in the signal to the controller. As soon as the controller detects the rise in temperature, relative to the set point, the controller output increases according to proportional action. The increase in signal to the converter causes the output pressure from the converter to increase and open the valve wider in order to admit a greater flow of cooling water. The increased flow of cooling water will eventually reduce the output temperature and move it toward the set point. From this qualitative description, we see that the flow of signals from one component to the next is such that the temperature of the heat exchanger should return toward the set point. In a well-tuned control system, the response of the temperature will oscillate around the set point before coming to steady state. We shall give considerable attention to the transient response of a control system in the remainder of this book. Further discussion will also be given on control valves in Chap. 20 and on controllers in Chap. 35.

For convenience in describing various control laws (or algorithms) in the next part of this chapter, the transducer, controller, and converter will be lumped into one block as shown in Fig. 10.3.

This concludes our brief introduction to valves and controllers. We now present transfer functions for such devices. These transfer functions, especially for controllers, are based on ideal devices that can be only approximated in practice. The degree of approximation is sufficiently good to warrant use of these transfer functions to describe the dynamic behavior of controller mechanisms for ordinary design purposes.

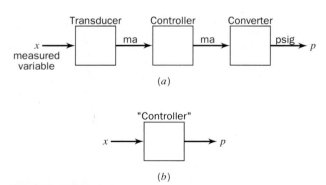

FIGURE 10-3
Equivalent block for transducer, controller, and converter.

IDEAL TRANSFER FUNCTIONS

Control Valve

A pneumatic valve always has some dynamic lag, which means that the stem motion does not respond instantaneously to a change in the applied pressure from the controller. From experiments conducted on pneumatic valves, it has been found that the relationship between flow and valve-top pressure for a linear valve can often be represented by a first-order transfer function; thus

$$\frac{Q(s)}{P(s)} = \frac{K_v}{\tau_v s + 1}$$

where K_v is the steady-state gain, i.e., the constant of proportionality between steady-state flow rate and valve-top pressure, and τ_v is the time constant of the valve.

In many practical systems, the time constant of the valve is very small when compared with the time constants of other components of the control system, and the transfer function of the valve can be approximated by a constant

$$\frac{Q(s)}{P(s)} = K_v$$

Under these conditions, the valve is said to contribute negligible dynamic lag.

To justify the approximation of a fast valve by a transfer function, which is simply K_v, consider a first-order valve and a first-order process connected in series, as shown in Fig. 10.4.

According to the discussion of Chap. 7, if we assume no interaction, the transfer function from $P(s)$ to $Y(s)$ is

$$\frac{Y(s)}{P(s)} = \frac{K_v K_P}{(\tau_v s + 1)(\tau_P s + 1)}$$

The assumption of no interaction is generally valid for this case.

For a unit-step change in P,

$$Y = \frac{1}{s} \frac{K_v K_P}{(\tau_v s + 1)(\tau_P s + 1)}$$

the inverse of which is

$$Y(t) = (K_v K_P)\left[1 - \frac{\tau_v \tau_P}{\tau_v - \tau_P}\left(\frac{1}{\tau_P}e^{-t/\tau_v} - \frac{1}{\tau_v}e^{-t/\tau_P}\right)\right]$$

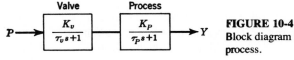

Valve **Process**

$$\frac{K_v}{\tau_v s + 1}$$ $$\frac{K_P}{\tau_P s + 1}$$

$P \longrightarrow$ $\longrightarrow Y$

FIGURE 10-4
Block diagram for a first-order valve and a first-order process.

If $\tau_v \ll \tau_P$, this equation is approximately

$$Y(t) = K_v K_P (1 - e^{-t/\tau_P})$$

The last expression is the unit-step response of the transfer function

$$\frac{Y(s)}{P(s)} = K_v \frac{K_P}{\tau_P s + 1}$$

so that the combination of process and valve is essentially first-order. This clearly demonstrates that, when the time constant of the valve is much smaller than that of the process, the valve transfer function can be taken as K_v.

A typical pneumatic valve has a time constant of the order of 1 sec. Many industrial processes behave as first-order systems or as a series of first-order systems having time constants that may range from a minute to an hour. For these systems we have shown that the lag of the valve is negligible, and we shall make frequent use of this approximation.

Controllers

In this section, we shall present the transfer functions for the controllers frequently used in industrial processes. Because the transducer and the converter will be lumped together with the controller for simplicity, the result is that the input will be the measured variable x (e.g. temperature, level, etc.) and the output will be a pneumatic signal p. (See Fig. 10.3) Actually this form (x as input and p as output) applies to a pneumatic controller. For convenience, we shall refer to the lumped components as the controller in the following discussion, even though the actual electronic controller is but one of the components.

PROPORTIONAL CONTROL. The proportional controller produces an output signal (pressure in the case of a pneumatic controller, current or voltage for an electronic controller) that is proportional to the error ϵ. This action may be expressed as

$$p = K_c \epsilon + p_s \tag{10.1}$$

where p = output signal from controller, psig or ma
$\quad\quad\;\; K_c$ = gain, or sensitivity
$\quad\quad\;\; \epsilon$ = error = set point − measured variable
$\quad\quad\;\; p_s$ = a constant

The error ϵ, which is the difference between the set point and the signal from the measuring element, may be in any suitable units. However, the units of set point and measured variable must be the same, since the error is the difference between these quantities.

In a controller having adjustable gain, the value of the gain K_c can be varied by moving a knob in the controller. The value of p_s is the value of the output signal when ϵ is zero, and in most controllers p_s can be adjusted to obtain the required output signal when the control system is at steady state and $\epsilon = 0$.

To obtain the transfer function of Eq. (10.1), we first introduce the deviation variable

$$P = p - p_s$$

into Eq. (10.1). At time $t = 0$, we assume the error ϵ_s to be zero. Then ϵ is already a deviation variable. Equation (10.1) Becomes

$$P(t) = K_c \epsilon(t) \tag{10.2}$$

Taking the transform of Eq. (10.2) gives the transfer function of an ideal proportional controller

$$\frac{P(s)}{\epsilon(s)} = K_c \tag{10.3}$$

The term *proportional band* is commonly used among process control engineers in place of the term *gain*. *Proportional band* (pb) is defined as the error (expressed as a percentage of the range of measured variable) required to move the valve from fully closed to fully open. A frequently used synonym is *bandwidth*. These terms will be most easily understood by considering the following example.

Example 10.1. A pneumatic proportional controller is used to control temperature within the range of 60 to 100°F. The controller is adjusted so that the output pressure goes from 3 psi (valve fully open) to 15 psi (valve fully closed) as the measured temperature goes for 71 to 75° F with the set point held constant. Find the gain and the proportional band.

$$\text{Proportional band} = \frac{(75°F - 71°F)}{(100°F - 60°F)} \times 100$$

$$= 10\%$$

$$\text{Gain} = \frac{\Delta P}{\Delta \epsilon} = \frac{(15 \text{ psi} - 3 \text{ psi})}{(75°F - 71°F)} = 3 \text{ psi/}°F$$

Now assume that the proportional band of the controller is changed to 75 percent. Find the gain and the temperature change necessary to cause a valve to go from fully open to fully closed.

$$\Delta T = (\text{proportional band}) (\text{range})$$

$$= 0.75(40°F)$$

$$= 30°F$$

$$\text{Gain} = \frac{12 \text{ psi}}{30°F} = 0.4 \text{ psi/}°F$$

From this example, we see that proportional gain corresponds inversely with proportional band; thus

$$\text{Proportional gain} \propto 1/\text{proportional band}$$

The gain K_c has the units of psi/unit of measured variable (e.g. psi/°F in Example 10.1). If the actual controller of Fig. 10.3*a* is considered, both the input and the output units are in milliamperes. In this case the gain will be dimensionless (i.e., ma/ma). Furthermore, the relation between proportional band (pb) in percentage and K_c will be

$$K_c = 100/[pb(\%)]$$

ON-OFF CONTROL. A special case of proportional control is on-off control. If the gain K_c is made very high, the valve will move from one extreme position to the other if the pen deviates only slightly from the set point. This very sensitive action is called on-off action because the valve is either fully open (on) or fully closed (off); i.e., the valve acts like a switch. This is a very simple controller and is exemplified by the thermostat used in a home-heating system. The bandwidth of an on-off controller is approximately zero.

For various reasons, one of which was suggested in Chap. 1, it is often desirable to add other modes of control to the basic proportional action. These modes, integral and derivative action, are discussed below with the objective of obtaining the ideal transfer functions of the expanded controllers. The reasons for introducing these modes will be discussed briefly at the end of this chapter and in more detail in later chapters.

PROPORTIONAL-INTEGRAL (PI) CONTROL.   This mode of control is described by the relationship

$$p = K_c\epsilon + \frac{K_c}{\tau_I} \int_0^t \epsilon \, dt + p_s \tag{10.4}$$

where K_c = gain
$\quad\quad \tau_I$ = integral time, min
$\quad\quad p_s$ = constant

In this case, we have added to the proportional action term, $K_c\epsilon$, another term that is proportional to the integral of the error. The values of K_c and τ_I may be varied by two knobs in the controller.

To visualize the response of this controller, consider the response to a unit-step change in error, as shown in Fig. 10.5. This unit-step response is most directly obtained by inserting $\epsilon = 1$ into Eq. (10.4), which yields

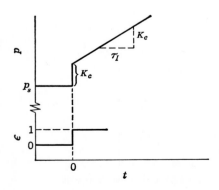

FIGURE 10-5
Response of a PI controller to a unit-step change in error.

$$p(t) = K_c + \frac{K_c}{\tau_I}t + p_s \qquad (10.5)$$

Notice that p changes suddenly by an amount K_c, and then changes linearly with time at a rate K_c/τ_I.

To obtain the transfer function of Eq. (10.4), we again introduce the deviation variable $P = p - p_s$ into Eq. (10.4) and then take the transform to obtain

$$\frac{P(s)}{\epsilon(s)} = K_c\left(1 + \frac{1}{\tau_I s}\right) \qquad (10.6)$$

Some manufacturers prefer to use the term *reset rate*, which is defined as the reciprocal of τ_I. The integral adjustment knob on a controller may be marked in terms of integral time or reset rate. The calibration of the proportional and integral knobs is often checked by observing the jump and slope of the step response shown in Fig. 10.5.

PROPORTIONAL-DERIVATIVE (PD) CONTROL. This mode of control may be represented by

$$p = K_c\epsilon + K_c\tau_D\frac{d\epsilon}{dt} + p_s \qquad (10.7)$$

where K_c = gain
τ_D = derivative time, min
p_s = constant

In this case, we have added to the proportional term another term, $K_c\tau_D \, d\epsilon/dt$, which is proportional to the derivative of the error. The values of K_c and τ_D may be varied separately by knobs on the controller. Other terms that are used to describe the derivative action are *rate control* and *anticipatory control*.

The action of this controller can be visualized by considering the response to a linear change in error as shown in Fig. 10.6. This response is obtained by introducing the linear function $\epsilon(t) = At$ into Eq. (10.7) to obtain

$$p(t) = AK_ct + AK_c\tau_D + p_s$$

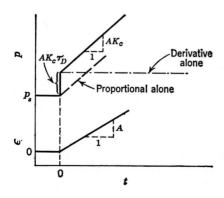

FIGURE 10-6
Response of a PD controller to a linear input in error.

Notice that p changes suddenly by an amount $AK_c\tau_D$ as a result of the derivative action and then changes linearly at a rate AK_c. The effect of derivative action in this case is to anticipate the linear change in error by adding additional output $AK_c\tau_D$ to the proportional action.

To obtain the transfer function from Eq. (10.7), we introduce the deviation variable $P = p - p_s$ and then take the transform to obtain

$$\frac{P(s)}{\epsilon(s)} = K_c(1 + \tau_D s) \tag{10.8}$$

PROPORTIONAL-INTEGRAL-DERIVATIVE (PID) CONTROL. This mode of control is combination of the previous modes and is given by the expression

$$p = K_c\epsilon + K_c\tau_D\frac{d\epsilon}{dt} + \frac{K_c}{\tau_I}\int_0^t \epsilon \, dt + p_s \tag{10.9}$$

In this case, the controller contains three knobs for adjusting K_c, τ_D, and τ_I. The transfer function for this controller can be obtained from the Laplace transform of Eq. (10.9); thus

$$\frac{P(s)}{\epsilon(s)} = K_c\left(1 + \tau_D s + \frac{1}{\tau_I s}\right) \tag{10.10}$$

Motivation for Addition of Integral and Derivative Control Modes

Having introduced ideal transfer functions for integral and derivative modes of control, we now wish to indicate the practical motivation for use of these modes. The curves of Fig. 10.7 show the behavior of a typical, feedback control system using different kinds of control when it is subjected to a permanent disturbance. This may be visualized in terms of the tank-temperature control system of Chap. 1 after a step change in T_i. The value of the controlled variable is seen to rise

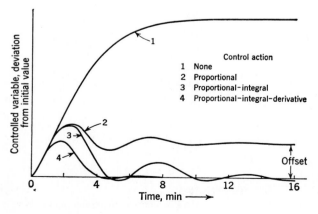

Control action

1 None
2 Proportional
3 Proportional-integral
4 Proportional-integral-derivative

Offset

FIGURE 10-7
Response of a typical control system showing the effects of various modes of control.

at time zero owing to the disturbance. With no control, this variable continues to rise to a new steady-state value. With control, after some time the control system begins to take action to try to maintain the controlled variable close to the value that existed before the disturbance occurred.

With proportional action only, the control system is able to arrest the rise of the controlled variable and ultimately bring it to rest at a new steady-state value. The difference between this new steady-state value and the original value is called *offset*. For the particular system shown, the offset is seen to be only 22 percent of the ultimate change that would have been realized for this disturbance in the absence of control.

As shown by the PI curve, the addition of integral action eliminates the offset; the controlled variable ultimately returns to the original value. This advantage of integral action is balanced by the disadvantage of a more oscillatory behavior.

The addition of derivative action to the PI action gives a definite improvement in the response. The rise of the controlled variable is arrested more quickly, and it is returned rapidly to the original value with little or no oscillation. Discussion of the PD mode is deferred to a later chapter.

The selection among the control systems whose responses are shown in Fig. 10.7 depends on the particular application. If an offset of 22 percent is tolerable, proportional action would likely be selected. If no offset were tolerable, integral action would be added. If excessive oscillations had to be eliminated, derivative action might be added. The addition of each mode means, as we shall see in later chapters, more difficult controller adjustment. Our goal in forthcoming chapters will be to present the material that will enable the reader to develop curves such as those of Fig. 10.7 and thereby to design efficient, economic control systems.

SUMMARY

In this chapter we have presented a brief discussion of control valves and controllers. In addition, we have presented ideal transfer functions to represent their dynamic behavior and some typical results of using these controllers.

The ideal transfer functions actually describe the action of many types of controllers, including pneumatic, electronic, computer-based, hydraulic, mechanical, and electrical systems. Hence, the mathematical analyses of control systems to be presented in later chapters, which are based upon first- and second-order systems, transportation lags, and ideal controllers, generalize to many branches of the control field. After studying this text on process control, the reader should be able to apply the knowledge to, for example, problems in mechanical control systems. All that is required is a preliminary study of the physical nature of the systems involved.

PROBLEMS

10.1. A pneumatic PI controller has an output pressure of 10 psi when the set point and pen point are together. The set point and pen point are suddenly displaced by 0.5 in. (i.e., a step change in error is introduced) and the following data are obtained:

Time, sec	psig
0−	10
0+	8
20	7
60	5
90	3.5

Determine the actual gain (psig per inch displacement) and the integral time.

10.2. A unit-step change in error is introduced into a PID controller. If $K_c = 10$, $\tau_I = 1$, and $\tau_D = 0.5$, plot the response of the controller, $P(t)$.

10.3. An ideal PD controller had the transfer function

$$\frac{P}{\epsilon} = K_c(\tau_D s + 1)$$

An actual PD controller had the transfer function

$$\frac{P}{\epsilon} = K_c \frac{\tau_D s + 1}{(\tau_D/\beta)s + 1}$$

where β is a large constant in an industrial controller.

   If a unit-step change in error is introduced into a controller having the second transfer function, show that

$$P(t) = K_c(1 + Ae^{-\beta t/\tau_D})$$

where A is a function of β which you are to determine. For $\beta = 5$ and $K_c = 0.5$, plot $P(t)$ versus t/τ_D. As $\beta \to \infty$, show that the unit-step response approaches that for the ideal controller.

10.4. A PID controller is at steady state with an output pressure of 9 psig. The set point and pen point are initially together. At time $t = 0$, the set point is moved away from the pen point at a rate of 0.5 in./min. The motion of the set point is in the direction of *lower* readings. If the knob settings are

$$K_c = 2 \text{ psig/in. of pen travel}$$
$$\tau_i = 1.25 \text{ min}$$
$$\tau_D = 0.4 \text{ min}$$

plot the output pressure versus time.

10.5. The input (ϵ) to a PI controller is shown in Fig. P10.5. Plot the output of the controller if $K_c = 2$ and $\tau_I = 0.50$ min.

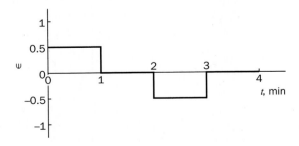

FIGURE P10-5

BLOCK DIAGRAM OF A CHEMICAL-REACTOR CONTROL SYSTEM

To tie together the principles developed thus far and to illustrate further the procedure for reduction of a physical control system to a block diagram, we consider in this chapter the two-tank chemical-reactor control system of Fig. 11.1. This entire chapter serves as an example and may be omitted by the reader with no loss in continuity.

Description of System

A liquid stream enters tank 1 at a volumetric flow rate F cfm and contains reactant A at a concentration of c_0 moles A/ft^3. Reactant A decomposes in the tanks according to the irreversible chemical reaction

$$A \rightarrow B$$

The reaction is first-order and proceeds at a rate

$$r = kc$$

where r = moles A decomposing/$(\text{ft}^3)(\text{time})$

$\quad\quad c$ = concentration of A, moles A/ft^3

$\quad\quad k$ = velocity constant, a function of temperature

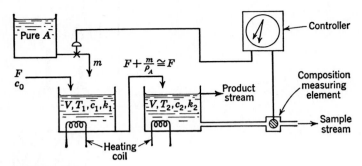

FIGURE 11-1
Control of a stirred-tank chemical reactor.

The reaction is to be carried out in a series of two stirred tanks. The tanks are maintained at different temperatures. The temperature in tank 2 is to be greater than the temperature in tank 1, with the result that k_2, the velocity constant in tank 2, is greater than that in tank 1, k_1. We shall neglect any changes in physical properties due to chemical reaction.

The purpose of the control system is to maintain c_2, the concentration of A leaving tank 2, at some desired value in spite of variation in inlet concentration c_0. This will be accomplished by adding a stream of pure A to tank 1 through a control valve.

Reactor Transfer Functions

We begin the analysis by making a material balance on A around tank 1; thus

$$V\frac{dc_1}{dt} = Fc_0 - \left(F + \frac{m}{\rho_A}\right)c_1 - k_1Vc_1 + m \tag{11.1}$$

where m = molar flow rate of pure A through the valve, lb moles/min
ρ_A = density of pure A, lb moles/ft^3
V = holdup volume of tank, a constant, ft^3

It is assumed that the volumetric flow of A through the valve m/ρ_A is much less than the inlet flow rate F with the result that Eq. (11.1) can be written

$$V\frac{dc_1}{dt} + (F + k_1V)c_1 = Fc_0 + m \tag{11.2}$$

This last equation may be written in the form

$$\tau_1\frac{dc_1}{dt} + c_1 = \frac{1}{1 + k_1V/F}c_0 + \frac{1}{F(1 + k_1V/F)}m \tag{11.3}$$

where

$$\tau_1 = \frac{V}{F + k_1V}$$

At steady state, $dc_1/dt = 0$, and Eq. (11.3) becomes

$$c_{1_s} = \frac{1}{1 + k_1 V/F} c_{0_s} + \frac{1}{F(1 + k_1 V/F)} m_s \tag{11.4}$$

where s refers to steady state.

Subtracting Eq. (11.4) from (11.3) and introducing the deviation variables

$$C_1 = c_1 - c_{1_s}$$
$$C_0 = c_0 - c_{0_s}$$
$$M = m - m_s$$

give

$$\tau_1 \frac{dC_1}{dt} + C_1 = \frac{1}{1 + k_1 V/F} C_0 + \frac{1}{F(1 + k_1 V/F)} M \tag{11.5}$$

Taking the transform of Eq. (11.5) yields the transfer function of the first reactor:

$$C_1(s) = \frac{1/(1 + k_1 V/F)}{\tau_1 s + 1} C_0(s) + \frac{1/[F(1 + k_1 V/F)]}{\tau_1 s + 1} M(s) \tag{11.6}$$

A material balance on A around tank 2 gives

$$V \frac{dc_2}{dt} = F(c_1 - c_2) - k_2 V c_2 \tag{11.7}$$

As with tank 1, this last equation can be written in terms of deviation variables and arranged to give

$$\tau_2 \frac{dC_2}{dt} + C_2 = \frac{1}{1 + k_2 V/F} C_1 \tag{11.8}$$

where

$$\tau_2 = \frac{V}{F + k_2 V}$$
$$C_2 = c_2 - c_{2_s}$$

Taking the transform of Eq. (11.8) gives the transfer function for the second reactor:

$$C_2(s) = \frac{1/(1 + k_2 V/F)}{\tau_2 s + 1} C_1(s) \tag{11.9}$$

To obtain some numerical results, we shall assume the following data to apply to the system:

$$\text{Molecular weight of } A = 100 \text{ lb/lb mole}$$
$$\rho_A = 0.8 \text{ lb mole/ft}^3$$
$$c_{0_s} = 0.1 \text{ lb mole } A/\text{ft}^3$$
$$F = 100 \text{ cfm}$$
$$m_s = 1.0 \text{ lb mole/min}$$

$$k_1 = \tfrac{1}{6} \text{ min}^{-1}$$

$$k_2 = \tfrac{2}{3} \text{ min}^{-1}$$

$$V = 300 \text{ ft}^3$$

Substituting these constants into the parameters of the problem yields the following values:

$$\tau_1 = 2 \text{ min}$$

$$\tau_2 = 1 \text{ min}$$

$$c_{1_s} = 0.0733 \text{ lb mole } A/\text{ft}^3$$

$$c_{2_s} = 0.0244 \text{ lb mole } A/\text{ft}^3$$

$$m_s/\rho_A = 1.25 \text{ cfm}$$

Control Valve

Assume that the control valve selected for the process has the following characteristics: The flow of A through the valve varies linearly from zero to 2 cfm as the valve-top pressure varies from 3 to 15 psig. The time constant τ_v of the valve is so small compared with the other time constants in the system that its dynamics can be neglected.

From the data given, the valve sensitivity is computed as

$$K_v = \frac{2 - 0}{15 - 3} = \tfrac{1}{6} \text{ cfm/psi}$$

Since $m_s/\rho_A = 1.25$ cfm, the normal operating pressure on the valve is

$$p_s = 3 + \frac{1.25}{2}(15 - 3) = 10.5 \text{ psi} \tag{11.10}$$

The equation for the valve is therefore

$$m = [1.25 + K_v(p - 10.5)]\rho_A \tag{11.11}$$

In terms of deviation variables, this can be written

$$M = K_v \rho_A P \tag{11.12}$$

where

$$M = m - 1.25\rho_A$$

$$P = p - 10.5$$

Taking the transform of Eq. (11.12) gives

$$\frac{M(s)}{P(s)} = K_v \rho_A \tag{11.13}$$

as the valve transfer function.

Measuring Element

For illustration, assume that the measuring element converts concentration of A to a pneumatic signal. Specifically, the output of the measuring element varies from 3 to 15 psig as the concentration of A varies from 0.01 to 0.05 lb mole A/ft^3. We shall assume that the concentration measuring device is linear and has negligible lag. The sensitivity (or gain) of the measuring device is therefore

$$K_m \frac{15 - 3}{0.05 - 0.01} = 300 \text{ psi/(lb mole/ft}^3)$$

Since c_{2_s} is 0.0244 lb mole/ft^3, the normal signal from the measuring device is

$$\frac{0.0244 - 0.01}{0.05 - 0.01}(15 - 3) + 3.0 = 4.32 + 3.0 = 7.32 \text{ psig}$$

The equation for the measuring device is therefore

$$b = 7.32 + K_m(c_2 - 0.0244) \tag{11.14}$$

where b is the output pressure (psig) from the measuring device. In terms of deviation variables, Eq. (11.14) becomes

$$B = K_m C_2 \tag{11.15}$$

where $B = b - 7.32$ and $C_2 = c_2 - c_{2_s}$.

The transfer function for the measuring device is therefore

$$\frac{B(s)}{C_2(s)} = K_m \tag{11.16}$$

A measuring device that changes the units between input and output signals is called a transducer; in the present case, the concentration signal is transduced to a pneumatic signal.

Controller

For convenience, we shall assume the controller to have proportional action, in which case the relation between controller output pressure and error is

$$p = p_s + K_c(c_R - b) = p_s + K_c \epsilon \tag{11.17}$$

where c_R = desired pneumatic signal (or set point), psig
K_c = controller sensitivity, psig/psig
ϵ = error = $c_R - b$, psig

In terms of deviation variables, Eq. (11.17) becomes

$$P = K_c \epsilon \tag{11.18}$$

The transform of this equation gives the transfer function of the controller

$$\frac{P(s)}{\epsilon(s)} = K_c \tag{11.19}$$

Assuming the set point and the signal from the measuring device to be the same when the system is at steady state under normal conditions, we have for the reference value of the set point

$$c_{R_s} = b = 7.32 \text{ psig}$$

The corresponding deviation variable for the set point is

$$C_R = c_R - c_{R_s}$$

Transportation Lag

A portion of the liquid leaving tank 2 is continuously withdrawn through a sample line, containing a concentration-measuring element, at a rate of 0.1 cfm. The measuring element must be remotely located from the process, because rigid ambient conditions must be maintained for accurate concentration measurements. The sample line has a length of 50 ft, and the cross-sectional area of the line is 0.001 ft^2.

The sample line can be represented by a transportation lag with parameter

$$\tau_d = \frac{\text{volume}}{\text{flow rate}} = \frac{(50)(0.001)}{0.1} = 0.5 \text{ min}$$

The transfer function for the sample line is, therefore,

$$e^{-\tau_d s} = e^{-0.5s}$$

Block Diagram

We have now completed the analysis of each component of the control system and have obtained a transfer function for each. These transfer functions can now be combined so that the overall system is represented by the block diagram in Fig. 11.2.

In Fig. 11.2, a block containing the transfer function K_m is placed at the positive inlet of the comparator in order to relate the set point in concentration units to a pneumatic signal, which matches the units of the feedback signal B. If the pneumatic controller in Fig. 11.2 were replaced by an electronic or computer-

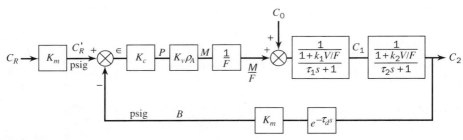

FIGURE 11-2
Block diagram for a chemical-reactor control system.

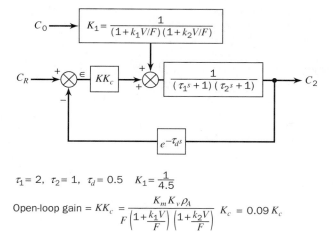

$\tau_1 = 2, \quad \tau_2 = 1, \quad \tau_d = 0.5 \quad K_1 = \dfrac{1}{4.5}$

$\text{Open-loop gain} = KK_c = \dfrac{K_m K_v \rho_A}{F\left(1 + \dfrac{k_1 V}{F}\right)\left(1 + \dfrac{k_2 V}{F}\right)} \quad K_c = 0.09 \, K_c$

FIGURE 11-3
Equivalent block diagram for a chemical-reactor control system (C_R is now in concentration units).

based controller, the block for the controller in Fig. 11.2 would be replaced by two blocks; one for the electronic controller and one for the converter, which converts the controller output (ma) to the pneumatic signal (psig). An equivalent diagram is shown in Fig. 11.3 in which some of the blocks have been combined.

Numerical quantities for the parameters in the transfer functions are given in Fig. 11.3. It should be emphasized that the block diagram is written for deviation variables. The true steady-state values, which are not given by the diagram, must be obtained from the analysis of the problem.

The example analyzed in this chapter will be used later in discussion of control system design. The design problem will be to select a value of K_c that gives satisfactory control of the composition C_2 despite the rather long transportation lag involved in getting information to the controller. In addition, we shall want to consider possible use of other modes of control for the system.

PROBLEMS

11.1. In the process shown in Fig. P11.1, the concentration of salt leaving the second tank is controlled using a proportional controller by adding concentrated solution through a control valve. The following data apply:

1. The controlled concentration is to be 0.1 lb salt/ft^3 solution. The inlet concentration c_i is always less than 0.1 lb/ft^3.
2. The concentration of concentrated salt solution is 30 lb salt/ft^3 solution.
3. Transducer: the output of the transducer varies linearly from 3 to 15 psig as the concentration varies from 0.05 to 0.15 lb/ft^3.
4. Controller: the controller is a pneumatic, direct-acting, proportional controller.
5. Control valve: as valve-top pressure varies from 3 to 15 psig, the flow through the control valve varies linearly from 0 to 0.005 cfm.

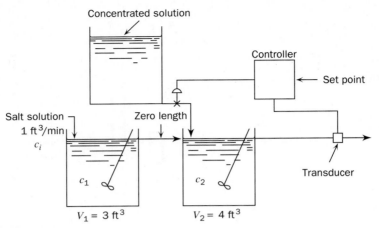

FIGURE P11-1

6. It takes 30 sec for the solution leaving the second tank to reach the transducer at the end of the pipe.

Draw a block diagram of the control system. Place in each block the appropriate transfer function. Calculate all the constants and give the units.

CHAPTER 12

CLOSED-LOOP TRANSFER FUNCTIONS

Standard Block-Diagram Symbols

In Chap. 9, a block diagram was developed for the control of a stirred-tank heater (Fig. 9.2). In Fig. 12.1, the block diagram has been redrawn and incorporates some standard symbols for the variables and transfer functions, which are widely used in the control literature. These symbols are defined as follows:

$$R = \text{set point or desired value}$$
$$C = \text{controlled variable}$$
$$\epsilon = \text{error}$$
$$B = \text{variable produced by measuring element}$$
$$M = \text{manipulated variable}$$
$$U = \text{load variable or disturbance}$$
$$G_c = \text{transfer function of controller}$$
$$G_1 = \text{transfer function of final control element}$$
$$G_2 = \text{transfer function of process}$$
$$H = \text{transfer function of measuring element}$$

In some cases, the blocks labeled G_c and G_1 will be lumped together into a single block as was done in Chap. 9. The series of blocks between the comparator and

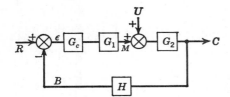

FIGURE 12-1
Standard control system nomenclature.

the controlled variable, which consist of G_c, G_1, and G_2, is referred to as the *forward path*. The block H between the controlled variable and the comparator is called the *feedback path*. The use of G for a transfer function in the forward path and H for one in the feedback path is a common convention.

The product GH, which is the product of all transfer functions ($G_c G_1 G_2 H$) in the loop, is called the *open-loop transfer function*. We call GH the open-loop transfer function because it relates the measured variable B to the set point R if the feedback loop (of Fig. 12.1) is disconnected (i.e., opened) from the comparator. The subject of this chapter is the closed-loop transfer function, which relates two variables when the loop of Fig. 12.1 is closed.

In more complex systems, the block diagram may contain several feedback paths and several loads. An example of a multiloop system, which is shown in Fig. 12.2, is cascade control. Several multiloop systems of industrial importance are presented in Chap. 18.

Overall Transfer Function for Single-Loop Systems

Once a control system has been described by a block diagram, such as the one shown in Fig. 12.1, the next step is to determine the transfer function relating C to R or C to U. We shall refer to these transfer functions as *overall* transfer functions because they apply to the entire system. These overall transfer functions are used to obtain considerable information about the control system, as will be demonstrated in the succeeding chapters. For the present it is sufficient to note that they are useful in determining the response of C to any change in R and U.

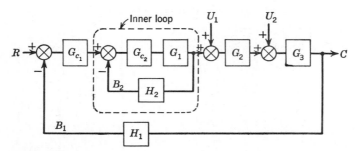

FIGURE 12-2
Block diagram for a multiloop, multiload system.

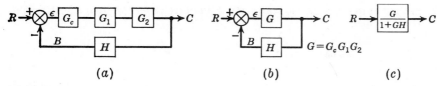

FIGURE 12-3
Block-diagram reduction to obtain overall transfer function.

The response to a change in set point R, obtained by setting $U = 0$, represents the solution to the servo problem. The response to a change in load variable U, obtained by setting $R = 0$, is the solution to the regulator problem. A systematic approach for obtaining the overall transfer function for set-point change and load change will now be presented.

Overall Transfer Function for Change in Set Point

For this case, $U = 0$ and Fig. 12.1 may be simplified or reduced as shown in Fig. 12.3. In this reduction, we have made use of a simple rule of block-diagram reduction which states that a block diagram consisting of several transfer functions in series can be simplified to a single block containing a transfer function that is the product of the individual transfer functions.

This rule can be proved by considering two noninteracting blocks in series as shown in Fig. 12.4. This block diagram is equivalent to the equations

$$\frac{Y}{X} = G_A \qquad \frac{Z}{Y} = G_B$$

Multiplying these equations gives

$$\frac{Y}{X}\frac{Z}{Y} = G_A G_B$$

which simplifies to

$$\frac{Z}{X} = G_A G_B$$

Thus, the intermediate variable Y has been eliminated, and we have shown the overall transfer function Z/X to be the product of the transfer functions $G_A G_B$. This proof for two blocks can be easily extended to any number of blocks to give the rule for the general case. This rule was developed in Chap. 7 for the specific case of several noninteracting, first-order systems in series.

FIGURE 12-4
Two noninteracting blocks in series.

With this simplification the following equations can be written directly from Fig. 12.3b.

$$C = G\epsilon \tag{12.1}$$

$$B = HC \tag{12.2}$$

$$\epsilon = R - B \tag{12.3}$$

Since there are four variables and three equations, we can solve the equations simultaneously for C in terms of R as follows:

$$C = G(R - B)$$

$$C = G(R - HC)$$

$$C = GR - GHC$$

or finally

$$\frac{C}{R} = \frac{G}{1 + GH} \tag{12.4}$$

This is the overall transfer function relating C to R and may be represented by an equivalent block diagram as shown in Fig. 12.3c.

Overall Transfer Function for Change in Load

In this case $R = 0$, and Fig. 12.1 is drawn as shown in Fig. 12.5a. From the diagram we can write the following equations:

$$C = G_2(U + M) \tag{12.5}$$

$$M = G_cG_1\epsilon \tag{12.6}$$

$$B = HC \tag{12.7}$$

$$\epsilon = -B \tag{12.8}$$

Again the number of variables (C, U, M, B, ϵ) exceeds by one the number of equations, and we can solve for C in terms of U as follows:

$$C = G_2(U + G_cG_1\epsilon)$$

$$C = G_2[U + G_cG_1(-HC)]$$

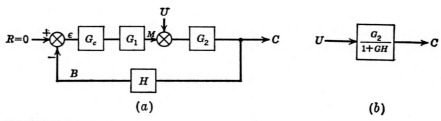

(a)   (b)

FIGURE 12-5
Block diagram for change in load.

or finally

$$\frac{C}{U} = \frac{G_2}{1 + GH} \tag{12.9}$$

where $G = G_c G_1 G_2$. Notice that the transfer functions for load change or set-point change have denominators that are identical, $1 + GH$.

The following simple rule serves to generalize these results for the single-loop feedback system shown in Fig. 12.1: the transfer function relating any pair of variables X, Y is obtained by the relationship

$$\frac{Y}{X} = \frac{\pi_f}{1 + \pi_l} \qquad \text{negative feedback} \tag{12.10}$$

where π_f = product of transfer functions in the path between the locations of the signals X and Y

π_l = product of all transfer functions in the loop (i.e., in Fig. 12.1, $\pi_l = G_c G_1 G_2 H$)

If this rule is applied to finding C/R in Fig. 12.1, we obtain

$$\frac{C}{R} = \frac{G_c G_1 G_2}{1 + G_c G_1 G_2 H} = \frac{G}{1 + GH}$$

which is the same as before. For positive feedback, the reader should show that the following result is obtained:

$$\frac{Y}{X} = \frac{\pi_f}{1 - \pi_l} \qquad \text{positive feedback} \tag{12.11}$$

Example 12.1. Determine the transfer functions C/R, C/U_1, and B/U_2 for the system show in Fig. 12.6. Also determine an expression for C in terms of R and U_1 for the situation when both set-point change and load change occur simultaneously.

Using the rule given by Eq. (12.10), we obtain by inspection the results

$$\frac{C}{R} = \frac{G_c G_1 G_2 G_3}{1 + G} \tag{12.12}$$

$$\frac{C}{U_1} = \frac{G_2 G_3}{1 + G} \tag{12.13}$$

$$\frac{B}{U_2} = \frac{G_3 H_1 H_2}{1 + G} \tag{12.14}$$

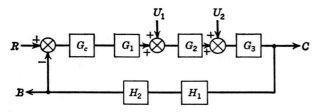

FIGURE 12-6
Block diagram for Example 12.1.

where $G = G_cG_1G_2G_3H_1H_2$. The reader should check one or more of these results by the direct method of solution of simultaneous equations.

For separate changes in R and U_1, we may obtain the response C from Eqs. (12.12) and (12.13); thus

$$C = \frac{G_cG_1G_2G_3}{1+G}R \tag{12.15}$$

and

$$C = \frac{G_2G_3}{1+G}U_1 \tag{12.16}$$

If both R and U_1 occur simultaneously, the principle of superposition requires that the overall response be the sum of the individual responses; thus

$$C = \frac{G_cG_1G_2G_3}{1+G}R + \frac{G_2G_3}{1+G}U_1 \tag{12.17}$$

Overall Transfer Function for Multiloop Control Systems

To illustrate how one obtains the overall transfer function for a multiloop system, consider the next example in which the method used is to reduce the block diagram to a single-loop diagram by application of the rules summarized by Eqs. (12.10) and (12.11).

Example 12.2. Determine the transfer function C/R for the system shown in Fig. 12.7. This block diagram represents a cascade control system, which will be discussed later.

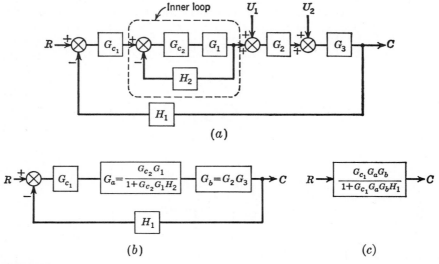

FIGURE 12-7
Block diagram reduction: (a) original diagram, (b) first reduction, (c) final single-block diagram.

Obtaining the overall transfer function C/R for the system represented by Fig. 12.7a is straightforward if we first reduce the inner loop (or minor loop) involving G_{c_2}, G_1, and H_2 to a single block, as we have just done in the case of Fig. 12.1. For convenience, we may also combine G_2 and G_3 into a single block. These reductions are shown in Fig. 12.7b. Figure 12.7b is a single-loop block diagram that can be reduced to one block as shown in Fig. 12.7c.

It should be clear without much detail that to find any other transfer function such as C/U_1 in Fig. 12.7a, we proceed in the same manner, i.e., first reduce the inner loop to a single-block equivalent.

SUMMARY

In this chapter, we have illustrated the procedure for reducing the block diagram of a control system to a single block that relates one input to one output variable. This procedure consists of writing, directly from the block diagram, a sufficient number of linear algebraic equations and solving them simultaneously for the transfer function of the desired pair of variables. For single-loop control systems, a simple rule was developed for finding the transfer function between any desired pair of input-output variables. This rule is also useful in reducing a multiloop system to a single-loop system.

It should be emphasized that regardless of the pair of variables selected, the denominator of the closed-loop transfer function will always contain the same term, $1 + G$, where G is the open-loop transfer function of the single-loop control system. In the succeeding chapters, frequent use will be made of the material in this chapter to determine the overall response of control systems.

PROBLEMS

12.1. Determine the transfer function $Y(s)/X(s)$ for the block diagrams shown in Fig. P12.1. Express the results in terms of G_a, G_b, and G_c.

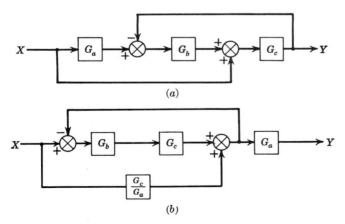

(a)

(b)

FIGURE P12-1

12.2. Find the transfer function $Y(s)/X(s)$ of the system shown in Fig. P12.2.

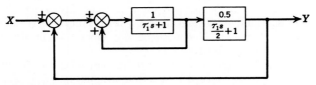

FIGURE P12-2

12.3. For the control system shown in Fig. P12.3 determine the transfer function $C(s)/R(s)$.

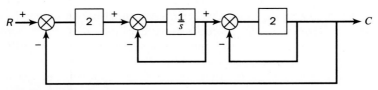

FIGURE P12-3

12.4. Derive the transfer function Y/X for the control system shown in Fig. P12.4.

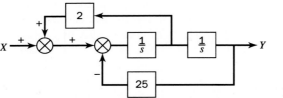

FIGURE P12-4

13

TRANSIENT RESPONSE OF SIMPLE CONTROL SYSTEMS

In this chapter the results of all the previous chapters will be applied to determining the transient response of a simple control system to changes in set point and load.* Considerable use will be made of the results of Chaps. 5 through 8 (Part II) because the overall transfer functions for the examples presented here reduce to first- and second-order systems.

Consider the control system for the heated, stirred tank that has been discussed in Chaps. 1 and 9 and is represented by Fig. 13.1. The reader may want to refer to Chap. 9 for a description of this control system.

In Fig. 13.1a, the sketch of the apparatus is drawn in such a way that the source of heat (electricity or steam) is not specified. To make this problem more realistic, we have shown in Fig. 13.1b that the source of heat is steam that is discharged directly into the water and in Fig. 13.1c the source of heat is electrical. In the latter drawing, a device known as a power controller provides electrical power to a resistance heater proportional to the signal from the controller.

---

*The reader who is interested in the simulation of control systems by digital computer is advised to study Chap. 34 at this point.

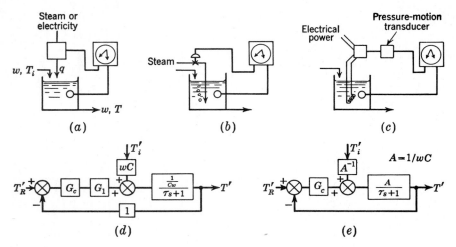

FIGURE 13-1
Block diagram of temperature-control system.

The block diagram is shown in Fig. 13.1*d*. The block representing the process is taken directly from Fig. 9.3. To reduce the number of symbols $1/wC$ has been replaced by A in Fig. 13.1*e*.

Throughout this chapter, we shall assume that the valve does not have any dynamic lag, for which case the transfer function of the valve (G_1 in Fig. 13.1) will be taken as a constant K_v. This assumption was shown to be reasonable in Chap. 10. To simplify the discussion further, K_v has been taken as 1. (If K_v were other than 1, we may simply replace G_c by G_cK_v in the ensuing discussion.)

In the first part of the chapter, we shall also assume that there is no dynamic lag in the measuring element ($\tau_m = 0$), so that it may be represented by a transfer function that is simply the constant 1. A bare thermocouple will have a response that is so fast that for all practical purposes it can be assumed to follow the slowly changing bath temperature without lag. When the feedback transfer function is unity, the system is called a *unity-feedback* system.

Introducing these assumptions leads to the simplified block diagram of Fig. 13.1*e*, for which we shall obtain overall transfer functions for changes in set point and load when proportional and proportional-integral control are used.

Proportional Control for Set-Point Change (Servo Problem)

For proportional control, $G_c = K_c$. Using the methods developed in the previous chapter, the overall transfer function in Fig. 13.1*e* is

$$\frac{T'}{T'_R} = \frac{K_cA/(\tau s + 1)}{1 + K_cA/(\tau s + 1)} = \frac{K_cA}{\tau s + 1 + K_cA} \tag{13.1}$$

This may be rearranged in the form of a first-order lag to give

$$\frac{T'}{T'_R} = \frac{A_1}{\tau_1 s + 1} \tag{13.2}$$

where $\tau_1 = \dfrac{\tau}{1 + K_cA}$

$$A_1 = \frac{K_cA}{1 + K_cA} = \frac{1}{1 + 1/K_cA}$$

According to this result, the response of the tank temperature to change in set point is first-order. The time constant for the control system, τ_1, is less than that of the stirred tank itself, τ. This means that one of the effects of feedback control is to speed up the response. We may use the results of Chap. 5 to find the response to a variety of inputs.

The response of the system to a unit-step change in set point T'_R is shown in Fig. 13.2. (We have selected a unit change in set point for convenience; responses to steps of other magnitudes are obtained by superposition.) For this case of a unit-step change in set point, T' approaches $A_1 = K_cA/(1 + K_cA)$, a fraction of unity. The desired change is, of course, 1. Thus, the ultimate value of the temperature $T'(\infty)$ does not match the desired change. This discrepancy is called *offset* and is defined as

$$\text{Offset} = T'_R(\infty) - T'(\infty) \tag{13.3}$$

In terms of the particular control system parameters

$$\text{Offset} = 1 - \frac{K_cA}{1 + K_cA} = \frac{1}{1 + K_cA} \tag{13.4}$$

This discrepancy between set point and tank temperature at steady state is characteristic of proportional control. In some cases offset cannot be tolerated. However, notice from Eq. (13.4) that the offset decreases as K_c increases, and in theory the offset could be made as small as desired by increasing K_c to a sufficiently large value. To give a full answer to the problem of eliminating offset by high controller gain requires a discussion of stability and the response of the system when other lags, which have been neglected, are included in the system. Both these subjects are to be covered later. For the present we shall simply say that whether or not proportional control is satisfactory depends on the amount of offset that can be tolerated, the speed of response of the system, and the amount of gain that can be provided by the controller without causing the system to go unstable.

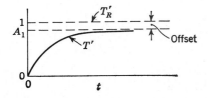

FIGURE 13-2
Unit-step response for set-point change (P control).

Proportional Control for Load Change (Regulator Problem)

The same control system shown in Fig. 13.1e is to be considered. This time the set point remains fixed; that is, $T_R' = 0$. We are interested in the response of the system to a change in the inlet stream temperature, i.e., to a load change.

Using the methods of Chap. 12, the overall transfer function becomes

$$\frac{T'}{T_i'} = \frac{AA^{-1}/(\tau s + 1)}{1 + K_c A/(\tau s + 1)} = \frac{1}{\tau s + 1 + K_c A} \tag{13.5}$$

This may be arranged in the form of the first-order lag; thus

$$\frac{T'}{T_i'} = \frac{A_2}{\tau_1 s + 1} \tag{13.6}$$

where $A_2 = \dfrac{1}{1 + K_c A}$

$\tau_1 = \dfrac{\tau}{1 + K_c A}$

As for the case of set-point change, we have an overall response that is first-order. The overall time constant τ_1 is the same as for set-point changes. The response of the system to a unit-step change in inlet temperature T_i' is shown in Fig. 13.3. It may be seen that T' approaches $1/(1 + K_c A)$. To demonstrate the benefit of control, we have shown the response of the tank temperature (open-loop response) to a unit-step change in inlet temperature if no control were present; that is, $K_c = 0$. In this case, the major advantage of control is in reduction of offset. From Eq. (13.3), the offset becomes

$$\text{Offset} = T_R'(\infty) - T'(\infty) = 0 - \frac{1}{1 + K_c A}$$

$$= -\frac{1}{1 + K_c A} \tag{13.7}$$

As for the case of a step change in set point, the offset is reduced as controller gain K_c is increased.

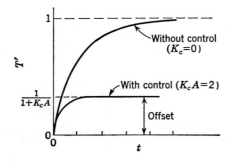

FIGURE 13-3
Unit-step response for load change (P control).

Proportional-Integral Control for Load Change

In this case, we replace G_c in Fig. 13.1e by $K_c(1 + 1/\tau_I s)$. The overall transfer function for load change is therefore

$$\frac{T'}{T'_i} = \frac{AA^{-1}/(\tau s + 1)}{1 + [K_cA/(\tau s + 1)](1 + 1/\tau_I s)} \tag{13.8}$$

Rearranging this gives

$$\frac{T'}{T'_i} = \frac{\tau_I s}{(\tau s + 1)(\tau_I s) + K_cA(\tau_I s + 1)}$$

or

$$\frac{T'}{T'_i} = \frac{\tau_I s}{\tau\tau_I s^2 + (K_cA\tau_I + \tau_I)s + K_cA}$$

Since the denominator contains a quadratic expression, the transfer function may be written in the standard form of the transportation lag to give

$$\frac{T'}{T'_i} = \frac{(\tau_I/K_cA)s}{(\tau\tau_I/K_cA)s^2 + \tau_I(1 + 1/K_cA)s + 1}$$

or

$$\frac{T'}{T'_i} = \frac{A_1 s}{\tau_1^2 s^2 + 2\zeta\tau_1 s + 1} \tag{13.9}$$

where $A_1 = \dfrac{\tau_I}{K_cA}$

$$\tau_1 = \sqrt{\frac{\tau\tau_I}{K_cA}}$$

$$\zeta = \frac{1}{2}\sqrt{\frac{\tau_I}{\tau}}\frac{1 + K_cA}{\sqrt{K_cA}}$$

For a unit-step change in load, $T'_i = 1/s$. Combining this with Eq. (13.9) gives

$$T' = \frac{A_1}{\tau_1^2 s^2 + 2\zeta\tau_1 s + 1} \tag{13.10}$$

Equation (13.10) shows that the response of the tank temperature is equivalent to the response of a second-order system to an impulse function of magnitude A_1. Since we have studied the impulse response of a second-order system in Chap. 8, the solution to the present problem is already known. This justifies in part our previous work on transients. Using Eq. (8.31), the impulse response for this system may be written for $\zeta < 1$ as

$$T' = A_1\left(\frac{1}{\tau_1}\frac{1}{\sqrt{1 - \zeta^2}}e^{-\zeta t/\tau_1}\sin\sqrt{1 - \zeta^2}\frac{t}{\tau_1}\right) \tag{13.11}$$

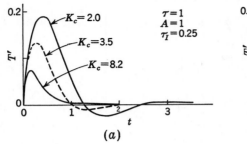
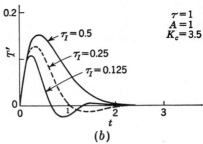

FIGURE 13-4
Unit-step response for load change (PI control).

Although the response of the system can be determined from Eq. (13.11) or Fig. 8.5, the effect of varying K_c and τ_I on the system response can be seen more clearly by plotting response curves, such as those shown in Fig. 13.4. From Fig. 13.4a, we see that an increase in K_c, for a fixed value of τ_I, improves the response by decreasing the maximum deviation and by making the response less oscillatory. The formula for ζ in Eq. (13.9) shows that ζ increases with K_c, which indicates that the response is less oscillatory. Figure 13.4b shows that, for a fixed value of K_c, a decrease in τ_I decreases the maximum deviation and period. However, a decrease in τ_I causes the response to become more oscillatory, which means that ζ decreases. This effect of τ_I on the oscillatory nature of the response is also given by the formula for ζ in Eq. (13.9).

For this case, the offset as defined by Eq. (13.3) is zero; thus

$$\text{Offset} = T'_R(\infty) - T'(\infty)$$
$$= 0 - 0 = 0$$

One of the most important advantages of PI control is the elimination of offset.

Proportional-Integral Control for Set-Point Change

Again, the controller transfer function is $K_c(1 + 1/\tau_I s)$, and we obtain from Fig. 13.1e the transfer function

$$\frac{T'}{T'_R} = \frac{K_c A(1 + 1/\tau_I s)[1/(\tau s + 1)]}{1 + K_c A(1 + 1/\tau_I s)[1/(\tau s + 1)]} \tag{13.12}$$

This equation may be reduced to the standard quadratic form to give

$$\frac{T'}{T'_R} = \frac{\tau_I s + 1}{\tau_1^2 s^2 + 2\zeta \tau_1 s + 1} \tag{13.13}$$

where τ_1 and ζ are the same functions of the parameters as in Eq. (13.9). Introducing a unit-step change ($T'_R = 1/s$) into Eq. (13.13) gives

$$T' = \frac{1}{s} \frac{\tau_I s + 1}{\tau_1^2 s^2 + 2\zeta \tau_1 s + 1} \tag{13.14}$$

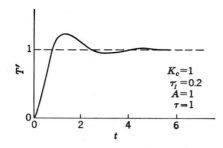

$K_c = 1$
$\tau_i = 0.2$
$A = 1$
$\tau = 1$

FIGURE 13-5
Unit-step response for set point change (PI control).

To obtain the response of T' in the time domain, Eq. (13.14) is expanded into two terms:

$$T' = \frac{\tau_I}{\tau_1^2 s^2 + 2\zeta\tau_1 s + 1} + \frac{1}{s}\frac{1}{\tau_1^2 s^2 + 2\zeta\tau_1 s + 1} \tag{13.15}$$

The first term on the right is equivalent to the response of a second-order system to an impulse function of magnitude τ_I. The second term is the unit-step response of a second-order system. It is convenient to use Figs. 8.2 and 8.5 to obtain the response for Eq. (13.15). For $\zeta < 1$, an analytic expression for T' is

$$T' = \frac{\tau_I}{\tau_1\sqrt{1-\zeta^2}}e^{-\zeta t/\tau_1}\sin\sqrt{1-\zeta^2}\frac{t}{\tau_1}$$

$$+ 1 - \frac{1}{\sqrt{1-\zeta^2}}e^{-\zeta t/\tau_1}\sin\left(\sqrt{1-\zeta^2}\frac{t}{\tau_1} + \tan^{-1}\frac{\sqrt{1-\zeta^2}}{\zeta}\right) \tag{13.16}$$

The last expression was obtained by combining Eqs. (8.17) and (8.31). A typical response for T' is shown in Fig. 13.5. The offset as defined by Eq. (13.3) is zero; thus

$$\text{Offset} = T'_R(\infty) - T'(\infty)$$
$$= 1 - 1 = 0$$

Again notice that the integral action in the controller has eliminated the offset.

Proportional Control of System with Measurement Lag

In the previous examples the lag in the measuring element was assumed to be negligible, for which case the feedback transfer function was taken as 1. We now consider the same control system, the stirred-tank heater of Fig. 13.1, with a first-order measuring element having a transfer function $1/(\tau_m s + 1)$. The block diagram for the modified system is now shown in Fig. 13.6. By the usual procedure, the transfer function for set-point changes may be written

$$\frac{T'}{T'_R} = \frac{A_1(\tau_m s + 1)}{\tau_2^2 s^2 + 2\zeta_2\tau_2 s + 1} \tag{13.17}$$

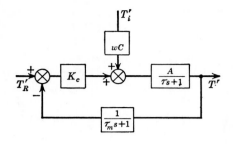

FIGURE 13-6
Control system with measurement lag.

where

$$A_1 = \frac{K_c A}{1 + K_c A}$$

$$\tau_2 = \sqrt{\frac{\tau \tau_m}{1 + K_c A}}$$

$$\zeta_2 = \frac{\tau + \tau_m}{2\sqrt{\tau \tau_m}} \frac{1}{\sqrt{1 + K_c A}}$$

We shall not obtain an expression for the transient response for this case, for it will be of the same form as Eq. (13.16). Adding the first-order measuring lag to the control system of Fig. 13.1 produces a second-order system even for proportional control. This means there will be an oscillatory response for an appropriate choice of the parameters τ, τ_m, K_c, and A. In order to understand the effect of gain K_c and measuring lag τ_m on the behavior of the system, response curves are shown in Fig. 13.7 for various combinations of K_c and τ_m for a fixed value of $\tau = 1$. In general, the response becomes more oscillatory, or less stable, as K_c or τ_m increases.

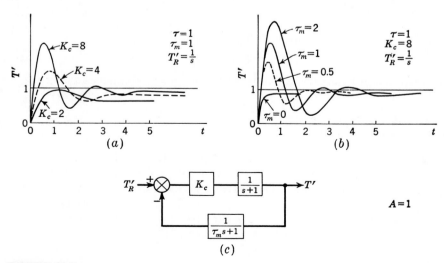

FIGURE 13-7
Effect of controller gain and measuring lag on system response for unit-step change in set point.

For a fixed value of $\tau_m = 1$, Fig. 13.7a shows that the offset is reduced as K_c increases; however, this improvement in steady-state performance is obtained at the expense of a poorer transient response. As K_c increases, the overshoot becomes excessive and the response becomes more oscillatory. In general, we shall find that a control system having proportional control will require a value of K_c that is based on a compromise between low offset and satisfactory transient response.

For a fixed value of controller gain ($K_c = 8$), Fig. 13.7b shows that an increase in measurement lag produces a poorer transient response in that the overshoot becomes greater and the response more oscillatory as τ_m increases. This behavior illustrates a general rule that the measuring element in a control system should respond quickly if satisfactory response is to be achieved.

SUMMARY

In this chapter, we have confined our attention to the response of simple control systems that were either first-order or second-order. This means that the transient response can be found by referring to Chaps. 5 and 8. However, if integral action were added to the controller in the system of Fig. 13.6, the overall transfer function would have a third-order polynomial in the denominator. Inversion would require factoring a cubic, which is generally a difficult task. Actually, systems with denominator polynomials of order greater than two are the rule rather than the exception. Hence, we shall develop in forthcoming chapters convenient techniques for studying the response of higher-order control systems. These techniques will be of direct use in control system design.

In Chap. 1, PI control of a heated, stirred tank with measurement lag was discussed. It was indicated that incorrect selection of controller parameters could lead to a response with increasing amplitude. These unstable responses can occur in all systems with third- or higher-order polynomials in the denominator of the overall transfer function. In the next chapter, we shall present a concrete definition of stability and begin the development of methods for determining stability in control systems.

PROBLEMS

13.1. The set point of the control system shown in Fig. P13.1 is given a step change of 0.1 unit. Determine:
   (a) The maximum value of C and the time at which it occurs.
   (b) The offset.
   (c) The period of oscillation.
   Draw a sketch of $C(t)$ as a function of time.

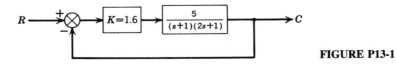

FIGURE P13-1

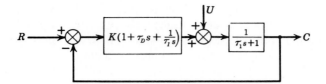

13.2. The control system shown in Fig. P13.2 contains a three-mode controller.
  (*a*) For the closed loop, develop formulas for the natural period of oscillation τ and the damping factor ζ in terms of the parameters K, τ_D, τ_I, and τ_1.

  For the following parts, $\tau_D = \tau_I = 1$ and $\tau_1 = 2$,

  (*b*) Calculate ζ when K is 0.5 and when K is 2.
  (*c*) Do ζ and τ approach limiting values as K increases, and if so, what are these values?
  (*d*) Determine the offset for a unit-step change in load if K is 2.
  (*e*) Sketch the response curve (C versus t) for a unit-step change in load when K is 0.5 and when K is 2.
  (*f*) In both cases of part (*e*) determine the maximum value of C and the time at which it occurs.

13.3. The location of a load change in a control loop may affect the system response. In the block diagram shown in Fig. P13.3, a unit-step change in load enters at either location 1 or location 2.
  (*a*) What is the frequency of the transient response when the load enters at location 1 and when the load enters at location 2?
  (*b*) What is the offset when the load enters at location 1 and when it enters at location 2?
  (*c*) Sketch the transient response to a step change in U_1 and to a step change in U_2.

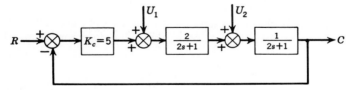

FIGURE P13-3

13.4. Consider the liquid-level control system shown in Fig. P13.4. The tanks are non-interacting. The following information is known:

  1. The resistances on the tanks are linear. These resistances were tested separately, and it was found that, if the steady-state flow rate q cfm is plotted against steady-state tank level h ft, the slope of the line dq/dh is 2 ft^2/min.
  2. The cross-sectional area of each tank is 2 ft^2.
  3. The control valve was tested separately, and it was found that a change of 1 psi in pressure to the valve produced a change in flow of 0.1 cfm.
  4. There is no dynamic lag in the valve or the measuring element.

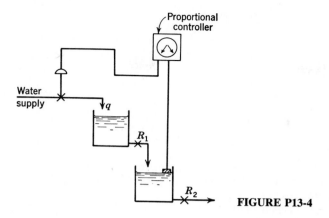

FIGURE P13-4

(a) Draw a block diagram of this control system, and in each block give the transfer function, with numerical values of the parameters.

(b) Determine the controller gain K_c for a critically damped response.

(c) If the tanks were connected so that they were interacting, what is the value of K_c needed for critical damping?

(d) Using 1.5 times the value of K_c determined in part (c), determine the response of the level in tank 2 to a step change in set point of 1 in. of level.

13.5. A PD controller is used in a control system having a first-order process and a measurement lag as shown in Fig. P13.5.

(a) Find expressions for ζ and τ for the closed-loop response.

(b) If $\tau_1 = 1$ min, $\tau_m = 10$ sec, find K_c so that $\zeta = 0.7$ for the two cases: (1) $\tau_D = 0$, (2) $\tau_D = 3$ sec.

(c) Compare the offset and period realized for both cases, and comment on the advantage of adding the derivative mode.

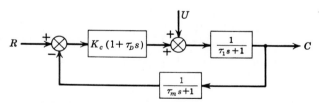

FIGURE P13-5

13.6. The thermal system shown in Fig. P13.6 is controlled by a PD controller.

$$\text{Data: } w = 250 \text{ lb/min}$$
$$\rho = 62.5 \text{ lb/ft}^3$$
$$V_1 = 4 \text{ ft}^3$$
$$V_2 = 5 \text{ ft}^3$$
$$V_3 = 6 \text{ ft}^3$$
$$C = 1 \text{ Btu/(lb)(°F)}$$

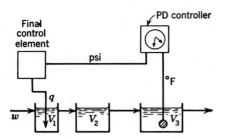

FIGURE P13-6

A change of 1 psi from the controller changes the flow rate of heat q by 500 Btu/min. The temperature of the inlet stream may vary. There is no lag in the measuring element.

(a) Draw a block diagram of the control system with the appropriate transfer function in each block. Each transfer function should contain numerical values of the parameters.

(b) From the block diagram, determine the overall transfer function relating the temperature in tank 3 to a change in set point.

(c) Find the offset for a unit-step change in inlet temperature if the controller gain K_c is 3 psi per °F of temperature error and the derivative time is 0.5 min.

13.7. (a) For the control system shown in Fig. P13.7, obtain the closed-loop transfer function C/U.

(b) Find the value of K_c for which the closed-loop response has a ζ of 2.3.

(c) Find the offset for a unit-step change in U if $K_c = 4$.

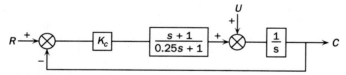

FIGURE P13-7

13.8. For the control system shown in Fig. P13.8, determine:

(a) $C(s)/R(s)$
(b) $C(\infty)$
(c) offset
(d) $C(0.5)$
(e) whether the closed-loop response is oscillatory

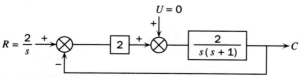

FIGURE P13-8

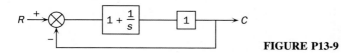

FIGURE P13-9

13.9. For the control system shown in Fig. P13.9, determine an expression for $C(t)$ if a unit-step change occurs in R. Sketch the response $C(t)$ and compute $C(2)$.

13.10. Compare the responses to a unit-step change in set point for the system shown in Fig. P13.10 for both negative feedback and positive feedback. Do this for K_c of 0.5 and 1.0. Compare these responses by sketching $C(t)$.

FIGURE P13-10

CHAPTER
14

STABILITY

CONCEPT OF STABILITY

In the previous chapter, the overall response of the control system was no higher than second-order. For these systems, the step response must resemble those of Fig. 5.6 or of Fig. 8.2. Hence, the system is inherently *stable*. In this chapter we shall consider the problem of stability in a control system (Fig. 14.1) only slightly more complicated than any studied previously. This system might represent proportional control of two stirred-tank heaters with measuring lag. In this discussion, only set-point changes are to be considered. From the methods developed in Chap. 12 for determining the overall transfer function, we have from Fig. 14.1.

$$\frac{C}{R} = \frac{K_c G}{1 + K_c G H} \tag{14.1}$$

In terms of the particular transfer functions shown in Fig. 14.1, C/R becomes, after some rearrangement,

$$\frac{C}{R} = \frac{K_c(\tau_3 s + 1)}{(\tau_1 s + 1)(\tau_2 s + 1)(\tau_3 s + 1) + K_c} \tag{14.2}$$

The denominator of Eq. (14.2) is third-order. For a unit-step change in R, the transform of the response is

$$C = \frac{1}{s} \frac{K_c(\tau_3 s + 1)}{(\tau_1 s + 1)(\tau_2 s + 1)(\tau_3 s + 1) + K_c} \tag{14.3}$$

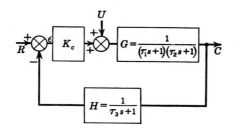

FIGURE 14-1
Third-order control system.

To obtain the transient response $C(t)$, it is necessary to find the inverse of Eq. (14.3). This requires obtaining the roots of the denominator of Eq. (14.2), which is third-order. We can no longer find these roots as easily as we did for the second-order systems by use of the quadratic formula. However, in principle they can always be obtained by algebraic methods.

It is apparent that the roots of the denominator depend on the particular values of the time constants and K_c. These roots determine the nature of the transient response, according to the rules presented in Fig. 3.1 and Table 3.1. It is of interest to examine the nature of the response for the control system of Fig. 14.1 as K_c is varied, assuming the time constants τ_1, τ_2, and τ_3 to be fixed. To be specific, consider the step response for $\tau_1 = 1$, $\tau_2 = \frac{1}{2}$, and $\tau_3 = \frac{1}{3}$ for several values of K_c. Without going into the detailed calculations at this time, the results of inversion of Eq. (14.3) are shown as response curves in Fig. 14.2. From these response curves, it is seen that, as K_c increases, the system response becomes more oscillatory. In fact, beyond a certain value of K_c, the successive amplitudes of the response grow rather than decay; this type of response is called *unstable*. Evidently, for some values of K_c, there is a pair of roots corresponding to s_4 and s_4^* of Fig. 3.1. As control system designers, we are clearly interested in being able to determine quickly the values of K_c that give unstable responses, such as that corresponding to $K_c = 12$ in Fig. 14.2.

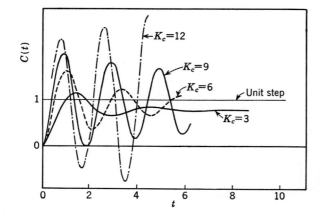

FIGURE 14-2
Response of control system of Fig. 14-1 for a unit-step change in set point.

If the order of Eq. (14.2) had been higher than three, the calculations necessary to obtain Fig. 14.2 would have been even more difficult. In the next chapter, on root-locus methods, a powerful graphical tool for finding the necessary roots will be developed. In this chapter, the focus is on developing a clearer understanding of the concept of stability. In addition, we shall develop a quick test for detecting roots having positive real parts, such as s_4 and s_4^* in Fig. 3.1.

Definition of Stability (Linear Systems)

For our purposes, a stable system will be defined as one for which the output response is bounded for all bounded inputs. A system exhibiting an unbounded response to a bounded input is unstable. This definition, although somewhat loose, is adequate for most of the linear systems and simple inputs that we shall study.

A bounded input function is a function of time that always falls within certain bounds during the course of time. For example, the step function and sinusoidal function are bounded inputs. The function $f(t) = t$ is obviously unbounded.

Although the definition of an unstable system states that the output becomes unbounded, this is true only in the mathematical sense. An actual physical system always exhibits bounds or restraints. A linear mathematical model (set of linear differential equations describing the system) from which stability information is obtained is meaningful only over a certain range of variables. For example, a linear control valve gives a linear relation between flow and valve-top pressure only over the range of pressure (or flow) corresponding to values between which the valve is shut tight or wide open. When the valve is wide open, for example, further change in pressure to the diaphragm will not increase the flow. We often describe such a limitation by the term *saturation*. A physical system, when unstable, may not follow the response of its linear mathematical model beyond certain physical bounds but rather may saturate. However, the prediction of stability by the linear model is of utmost importance in a real control system since operation with the valve shut tight or wide open is clearly unsatisfactory control.

STABILITY CRITERION

The purpose of this section is to translate the stability definition into a more simple criterion, one that can be used to ascertain the stability of control systems of the form shown in Fig. 14.3.

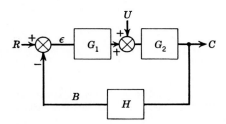

FIGURE 14-3
Basic single-loop control system.

CHARACTERISTIC EQUATION. From the block diagram of the control system (Fig. 14.3), we obtain by the methods of Chap. 12

$$C = \frac{G_1 G_2}{1 + G_1 G_2 H} R + \frac{G_2}{1 + G_1 G_2 H} U \tag{14.4}$$

In order to simplify the nomenclature, let $G = G_1 G_2 H$. We call G the *open-loop transfer function* because it relates the measured variable B to the set point R if the feedback loop of Fig. 14.3 is disconnected from the comparator (i.e., if the loop is opened). In terms of the open-loop transfer function G, Eq. (14.4) becomes

$$C = \frac{G_1 G_2}{1 + G} R + \frac{G_2}{1 + G} U \tag{14.5}$$

In principle, for given forcing functions $R(s)$ and $U(s)$, Eq. (14.5) may be inverted to give the control system response.

To determine under what conditions the system represented by Eq. (14.5) is stable, it is necessary to test the response to a bounded input. Suppose a unit-step change in set point is applied. Then

$$C(s) = \frac{G_1 G_2}{1 + G} \frac{1}{s} = \frac{G_1 G_2 F(s)}{s(s - r_1)(s - r_2) \ldots (s - r_n)} \tag{14.6}$$

where $r_1, r_2, \ldots, r_n$ are the n roots of the equation

$$1 + G(s) = 0 \tag{14.7}$$

and $F(s)$ is a function that arises in the rearrangement to the right-hand form of Eq. (14.6). Equation (14.7) is called the *characteristic equation* for the control system of Fig. 14.3. For example, for the control system of Fig. 14.1 the step response is

$$C(s) = \frac{G_1 G_2}{s(1 + G)}$$

$$= \frac{K_c}{(\tau_1 s + 1)(\tau_2 s + 1)} \Bigg/ s \left[1 + \frac{K_c}{(\tau_1 s + 1)(\tau_2 s + 1)(\tau_3 s + 1)} \right]$$

which may be rearranged to

$$C(s) = \frac{K_c(\tau_3 s + 1)}{s[\tau_1 \tau_2 \tau_3 s^3 + (\tau_1 \tau_2 + \tau_1 \tau_3 + \tau_2 \tau_3)s^2 + (\tau_1 + \tau_2 + \tau_3)s + (1 + K_c)]}$$

This is equivalent to

$$C(s) = \frac{K_c(\tau_3 s + 1)/\tau_1 \tau_2 \tau_3}{s(s - r_1)(s - r_2)(s - r_3)}$$

where r_1, r_2, and r_3 are the roots of the characteristic equation

$$\tau_1 \tau_2 \tau_3 s^3 + (\tau_1 \tau_2 + \tau_1 \tau_3 + \tau_2 \tau_3)s^2 + (\tau_1 + \tau_2 + \tau_3)s + (1 + K_c) = 0 \tag{14.8}$$

Evidently, for this case the function $F(s)$ in Eq. (14.6) is

$$F(s) = \frac{(\tau_1 s + 1)(\tau_2 s + 1)(\tau_3 s + 1)}{\tau_1 \tau_2 \tau_3}$$

In Chap. 3, the qualitative nature of the inverse transforms of equations such as Eq. (14.6) was discussed. It was shown that (see Fig. 3.1 and Table 3.1), if there are any of the roots $r_1, r_2, \ldots, r_n$ in the right half of the complex plane, the response $C(t)$ will contain a term that grows exponentially in time and the system is unstable. If there are one or more roots of the characteristic equation at the origin, there is an s^m in the denominator of Eq. (14.6) (where $m \geq 2$) and the response is again unbounded, growing as a polynomial in time. This condition specifies m as greater than or equal to 2, not 1, because one of the s terms in the denominator is accounted for by the fact that the input is a unit-step ($1/s$) in Eq. (14.6). If there is a pair of conjugate roots on the imaginary axis, the contribution to the overall step response is a pure sinusoid, which is bounded. However, if the bounded input is taken as $\sin \omega t$, where ω is the imaginary part of the conjugate roots, the contribution to the overall response is a sinusoid with an amplitude that increases as a polynomial in time.

It is evident from Eq. (14.5) that precisely the same considerations apply to a change in U. Therefore, the definition of *stability for linear systems* may be translated to the following criterion: a linear control system is unstable if any roots of its characteristic equation are on, or to the right of, the imaginary axis. Otherwise the system is stable.

It is important to note that the characteristic equation of a control system, which determines its stability, is the same for set-point or load changes. It depends only on $G(s)$, the open-loop transfer function. Furthermore, although the rules derived above were based on a step input, they are applicable to any input. This is true, first, by the definition of stability and, second, because if there is a root of the characteristic equation in the right half plane, it contributes an unbounded term in the response to any input. This follows from Eq. (14.5) after it is rearranged to the form of Eq. (14.6) for the particular input.

Therefore, the stability of a control system of the type shown in Fig. 14.3 is determined solely by its open-loop transfer function through the roots of the characteristic equation.

Example 14.1. In terms of Fig. 14.3, a control system has the transfer functions

$$G_1 = 10\frac{0.5s + 1}{s} \qquad \text{(PI controller)}$$

$$G_2 = \frac{1}{2s + 1} \qquad \text{(stirred tank)}$$

$$H = 1 \qquad \text{(measuring element without lag)}$$

We have suggested a physical system by the components placed in parentheses. Find the characteristic equation and its roots, and determine whether the system is stable.

The first step is to write the open-loop transfer function:

$$G = G_1 G_2 H = \frac{10(0.5s + 1)}{s(2s + 1)}$$

The characteristic equation is therefore

$$1 + \frac{10(0.5s + 1)}{s(2s + 1)} = 0$$

which is equivalent to

$$s^2 + 3s + 5 = 0$$

Solving by the quadratic formula gives

$$s = \frac{-3}{2} \pm \frac{\sqrt{9 - 20}}{2}$$

or

$$s_1 = \frac{-3}{2} + j \frac{\sqrt{11}}{2}$$

$$s_2 = \frac{-3}{2} - j \frac{\sqrt{11}}{2}$$

Since the real part of s_1 and s_2 is negative ($\frac{-3}{2}$), the system is stable.

ROUTH TEST FOR STABILITY

The Routh test is a purely algebraic method for determining how many roots of the characteristic equation have positive real parts; from this it can also be determined whether the system is stable, for if there are no roots with positive real parts, the system is stable. The test is limited to systems that have polynomial characteristic equations. This means that it cannot be used to test the stability of a control system containing a transportation lag. The procedure for application of the Routh test is presented without proof. The proof is available elsewhere (Routh, 1905) and is mathematically beyond the scope of this text.

The procedure for examining the roots is to write the characteristic equation in the form

$$a_0 s^n + a_1 s^{n-1} + a_2 s^{n-2} + \cdots + a_n = 0 \tag{14.9}$$

where a_0 is positive. (If a_0 is originally negative, both sides are multiplied by -1.) In this form, it is *necessary* that all the coefficients

$$a_0, a_1, a_2, \ldots, a_{n-1}, a_n$$

be positive if all the roots are to lie in the left half plane. If any coefficient is negative, the system is definitely unstable, and the Routh test is not needed to answer the question of stability. (However, in this case, the Routh test will tell us the number of roots in the right half plane.) If all the coefficients are positive, the system may be stable or unstable. It is then necessary to apply the following procedure to determine stability.

Routh Array

Arrange the coefficients of Eq. (14.9) into the first two rows of the Routh array, as follows:

Row				
1	a_0	a_2	a_4	a_6
2	a_1	a_3	a_5	a_7
3	b_1	b_2	b_3	
4	c_1	c_2	c_3	
5	d_1	d_2		
6	e_1	e_2		
7	f_1			
$n+1$	g_1			

The array has been filled in for $n = 7$ in order to simplify the discussion. For any other value of n, the array is prepared in the same manner. In general, there are $(n + 1)$ rows. For n even, the first row has one more element than the second row.

The elements in the remaining rows are found from the formulas

$$b_1 = \frac{a_1 a_2 - a_0 a_3}{a_1} \qquad b_2 = \frac{a_1 a_4 - a_0 a_5}{a_1} \cdots$$

$$c_1 = \frac{b_1 a_3 - a_1 b_2}{b_1} \qquad c_2 = \frac{b_1 a_5 - a_1 b_3}{b_1} \cdots$$

$$\cdots\cdots\cdots\cdots\cdots\cdots\cdots\cdots\cdots\cdots\cdots\cdots$$

The elements for the other rows are found from formulas that correspond to those just given. The elements in any row are always derived from the elements of the two preceding rows. During the computation of the Routh array, any row can be divided by a positive constant without changing the results of the test. (The application of this rule often simplifies the arithmetic.)

Having obtained the Routh array, the following theorems are applied to determine stability.

Theorems of the Routh Test

1. The necessary and sufficient condition for all the roots of the characteristic equation [Eq. (14.9)] to have negative real parts (stable system) is that all elements of the first column of the Routh array (a_0, a_1, b_1, c_1, etc.) be positive and nonzero.

2. If some of the elements in the first column are negative, the number of roots with a positive real part (in the right half plane) is equal to the number of sign changes in the first column.

3. If *one* pair of roots is on the imaginary axis, equidistant from the origin, and all other roots are in the left half plane, all the elements of the nth row will vanish and none of the elements of the preceding row will vanish. The location of the pair of imaginary roots can be found by solving the equation

$$Cs^2 + D = 0 \qquad (14.10)$$

where the coefficients C and D are the elements of the array in the $(n - 1)$th row as read from left to right, respectively. We shall find this last rule to be of value in the root-locus method presented in the next chapter.

The algebraic method for determining stability is limited in its usefulness in that all we can learn from it is whether a system is stable. It does not give us any idea of the degree of stability or the roots of the characteristic equation.

Example 14.2. Given the characteristic equation

$$s^4 + 3s^2 + 5s^2 + 4s + 2 = 0$$

determine the stability by the Routh criterion.

Since all the coefficients are positive, the system may be stable. To test this, form the following Routh array:

Row			
1	1	5	2
2	3	4	
3	$11/3$	$6/3$	
4	$26/11$	0	
5	2		

The elements in the array are found by applying the formulas presented in the rules; for example, b_1, which is the element in the first column, third row, is obtained by

$$b_1 = \frac{a_1 a_2 - a_0 a_3}{a_1}$$

or in terms of numerical values,

$$b_1 = \frac{(3)(5) - (1)(4)}{3} = \frac{15}{3} - \frac{4}{3} = \frac{11}{3}$$

Since there is no change in sign in the first column, there are no roots having positive real parts, and the system is stable.

In the appendix of Chap. 15, a BASIC program for computing the roots of a polynomial equation is given.

Example 14.3. (a) Using $\tau_1 = 1, \tau_2 = \frac{1}{2}, \tau_3 = \frac{1}{3}$, determine the values of K_c for which the control system in Fig. 14.1 is stable. (b) For the value of K_c for which the system is on the threshold of instability, determine the roots of the characteristic equation with the help of Theorem 3.

Solution. (a) The characteristic equation $1 + G(s) = 0$ becomes

$$1 + \frac{K_c}{(s + 1)[(s/2) + 1][(s/3) + 1]} = 0$$

Rearrangement of this equation for use in the Routh test gives

$$s^3 + 6s^2 + 11s + 6(1 + K_c) = 0 \qquad (14.11)$$

The Routh array is

Row		
1	1	11
2	6	$6(1 + K_c)$
3	$10 - K_c$	
4	$6(1 + K_c)$	

Since the proportional sensitivity of the controller (K_c) is a positive quantity, we see that the fourth entry in the first column, $6(1 + K_c)$, is positive. According to Theorem 1, all the elements of the first column must be positive for stability; hence

$$10 - K_c > 0$$
$$K_c < 10$$

It is concluded that the system will be stable only if $K_c < 10$, which agrees with Fig. 14.2.

(b) At $K_c = 10$, the system is on the verge of instability, and the element in the nth (third) row of the array is zero. According to Theorem 3, the location of the imaginary roots is obtained by solving

$$C s^2 + D = 0$$

where C and D are the elements in the $(n-1)$th row. For this problem, with $K_c = 10$, we obtain

$$6s^2 + 66 = 0$$
$$s = \pm j \sqrt{11}$$

Therefore, two of the roots on the imaginary axis are located at $\sqrt{11}$ and $-\sqrt{11}$.

The third root can be found by expressing Eq. (14.11) in factored form:

$$(s - s_1)(s - s_2)(s - s_3) = 0 \qquad (14.12)$$

where s_1, s_2, and s_3 are the roots. Introducing the two imaginary roots ($s_1 = j \sqrt{11}$ and $s_2 = -j \sqrt{11}$) into Eq. (14.12) and multiplying out the terms give

$$s^3 - s_3 s^2 + 11s - 11s_3 = 0$$

Comparing this equation with Eq. (14.11), we see that $s_3 = -6$. The roots of the characteristic equation are therefore $s_1 = j\sqrt{11}$, $s_2 = -j\sqrt{11}$, and $s_3 = -6$.

Example 14.4. Determine the stability of the system shown in Fig. 14.1 for which a PI controller is used. Use $\tau_1 = 1$, $\tau_2 = \frac{1}{2}$, $\tau_3 = \frac{1}{3}$, $K_c = 5$, and $\tau_I = 0.25$.

Solution. The characteristic equation is

$$1 + \frac{(K_c/\tau_1\tau_2\tau_3)(\tau_I s + 1)}{\tau_I s[s + (1/\tau_1)][s + (1/\tau_2)][s + (1/\tau_3)]} = 0$$

Using the parameters given above in this equation leads to

$$s^4 + 6s^3 + 11s^2 + 36s + 120 = 0$$

Notice that the order of the characteristic equation has increased from three to four as a result of adding integral action to the controller. The Routh array becomes

Row			
1	1	11	120
2	6	36	
3	5	120	
4	−108		
5	120		

Because there are two sign changes in the first column, we know from Theorem 2 of the Routh test that two roots have positive real parts. From the previous example we know that for $K_c = 5$ the system is stable with proportional control. With integral action present, however, the system is unstable for $K_c = 5$.

SUMMARY AND GUIDE FOR FURTHER STUDY

A definition of stability for a control system has been presented and discussed. This definition was translated into a simple mathematical criterion relating stability to the location of roots of the characteristic equation. Briefly, it was found that a control system is stable if all the roots of its characteristic equation lie in the left half of the complex plane. The Routh criterion, a simple algebraic test for detecting roots of a polynomial lying in the right half of the complex plane, was presented and applied to control system stability analysis. This criterion suffers from two limitations: (1) It is applicable only to systems with polynomial characteristic equations, and (2) it gives no information about the actual location of the roots and, in particular, their proximity to the imaginary axis.

This latter point is quite important, as can be seen from Fig. 14.2 and the results of Example 14.3. The Routh criterion tells us only that for $K_c < 10$ the system is stable. However, from Fig. 14.2 it is clear that the value $K_c = 9$

produces a response that is undesirable because it has a response time that is too long. In other words, the controlled variable oscillates too long before returning to steady state. It will be shown later that this happens because for $K_c = 9$ there is a pair of roots close to the imaginary axis.

In the next chapter tools will be developed for obtaining more information about the actual location of the roots of the characteristic equation. This will enable us to predict the form of the curves of Fig. 14.2 for various values of K_c. The advantage of these tools is that they are graphical and are easy to apply compared with standard algebraic solution of the characteristic equation.

There are two distinct approaches to this problem: root-locus methods and frequency-response methods. The former are discussed in Chap. 15 and the latter in Chaps. 16 and 17. These groups of chapters are written in parallel, and the reader may study one or both groups in either order. As a guide to making this decision, here are some general comments concerning the two approaches.

Root-locus methods allow rapid determination of the location of the roots of the characteristic equation as functions of parameters such as K_c of Fig. 14.1. However, they are difficult to apply to systems containing transportation lags. Also, they require a reasonably accurate knowledge of the theoretical process transfer function.

Frequency-response methods are an indirect solution to the location of the roots. They utilize the sinusoidal response of the open-loop transfer function to determine values of parameters such as K_c that keep these roots a "safe distance" from the right half plane. The actual transient response for a given value of K_c can be only crudely approximated. However, frequency-response methods are easily applied to systems containing transportation lags and may be used with only experimental knowledge of the unsteady-state process behavior.

A mastery of control theory requires knowledge of both methods because they are complementary. However, the reader may choose to study only frequency response and still be adequately prepared for most of the material in the remainder of this book. The choice of studying only root locus will be more restrictive in terms of preparation for subsequent chapters. In addition, much of the literature on process dynamics relies heavily on frequency-response methods.

PROBLEMS

14.1. Write the characteristic equation and construct the Routh array for the control system shown in Fig. P14.1. Is the system stable for (*a*) $K_c = 9.5$, (*b*) $K_c = 11$, (*c*) $K_c = 12$?

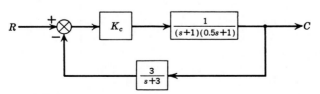

FIGURE P14-1

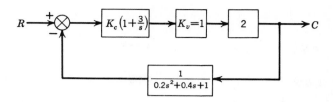

FIGURE P14-2

14.2. By means of the Routh test, determine the stability of the system shown in Fig. P14.2 when $K_c = 2$.

14.3. In the control system of Prob. 13.6, determine the value of gain (psi/° F) that just causes the system to be unstable if (a) $\tau_D = 0.25$ min, (b) $\tau_D = 0.5$ min.

14.4. Prove that, if one or more of the coefficients $(a_0, a_1, \ldots, a_n)$ of the characteristic equation [Eq. (14.9)] is negative or zero, then there is necessarily an unstable root. *Hint:* First show that a_1/a_0 is minus the sum of all the roots, a_2/a_0 is plus the sum of all possible products of two roots, a_j/a_0 is $(-1)^j$ times the sum of all possible products of j roots, etc.

14.5. Prove that the converse statement of Prob. 14.4, i.e., that an unstable root implies that one or more of the coefficients will be negative or zero, is untrue for all $n > 2$. *Hint:* To prove that a statement is untrue, it is only necessary to demonstrate a single counterexample.

14.6. Deduce an extension of the Routh criterion that will detect the presence of roots with real parts greater than $-\sigma$ for any specified $\sigma > 0$.

14.7. Show that any complex number s satisfying $|s| < 1$ yields a value of

$$z = \frac{1 + s}{1 - s}$$

that satisfies

$$\mathrm{Re}(z) > 0$$

(Hint: Let $s = x + jy; z = u + jv$. Rationalize the fraction, and equate real and imaginary parts of z and the rationalized fraction. Now consider what happens to the circle $x^2 + y^2 = 1$. To show that the *inside* of the circle goes over to the right half plane, consider a convenient point inside the circle.)

On the basis of this transformation, deduce an extension of the Routh criterion that will determine whether the system has roots inside the unit circle. Why might this information be of interest? How can the transformation be modified to consider circles of other radii?

14.8. Given the control diagram shown in Fig. P14.8, deduce by means of the Routh criterion those values of τ_I for which the output C is stable for all inputs R and U.

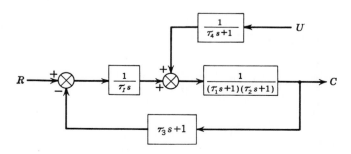

FIGURE P14-8

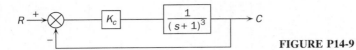

FIGURE P14-9

14.9. In the control system shown in Fig. P14.9, find the value of K_c for which the system is on the verge of instability. The controller is replaced by a PD controller, for which the transfer function is $K_c(\tau_D s + 1)$. If $K_c = 10$, determine the range of τ_D for which the system is stable.

14.10. (*a*) Write the characteristic equation for the control system shown in Fig. P14.10.
(*b*) Use the Routh Test to determine if the system is stable for $K_c = 4$.
(*c*) Determine the ultimate value of K_c, above which the system is unstable.

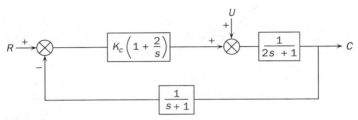

FIGURE P14-10

14.11. For the control system in Fig. P14.11, the characteristic equation is

$$s^4 + 4s^3 + 6s^2 + 4s + (1 + K) = 0$$

(*a*) Determine the value of K above which the system is unstable.
(*b*) Determine the value of K for which two of the roots are on the imaginary axis, and determine the values of these imaginary roots and the remaining two roots.

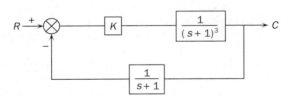

FIGURE P14-11

CHAPTER 15

ROOT
LOCUS

In the previous chapter on stability, Routh's criterion was introduced to provide an algebraic method for determining the stability of a simple feedback control system (Fig. 14.3) from the characteristic equation of the system [Eq. (14.7)]. This criterion also yields the number of roots of the characteristic equation that are located in the right half of the complex plane. In this chapter, we shall develop a graphical method for finding the actual values of the roots of the characteristic equation, from which we can obtain the transient response of the system to an arbitrary forcing function.

CONCEPT OF ROOT LOCUS

In the previous chapter, the response of the simple feedback control system, shown again in Fig. 15.1, was given by the expression

$$C = \frac{G_1 G_2}{1 + G} R + \frac{G_2}{1 + G} U \tag{15.1}$$

where $G = G_1 G_2 H$. The factor in the denominator, $1 + G$, when set equal to zero, is called the characteristic equation of the closed-loop system. The roots of the characteristic equation determine the form (or character) of the response $C(t)$ to any particular forcing function $R(t)$ or $U(t)$.

The *root-locus* method is a graphical procedure for finding the roots of $1 + G = 0$, as one of the parameters of G varies continuously. In our work, the parameter that will be varied is the gain (or sensitivity) K_c of the controller. We can illustrate the concept of a root-locus diagram by considering the example

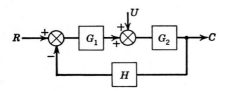

FIGURE 15-1
Simple feedback control system.

presented in Fig. 14.1, which is represented by the block diagram of Fig. 15.1
with

$$G_1 = K_c$$

$$G_2 = \frac{1}{(\tau_1 s + 1)(\tau_2 s + 1)}$$

$$H = \frac{1}{\tau_3 s + 1}$$

For this case, the open-loop transfer function is

$$G = \frac{K_c}{(\tau_1 s + 1)(\tau_2 s + 1)(\tau_3 s + 1)}$$

which may be written in the alternate form

$$G(s) = \frac{K}{(s - p_1)(s - p_2)(s - p_3)} \tag{15.2}$$

where $K = \dfrac{K_c}{\tau_1 \tau_2 \tau_3}$

$$p_1 = -\frac{1}{\tau_1} \qquad p_2 = -\frac{1}{\tau_2} \qquad p_3 = -\frac{1}{\tau_3}$$

The terms p_1, p_2, and p_3 are called the *poles* of the open-loop transfer function.
A *pole* of $G(s)$ is any value of s for which $G(s)$ approaches infinity. For example,
it is clear from Eq. (15.2) that, if $s = p_1$, the denominator of Eq. (15.2) is zero
and therefore $G(s)$ approaches infinity. Hence $p_1 = -1/\tau_1$ is a pole of $G(s)$.
    The characteristic equation for the *closed-loop* system is

$$1 + \frac{K}{(s - p_1)(s - p_2)(s - p_3)} = 0$$

This expression may be written

$$(s - p_1)(s - p_2)(s - p_3) + K = 0 \tag{15.3}$$

Using the same numerical values for the poles that were used at the beginning of
Chap. 14 ($-1, -2, -3$) gives

$$(s + 1)(s + 2)(s + 3) + K = 0 \tag{15.4}$$

where

$$K = 6K_c$$

Expanding the product of this equation gives

$$s^3 + 6s^2 + 11s + (K + 6) = 0 \tag{15.5}$$

which is third-order. For any particular value of controller gain K_c, we can obtain the roots of the characteristic equation [Eq. (15.5)]. For example, if $K_c = 4.41(K = 26.5)$, Eq. (15.5) becomes

$$s^3 + 6s^2 + 11s + 32.5 = 0$$

Solving* this equation for the three roots gives

$$r_1 = -5.10$$

$$r_2 = -0.45 - j2.5$$

$$r_3 = -0.45 + j2.5$$

By selecting other values of K, other sets of roots are obtained as shown in Table 15.1.

For convenience, we may plot the roots r_1, r_2, and r_3 on the complex plane as K changes continuously. Such a plot is called a root-locus diagram and is shown in Fig. 15.2. Notice that there are three loci or *branches* corresponding to the three roots and that they "emerge" or begin (for $K = 0$) at the poles of the open-loop transfer function $(-1, -2, -3)$. The direction of increasing K is indicated on the diagram by an arrow. Also the values of K are marked on each locus. The root-locus diagram for this system and others to follow is symmetrical with respect to the real axis, and only the portion of the diagram in the upper half plane need be drawn. This follows from the fact that the characteristic equation for a physical system contains coefficients that are real, and therefore complex roots of such an equation must appear in conjugate pairs.

The root-locus diagram has the distinct advantage of giving at a glance the character of the response as the gain of the controller is continuously changed. The diagram of Fig. 15.2 reveals two critical values of K; one is at K_2 where two of the roots become equal, and the other is at K_3 where two of the roots are pure imaginary. It should be clear from the discussion in Chap. 14 that the nature of the response $C(t)$ will depend only on the roots r_1, r_2, r_3. Thus, if the roots are all real, which occurs for $K < K_2$ in Fig. 15.2, the response will be nonoscillatory.

---

*The procedure for obtaining the roots of a higher-order equation, such as Eq. (15.5), is covered in any text on advanced algebra. In a later section of this chapter, we shall find the roots by a graphical technique called the root-locus method. There are also numerical methods for finding the roots. In Appendix 15A of this chapter, a BASIC computer program for computing the roots of a polynomial equation is given.

TABLE 15.1
Roots of the characteristic equation
$(s + 1)(s + 2)(s + 3) + K = 0$

$K = 6K_c$	r_1	r_2	r_3
0	−3	−2	−1
0.23	−3.10	−1.75	−1.15
0.39	−3.16	−1.42	−1.42
1.58	−3.45	$−1.28 − j0.75$	$−1.28 + j0.75$
6.6	−4.11	$−0.95 − j1.5$	$−0.95 + j1.5$
26.5	−5.10	$−0.45 − j2.5$	$−0.45 + j2.5$
60.0	−6.00	$0.0 − j3.32$	$0.0 + j3.32$
100.0	−6.72	$0.35 − j4$	$0.35 + j4$

If two of the roots are complex and have negative real parts $(K_2 < K < K_3)$, the response will include damped sinusoidal terms, which will produce an oscillatory response. If $K > K_3$, two of the roots are complex and have positive real parts, and the response is a growing sinusoid. Some of these types of response were shown in Fig. 14.2.

As another example of a root-locus diagram, let the proportional controller be replaced with a PI controller, for which case G_1 in Fig. 15.1 is

$$G_1 = K_c(1 + \frac{1}{\tau_I s})$$

For this case, the open-loop transfer function is

$$G(s) = \frac{K_c(\tau_I s + 1)}{\tau_I s(\tau_1 s + 1)(\tau_2 s + 1)(\tau_3 s + 1)}$$

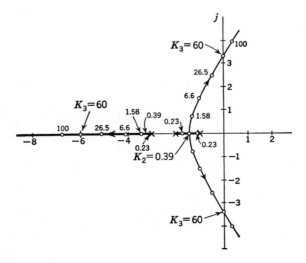

FIGURE 15-2
Root-locus diagram for
$(s + 1)(s + 2)(s + 3) + K = 0$.

which may be written in an alternate form

$$G(s) = \frac{K(s - z_1)}{s(s - p_1)(s - p_2)(s - p_3)} \qquad (15.6)$$

where $K = \frac{K_c}{\tau_1 \tau_2 \tau_3}$, $z_1 = -\frac{1}{\tau_I}$

$p_1 = -\frac{1}{\tau_1}$, $p_2 = -\frac{1}{\tau_2}$, $p_3 = -\frac{1}{\tau_3}$

The term z_1 is called a *zero* of the open-loop transfer function. A zero of $G(s)$ is any value of s for which $G(s)$ approaches zero. By comparing Eq. (15.6) with Eq.(15.2), we see that the addition of integral action contributes to the open-loop transfer function one *zero* at z_1 and one additional pole at the origin.

The characteristic equation corresponding to Eq. (15.6) is

$$1 + \frac{K(s - z_1)}{s(s - p_1)(s - p_2)(s - p_3)} = 0 \qquad (15.7)$$

This expression may be written

$$s(s - p_1)(s - p_2)(s - p_3) + K(s - z_1) = 0 \qquad (15.8)$$

As a specific example of the root-locus diagram corresponding to Eq. (15.8), let $\tau_1 = 1, \tau_2 = \frac{1}{2}, \tau_3 = \frac{1}{3}$, and $\tau_I = \frac{1}{4}$. These parameters are the same as those used in Example 14.4. The root-locus diagram is shown in Fig. 15.3.

Notice that for this case there are four loci corresponding to the four roots and that they emerge (at $K = 0$) from the open-loop poles $(0, -1, -2, -3)$. One of the loci moves toward the open-loop zero at -4 as K approaches infinity. The diagram in Fig. 15.3 should be compared with the one in Fig. 15.2 to see the effect of adding integral action to the control system. Notice that the value of $K = 3.84$, above which the roots move into the right half plane, is lower than the corresponding value of $K = 60$ for proportional control. The effect of

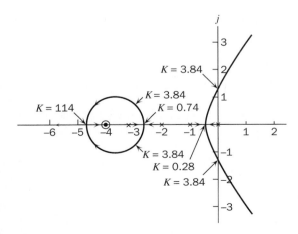

FIGURE 15-3
Root-locus diagram for
$s(s + 1)(s + 2)(s + 3) + K(s + 4) = 0; K = 6K_c$.

adding integral action has been to destablize the system in terms of the amount of proportional action that can be used before instability occurs.

A method for quickly sketching the root-locus diagram was developed by Evans (1954, 1948) and has been presented in many textbooks on control theory. In the next section, this method will be presented.

PLOTTING THE ROOT-LOCUS DIAGRAM

Having introduced the concept of root locus by two examples, here are some rules that were first introduced by Evans (1954, 1948) for plotting root-locus diagrams of characteristic equations of any order. Without these rules, the time and effort needed to plot root-locus diagrams would be too great to render them useful in engineering computations.

The first step in applying the root-locus technique to determine the roots of the characteristic equation of the closed-loop control system is to write the open-loop transfer function ($G = G_1G_2H$) in the standard form

$$G = K\frac{N}{D} \tag{15.9}$$

where K = constant

$$N = (s - z_1)(s - z_2) \cdots (s - z_m)$$

$$D = (s - p_1)(s - p_2) \cdots (s - p_n)$$

The term z_i is called a *zero* of the open-loop transfer function. The term p_i is called a *pole* of the open-loop transfer function. This term was defined earlier in this chapter. A *zero* of $G(s)$ is any value of s for which $G(s)$ equals zero. The factored terms $(s - z_i)$ and $(s - p_i)$ in N/D arise naturally in the open-loop transfer function. For example, in the control system considered at the beginning of this chapter, Eq. (15.2) was written in the standard form with

$$K = \frac{K_c}{\tau_1 \tau_2 \tau_3}$$

$$D = (s - p_1)(s - p_2)(s - p_3)$$

$$N = 1$$

The second example for PI control considered earlier [Eq. (15.6)] illustrates a situation where a term $(z - z_1)$ appears in N.

Using the form of G given by Eq. (15.9), the characteristic equation $1 + G = 0$ may be written in the alternate form

$$1 + K\frac{N}{D} = 0$$

or

$$D + KN = 0 \tag{15.10}$$

It is assumed in the remainder of this chapter that $n \geq m$, which is true for all physical systems. This being the case, the characteristic equation will be of nth order and have n roots, $r_1, r_2, \ldots, r_n$.

To develop the graphical method for determining the root locus, the characteristic equation is rewritten as

$$K \frac{N}{D} = -1 \tag{15.11}$$

In terms of the poles and zeros of the open-loop transfer function, Eq. (15.11) becomes

$$K \frac{(s - z_1)(s - z_2) \cdots (s - z_m)}{(s - p_1)(s - p_2) \cdots (s - p_n)} = -1 \tag{15.12}$$

Since the left-hand member is in general complex, we may write Eq. (15.12) in the equivalent form involving magnitude and phase angle; thus

$$K \frac{|s - z_1||s - z_2| \cdots |s - z_m|}{|s - p_1||s - p_2| \cdots |s - p_n|} = 1 \tag{15.13}$$

$$\begin{aligned} \angle(s - z_1) + \angle(s - z_2) + \cdots + \angle(s - z_m) \\ - [\angle(s - p_1) + \cdots + \angle(s - p_n)] = (2i + 1)\pi \end{aligned} \tag{15.14}$$

where i is any integer (positive or negative) or zero. Equations (15.13) and (15.14) may be used to find the root locus by trial and error as follows: The trace of the locus is found entirely from the *angle criterion* of Eq. (15.14), which is independent of K. After the locus is established, the gain K for any point on it may be obtained from Eq. (15.13), which we shall refer to as the *magnitude criterion*.

To understand the procedure for determining the root locus from the angle criterion [Eq. (15.14)], consider the simple example

$$K \frac{N}{D} = \frac{K(s - z_1)}{(s - p_1)(s - p_2)}$$

for which the poles and zeros are located as shown in Fig. 15.4. (It is convenient to indicate open-loop poles by $\times$ and open-loop zeros by $\bigcirc$ in root-locus diagrams.) To plot the root locus, a trial point (labeled s_c in Fig. 15.4) is selected and the vectors representing $(s_c - z_1)$, $(s_c - p_1)$, and $(s_c - p_2)$ are drawn. If the trial point is correct, all the angles associated with these vectors (labeled θ_1, θ_2, and α_1 in Fig. 15.4), when substituted into Eq. (15.14), will yield an odd multiple of π. For this example, the trial point s_c is correct if

$$\alpha_1 - \theta_1 - \theta_2 = (2i + 1)\pi$$

for some value of i. The trial point is moved until the angle criterion [Eq. (15.14)] is satisfied. After a sufficient number of trial points have been established as correct, the root locus is drawn by connecting them with a smooth curve. The gains K associated with various points on the locus are determined by use of the

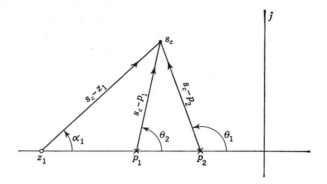

FIGURE 15-4
Use of the angle criterion to establish root locus.

magnitude criterion [Eq. (15.13)]. Again with reference to the example shown in Fig. 15.4, if we find the point s_c to be on the root locus by using the angle criterion, then the gain is obtained from Eq. (15.13); thus

$$\frac{K\,|\,s_c - z_1\,|}{|\,s_c - p_1\,||\,s_c - p_2\,|} = 1$$

Solving for K gives

$$K = \frac{|\,s_c - p_1\,||\,s_c - p_2\,|}{|\,s_c - z_1\,|}$$

It should be emphasized that the root-locus plot is symmetrical with respect to the real axis (i.e., complex roots occur as conjugate pairs). For this reason, the trial-and-error procedure for finding points on the loci need be done for only the upper half plane. The loci in the lower half plane can be drawn from symmetry.

In principle, the trial-and-error method will produce the root-locus plot; however, to save time it should be used only after applying the following rules, which give a rapid guide to the general location of the loci. These rules are proved in other texts [see Coughanowr and Koppel (1965)]. We state them below and then illustrate their use with examples. It will probably be expedient first to glance over the list of rules and then study them more carefully in conjunction with Examples 15.1 and 15.2.

Rules for Plotting Root-Locus Diagrams (Negative Feedback)

In the following rules $n \geq m$.

RULE 1. The number of loci or branches is equal to the number of open-loop poles, n.

RULE 2. The root loci begin at open-loop poles and terminate at open-loop zeros. The termination of $(n - m)$ of the loci will occur at the zeros at infinity along asymptotes to be described later. In the case of a qth-order pole,* q loci emerge from it. For a qth-order zero, q loci terminate there.

RULE 3. LOCUS ON REAL AXIS. The real axis is part of the root locus when the sum of the number of poles and zeros to the right of a point on the real axis is odd. It is necessary to consider only the real poles and zeros in applying this rule, for the complex poles and zeros always occur in conjugate pairs and their effects cancel in checking the angle criterion for points on the real axis. Furthermore, a qth-order pole (or zero) must be counted q times in applying the rule.

RULE 4. ASYMPTOTES. There are $(n - m)$ loci that approach (as $K \to \infty$) asymptotically $(n - m)$ straight lines, radiating from the *center of gravity* of the poles and zeros of the open-loop transfer function. The center of gravity is given by

$$\gamma = \frac{\displaystyle\sum_{j=1}^{n} p_j - \sum_{i=1}^{m} z_i}{n - m} \tag{15.15}$$

These asymptotic lines make angles of $\pi[(2k + 1)/(n - m)]$ with the real axis and are, therefore, equally spaced at angles $2\pi/(n - m)$ to each other ($k = 0, 1, 2 \ldots, n - m - 1$).

RULE 5. BREAKAWAY POINT. The point at which two root loci, emerging from adjacent poles (or moving toward adjacent zeros) on the real axis, intersect and then leave (or enter) the real axis is determined by the solution of the equation

$$\sum_{i=1}^{m} \frac{1}{s - z_i} = \sum_{j=1}^{n} \frac{1}{s - p_j} \tag{15.16}$$

These loci leave (or enter) the real axis at angles of $\pm\pi/2$. Equation (15.16) is solved by trial by checking it for various test points, $s = s_c$, on the real axis between the poles (or zeros) of interest. For real poles or zeros, the terms in the denominator of Eq. (15.16) are obtained by simply measuring distances along the real axis between the test point and the poles and zeros. If a pair of complex poles, $p_i = a_i \pm jb_i$, are present, add to the right side of Eq. (15.16) the term

$$\frac{2(s - a_i)}{(s - a_i)^2 + b_i^2}$$

---

*A pole p_a of order q is present in the open-loop transfer function if the denominator of G contains $(s - p_a)^q$. A zero z_a of order q is present if the numerator of G contains $(s - z_a)^q$.

(This term accounts for both poles of the complex pair.) This term is merely the result of simplifying the sum

$$\frac{1}{s - a_i - jb_i} + \frac{1}{s - a_i + jb_i}$$

For a pair of complex zeros, add a similar term to the left side of Eq. (15.16).

RULE 6. ANGLE OF DEPARTURE OR APPROACH. There are q loci emerging from each qth-order open-loop pole at angles determined by

$$\theta = \frac{1}{q}\left[(2k + 1)\pi + \sum_{i=1}^{m} \angle(p_a - z_i) - \sum_{\substack{j=1 \\ j \neq a}}^{n} \angle(p_a - p_j)\right] \tag{15.17}$$

$$k = 0, 1, 2, \ldots, q - 1$$

where p_a is a particular pole of order q. Each of the m loci that do not approach the asymptotes will terminate at one of the m zeros. They will approach their particular zeros at angles

$$\theta = \frac{1}{v}\left[(2k + 1)\pi + \sum_{j=1}^{n} \angle(z_b - p_j) - \sum_{\substack{i=1 \\ i \neq b}}^{m} \angle(z_b - z_i)\right] \tag{15.18}$$

$$k = 0, 1, 2, \ldots, v - 1$$

where z_b is a particular zero of order v. For simple poles (or zeros) on the real axis, the angle of departure (or approach) will be 0 or π.

An analog from potential theory is useful in plotting a root-locus diagram. It may be shown that the loci correspond to the paths taken by a positively charged particle in an electrostatic field which is established by poles (positive charges) and zeros (negative charges). In general, we may expect a locus to be repelled by a pole and attracted toward a zero.

Another general aid to plotting the loci is to be aware of the fact that for $n - m \geq 2$, the sum of the roots $(r_1 + r_2 + \cdots + r_n)$ is constant, real, and independent of K. This requires that motion of branches to the right be counterbalanced by the motion of other branches to the left.

Most of the open-loop transfer functions encountered in single-loop chemical process control systems will have all their poles on the real axis. In exceptional cases where the feedback path includes second-order measuring elements, such as a pressure transmitter, the open-loop transfer function will contain complex poles, but very often they will be located so far from the remaining dominant poles that they can be ignored.

These rules and guides will now be explained by applying them to specific examples.

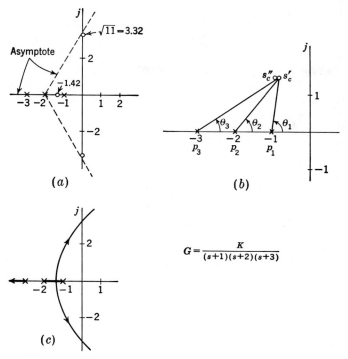

FIGURE 15-5
Root-locus construction for Example 15.1.

Example 15.1. Plot the root-locus diagram for the open-loop transfer function:*

$$G = \frac{K}{(s + 1)(s + 2)(s + 3)}$$

In general, our stepwise procedure will follow the same order in which the rules were presented.

1. Plot the open-loop poles as shown in Fig. 15.5a. The poles are indicated by ×. There are no open-loop zeros for this example.
2. (Rule 1) Since we have three poles, there are three branches.
3. (Rule 3) A portion of the locus is on the real axis between −1 and −2 and another portion is to the left of −3.

---

*To grasp more easily the graphical procedure for plotting the root locus, the reader should actually plot these examples according to the steps given in the solution. Also note that this is the same example that was treated by algebraic methods at the beginning of this chapter.

4. (Rule 4)   Since $n - m = 3$, we have three asymptotes and the center of gravity is $\gamma = (-3 - 2 - 1)/3 = -2$. Angles which the asymptotes make with the real axis are $\pi/3$, $3\pi/3$, and $5\pi/3$. These asymptotes are shown in Fig. 15.5a.

With these few steps completed, a rough sketch of the root-locus diagram can be made as follows: Since the real axis to the left of -3 is an asymptote and one branch emerges from the pole at -3, it should be clear that one entire branch is the real axis to the left of the -3. Furthermore, from the fact that two loci must emerge from the poles -1 and -2 and that the real axis between these poles is part of the locus, we see that two loci move toward each other along the real axis between -1 and -2 and eventually meet at some common point. Since the location of the asymptotes is known, it is therefore necessary that the two loci that meet on the real axis must break away and eventually follow the asymptotes. From these observations, we could sketch a root-locus diagram that closely resembles that of Fig. 15.5c. If the breakaway point and the crossings of the imaginary axis were known, the sketch could be made with considerable accuracy. We now continue the example by applying Rule 5 to find the breakaway point and the Routh test to find the crossings of the imaginary axis.

5. *Breakaway point.*   (Rule 5)   The roots emerging from -1 and -2 move toward each other until they meet, at which point the loci leave the real axis at angles of $\pm \pi/2$. The breakaway point is found from Eq. (15.16) as follows

$$0 = \frac{1}{s - p_1} + \frac{1}{s - p_2} + \frac{1}{s - p_3}$$

or

$$0 = \frac{1}{s + 1} + \frac{1}{s + 2} + \frac{1}{s + 3}$$

Solving this by trial and error gives

$$s = -1.42$$

6. To find the points at which the loci cross the imaginary axis, the Routh test (theorem 3) of Chap. 14 may be used. Writing the characteristic equation $D + KN = 0$ in polynomial form gives

$$D + KN = (s + 1)(s + 2)(s + 3) + K = 0$$

or

$$s^2 + 6s^2 + 11s + K + 6 = 0$$

from which we can write the Routh array:

Row		
1	1	11
2	6	$K + 6$
3	b_1	

The theorem states that, if one pair of roots are on the imaginary axis and all others in the left half plane, all the elements of the nth row must be zero. From this we obtain for the element b_1

$$b_1 = \frac{(6)(11) - (K + 6)}{6} = 0$$

Solving for K,

$$K = 60$$

A root on the imaginary axis is expressed as simply ja. Substituting $s = ja$ and $K = 60$ into the polynomial gives

$$-ja^3 - 6a^2 + 11aj + 66 = 0$$
$$(66 - 6a^2) + (11a - a^3)j = 0$$

Equating the real part or the imaginary part to zero gives

$$a = \pm\sqrt{11} = \pm 3.32$$

Therefore the loci intersect the imaginary axis at $+j\sqrt{11}$ and $-j\sqrt{11}$.

7. Having found these general features of the root-locus plot, we can sketch the root locus. If it is desirable to have a more accurate plot of the loci, the construction is continued by the trial-and-error method described earlier in this chapter.[†] To illustrate the method of finding roots, suppose the trial point, $s_c' = -0.75 + 1.5j$ of Fig. 15.5b, is selected. This point is checked by the angle criterion [Eq. (15.14), which for this example may be written

$$\angle(s + 1) + \angle(s + 2) + \angle(s + 3) = (2i + 1)\pi$$

or

$$\theta_1 + \theta_2 + \theta_3 = (2i + 1)\pi$$

From Fig. 15.5b, these angles are found to be

$$\theta_1 = 81° \qquad \theta_2 = 51° \qquad \theta_3 = 34°$$

and we have

$$81° + 51° + 34° = 166° \neq (2i + 1)\pi$$

[†] Several computer software packages are now available for plotting the root-locus diagram. For example, the program CC is especially useful for root-locus plotting. Details on CC and other software packages are given in Appendix 34A (of Chap. 34). Evans (1954, 1948), who developed the root-locus method, produced an instrument for plotting root-locus diagrams called the Spirule. The Spirule was essentially a drawing instrument that was used to add angles by rotating an arm with respect to a disk. The Spirule, which is no longer available, is now obsolete as a result of the availability of computer programs for plotting root-locus diagrams.

Shifting the trial point horizontally to the left will increase the sum of the angles. As a second trial point, $s_c'' = -0.95 + 1.5j$ gives for the sum of the angles

$$88° + 56° + 37° = 181° \cong \pi$$

This result is sufficiently close to π, which is $(2i + 1)\pi$ with $i = 0$, and we accept the point as one on the locus. In this manner, more points on the locus can be found and a curve drawn through them.

8. *Gain.* To determine the gain at various points along the loci, the magnitude criterion [Eq. (15.13)] is used. For example, if the gain at $s = -0.95 + j1.5$ (labeled s_c'' in Fig. 15.5b), is wanted, we measure the distances directly with a ruler; thus

$$|s - p_1| = 1.50$$
$$|s - p_2| = 1.82$$
$$|s - p_3| = 2.52$$

(It is important to measure the vector lengths in units that are consistent with those used on the axes of the graph.)

Substituting these values into Eq. (15.13) gives

$$\frac{K}{(1.50)(1.82)(2.52)} = 1$$

or $K = (1.50)(1.82)(2.52) = 6.8$. To find the point corresponding to $K = 6.8$ on the branch along the real axis to the left of p_3 requires a trial-and-error solution if the graphical approach is used. For example, if $s = -4.5$ is tried, we obtain

$$|s - p_1| = 3.5$$
$$|s - p_2| = 2.5$$
$$|s - p_3| = 1.5$$

from which we get

$$K = (1.5)(2.5)(3.5) = 13.1$$

We see that $s = -4.5$ does not correspond to a gain of 6.8. It is therefore necessary to try other values of s greater than -4.5 until the desired value of $K = 6.8$ is obtained. Although this procedure may seem very tedious, the actual calculations go quite quickly as the reader will discover while working out this example.

We also may find the root on the real axis more directly by applying the following theorem from algebra:

The sum of the roots $(r_1 + r_2 + \cdots + r_n)$ of the nth-order polynomial equation

$$a_0 x^n + a_1 x^{n-1} + \cdots + a_n = 0$$

is given by

$$(r_1 + r_2 + \cdots + r_n) = -\frac{a_1}{a_0}$$

In this case, we have just found the complex roots for $K = 6.8$ to be

$$r_2, r_3 = -0.95 \pm j1.5$$

The polynomial equation is

$$(s + 1)(s + 2)(s + 3) + K = 0$$

which can be expanded into

$$s^3 + 6s^2 + 11s + (K + 6) = 0$$

According to the theorem

$$r_1 + (-0.95 + j1.5) + (-0.95 - j1/5) = -\frac{6}{1}$$

or

$$6 = -[r_1 - 2(0.95)]$$

or

$$r_1 = -4.10$$

All the detailed steps needed to plot the root locus for this problem have been discussed. The complete locus is shown in Fig. 15.5c. This same plot is also shown in more detail in Fig. 15.2.

Example 15.2. Consider the block diagram for the control system shown in Fig. 15.6. This system may represent a two-tank, liquid-level system having a PID controller and a first-order measuring lag. The open-loop transfer function is

$$G = K_c \frac{1 + 2s/3 + 1/3s}{(20s + 1)(10s + 1)(0.5s + 1)}$$

Rearranging this into the standard form, KN/D, gives

$$G = \frac{K(s - z_1)(s - z_2)}{s(s - p_1)(s - p_2)(s - p_3)}$$

where $K = K_c/150$
$z_1 = -0.5$
$z_2 = -1$
$p_1 = -0.05$
$p_2 = -0.1$
$p_3 = -2$

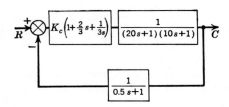

FIGURE 15-6
Block diagram for Example 15.2.

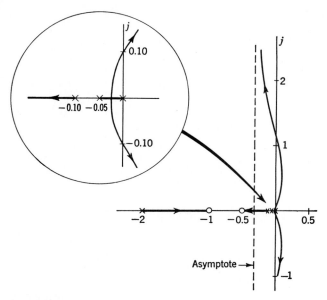

FIGURE 15-7
Root-locus diagram for Example 15.2.

In this case, there are four poles at 0, −0.05, −0.1, and −2 and two zeros at −0.5 and −1. These are plotted in Fig. 15.7. Note that the three-action controller contributes the pole at the origin and the zeros, −0.5 and −1. The steps for plotting the root-locus diagram are as follows:

1. Since there are four poles, there are four branches emerging from them.
2. Three portions of the root locus are on the real axis between 0 and −0.05, between −0.10 and −0.5, and between −1 and −2.
3. Since $n - m = 2$, there are two asymptotes, and the center of gravity is

$$\gamma = \frac{(-0.05 - 0.1 - 2) - (-0.5 - 1.0)}{2} = -0.325$$

The angles that the asymptotes make with the real axis are $\pm\pi/2$. These asymptotes are shown in Fig. 15.7.

At this stage, we can sketch part of the root-locus diagram. Since the locus is on the real axis between −0.1 and −0.5 and between −1 and −2, it should be evident that one branch moves from the pole at −2 to the zero at −1 and another branch moves from the pole at −0.1 to the zero at −0.5. The remaining two branches move from the poles at 0 and −0.05 toward each other along the real axis until they meet, at which point they must break away from the real axis and move in some way toward the vertical asymptotes that intersect the real axis at −0.325.

With the information now available, it is difficult to continue the sketch with confidence, for the breakaway point is so close to the origin that there is some likelihood that the loci will move into the right half plane before approaching the asymptote. If this should occur, each locus would have to cross the imaginary axis twice, in which case there would be an intermediate range of K over which the system is unstable. On either end of this range of K, the system is stable. This condition is called *conditional stability*. The possibility of the locus crossing the imaginary axis twice is suggested by the analog from potential theory that was mentioned earlier. This can be explained as follows: immediately after the locus leaves the real axis at the breakaway point, it has a tendency to move to the right half plane because the pole at -0.1 "repels" the locus. However, after the locus moves to a point sufficiently far from this repelling pole, it is attracted more strongly by the two zeros at -0.5 and -1 and has the tendency to return to the left half plane where we know it must eventually approach the vertical asymptote. Actually to determine whether or not the locus moves into the right half plane requires that the points at which the loci cross the imaginary axis be determined. This can be done by use of the Routh test as illustrated in Example 15.1. The details of the calculation will not be given here; however, the reader can show that there are two values of gain K which give a pair of roots of the characteristic equation that lie on the imaginary axis. These gains and corresponding roots are approximately

$$K = 0.004 \quad \text{or} \quad K_c = 0.6 \quad s = \pm j0.1$$

$$K = 2.4 \quad \text{or} \quad K_c = 360 \quad s = \pm j1.1$$

From these results, we conclude that the system will oscillate with constant amplitude with a frequency $\omega = 0.1$ rad/time when $K_c = 0.6$; it will also oscillate at constant amplitude with $\omega = 1.1$ when $K_c = 360$. The system is unstable for $0.6 < K_c < 360$. The system is stable for $K_c < 0.6$ and for $K_c > 360$. The complete root-locus diagram is sketched in Fig. 15.7.

SUMMARY

In this chapter, the rules for plotting root-locus diagrams have been presented and applied to several control systems. It should be emphasized that the basic advantage of this method is the speed and ease with which a rough sketch of the loci can be obtained. This sketch frequently gives much of the desired information on stability. A few further calculations of points on the locus are usually all that are necessary to obtain accurate, quantitative behavior of the roots.

The root locus for variation of parameters other than K_c, such as τ_D, has not been discussed here. The method of constructing this type of diagram is similar to that presented here and is discussed in detail in other texts [see Wilts (1960)].

Once the roots are available, the response of the system to any forcing function can be obtained by the usual procedures of partial fractions and inversion given in Chap. 3.

APPENDIX 15A

Table 15.1A gives a BASIC computer program for finding the roots of a polynomial equation by the Lin-Bairstow method [see Hovannessian and Pipes (1969)]. To use this program, arrange the polynomial equation in the form

$$a_n s^n + a_{n-1} s^{n-1} + a_{n-2} s^{n-2} + \ldots + a_o = 0$$

Before running the program, a DATA statement is used to list the order (n) and the coefficients of the polynomial equation as follows:

$$\text{DATA } n, a_n, a_{n-1}, \ldots, a_o$$

An example of the use of the root-finding program is shown in Table 15.2A; the example involves finding the roots of Eq. (15.5) for the case of $K = 26.5$.

TABLE 15.1A
BASIC program for finding roots of a polynomial equation

```
10 REM ROOTS OF POLYNOMIAL EQUATION
20 REM USING LIN-BAIRSTOW METHOD
30 REM AN*S**N + A(N-1)*S**(N-1) + A(N-2)*S**(N-2) + ...+ A0 = 0
40 REM DATA N, AN, A(N-1), ...,A0
50 REM REFERENCE: DIGITAL COMP METH IN ENGRG, HOVANNESSIAN, S.A.
     AND L. A. PIPES
100 DIM A(10),B(10),C(10),D(10)
110 READ N
120 PRINT "DEGREE"N
130 PRINT "COEFFICIENT"
140 FOR I = N TO 0 STEP -1
150 READ A(I)
160 PRINT A(I) ;
170 NEXT I
180 PRINT
190 PRINT
200 LET R=A(1)/A(2)
210 LET S=A(0)/A(2)
220 LET B(N)=A(N)
230 LET C(N)=0
240 LET D(N)=0
250 LET B(N-1)=A(N-1)-R*B(N)
260 LET C(N-1)=-B(N)
270 LET D(N-1)=0
280 FOR I=2 TO N-2
290 LET B(N-I)=A(N-I)-R*B(N-I+1)-S*B(N-I+2)
300 LET C(N-I)=-B(N-I+1)-R*C(N-I+1)-S*C(N-I+2)
310 LET D(N-I)=-B(N-I+2)-S*D(N-I+2)-R*D(N-I+1)
```

TABLE 15.1A *(Continued)*
BASIC program for finding roots of a polynomial equation

```
320 NEXT I
330 LET R1=A(1)-R*B(2)-S*B(3)
340 LET S1=A(0)-S*B(2)
350 LET T=-B(2)-R*C(2)-S*C(3)
360 LET U=-B(3)-S*D(3)-R*D(2)
370 LET V=-S*C(2)
380 LET W=-B(2)-S*D(2)
390 LET R2=(-R1*W+S1*U)/(T*W-U*V)
400 LET S2=(-T*S1+V*R1)/(T*W-U*V)
410 LET S=S+S2
420 LET R=R+R2
430 IF ABS(R2)<.00001 THEN 450
440 GOTO 220
450 LET G=R*R-4*S
460 IF G<0 THEN 490
470 PRINT "ROOTS";-R/2;"+OR-";SQR(G)/2
480 GOTO 500
490 PRINT "ROOTS";-R/2;"+OR-";SQR(-G)/2;"J"
500 LET N=N-2
510 PRINT
520 IF N=0 THEN 630
530 FOR I = N TO 0 STEP-1
540 LET A(I)=B(I+2)
550 NEXT I
560 IF N>2 THEN 200
570 IF N<2 THEN 610
580 LET R=A(N-1)/A(N)
590 LET S=A(N-2)/A(N)
600 GOTO 450
610 PRINT "ROOT",-A(N-1)/A(N)
620 DATA 3,1,6,11,32.5
630 END
```

TABLE 15.2A
Use of BASIC program of Table 15.1A for finding roots of Eq. (15.5):
$s^3 + 6s^2 + 11s + (K + 6) = 0$ with $K = 26.5$

```
RUN
DEGREE 3
COEFFICIENT
 1   6   11   32.5

ROOTS-.4534395 +OR- 2.485065 J

ROOT            -5.093121
```

PROBLEMS

15.1. Draw the root-locus diagram for the system shown in Fig. P15.1 where $G_c = K_c(1 + 0.5s + 1/s)$.

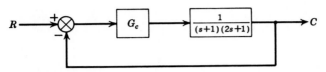

FIGURE P15-1

15.2. Draw the root-locus diagram for the system shown in Fig. P13.4 for (a) $\tau_I = 0.4$ min and (b) $\tau_I = 0.2$ min. (The proportional controller is replaced by a PI controller.) Determine the controller gain that just causes the system to become unstable. The values of parameters of the system are:

$$K_v = \text{valve constant } 0.070 \text{ cfm/psi}$$
$$K_m = \text{transducer constant } 6.74(\text{in. pen travel})/(\text{ft of tank level})$$
$$R_2 = 0.55 \text{ ft level/cfm}$$
$$\tau_1 = \text{time constant of tank 1} = 2.0 \text{ min}$$
$$\tau_2 = \text{time constant of tank 2} = 0.5 \text{ min}$$

The controller gain K_c has the units of pounds per square inch per inch of pen travel.

15.3. Sketch the root-locus diagram for the system shown in Fig. P14.2. If the system is unstable at higher values of K_c, find the roots on the imaginary axis and the corresponding value of K_c.

15.4. Sketch the root loci for the following equations:

(a) $1 + \dfrac{K}{(s + 1)(2s + 1)} = 0$

(b) $1 + \dfrac{K}{s(s + 1)(2s + 1)} = 0$

(c) $1 + \dfrac{K(4s + 1)}{s(s + 1)(2s + 1)} = 0$

(d) $1 + \dfrac{K(1.5s + 1)}{s(s + 1)(2s + 1)} = 0$

(e) $1 + \dfrac{K(0.5s + 1)}{s(s + 1)(2s + 1)} = 0$

On your sketch you should locate quantitatively all poles, zeros, and asymptotes. In addition show the parameter that is being varied along the locus and the direction in which the loci travel as this parameter is increased.

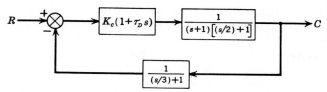

FIGURE P15-5

15.5. For the control system shown in Fig. P15.5.

Case 1: $\tau_D = \frac{2}{3}$

Case 2: $\tau_D = \frac{1}{9}$

(a) Sketch the root-locus diagram in each case.

(b) If the system can go unstable, find the value of K_c that just causes instability.

(c) Using Theorem 3 (Chap. 14) of the Routh test, find the locations (if any) at which the loci cross into the unstable region.

15.6. Draw the root-locus diagram for the control system shown in Fig. P15.6.

(a) Determine the value of K_c needed to obtain a root of the characteristic equation of the closed-loop response which has an imaginary part 0.75.

(b) Using the value of K_c found in part (a), determine all the other roots of the characteristic equation from the root-locus diagram.

(c) If a unit impulse is introduced into the set point, determine the response of the system, $C(t)$.

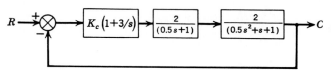

FIGURE P15-6

15.7. Plot the root-locus diagram for the system shown in Fig. P15.7. We may consider this system to consist of a process having negligible lag, an underdamped, second-order measuring element, and a PD controller. This system may approximate the control of flow rate, in which case the block labeled K_p would represent a valve having no dynamic lag. The feedback element would represent a flow measuring device, such as a mercury manometer placed across an orifice plate. Mercury manometers are known to have underdamped, second-order dynamics. Plot the diagram for $\tau_D = 1/3$.

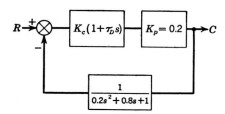

FIGURE P15-7

15.8. Draw the root-locus diagram for the proportional control of a plant having the transfer function $2/[(s + 1)^3]$. Determine the roots on the imaginary axis and the corresponding value of K_c.

15.9. (a) Show how you would adopt the usual root-locus method for variation in controller gain to the problem of obtaining the root-locus diagram for variation in τ_D for the control system shown in Fig. P15.9 for $K_c = 2$.

  (b) Plot the root-locus diagram for variation in τ_D with $K_c = 2$.

  (c) Determine the response of the system $C(t)$ for a unit-step change in R for $\tau_D = 0.5$, and $K_c = 2$. Sketch the response. What is the ultimate value of $C(t)$?

  Hint: Rearrange the open-loop transfer function to be in the form

  $$G(s) = \frac{\tau_D s}{s^2 + 1.5s + 1.5}$$

  Then apply the usual root-locus rules with τ_D taking the place of K_c.

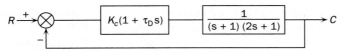

FIGURE P15-9

PART
IV

FREQUENCY
RESPONSE

INTRODUCTION
TO FREQUENCY
RESPONSE

Chapters 5 and 8 discussed briefly the response of first- and second-order systems to sinusoidal forcing functions. These frequency responses were derived by using the standard Laplace transform technique. In this chapter, a convenient graphical technique will be established for obtaining the frequency response of linear systems. The motivation for doing so will become apparent in the following chapter, where it will be found that frequency response is a valuable tool in the analysis and design of control systems.

   Many of the calculations in this chapter make use of complex numbers. The reader should review the two forms of complex numbers (rectangular and polar) and the basic operations used on complex numbers.

SUBSTITUTION RULE

A Fortunate Circumstance

Consider a simple first-order system with transfer function

$$G(s) = \frac{1}{\tau s + 1} \tag{16.1}$$

   Substituting the quantity $j\omega$ for s in Eq. (16.1) gives

$$G(j\omega) = \frac{1}{j\omega\tau + 1}$$

We may convert this expression to polar form by multiplying numerator and denominator by the conjugate of $(j\omega\tau + 1)$; the result is:

$$G(j\omega) = \frac{-j\omega\tau + 1}{(j\omega\tau + 1)(-j\omega\tau + 1)} = \frac{1}{1 + \omega^2\tau^2} - j\frac{\omega\tau}{1 + \omega^2\tau^2} \qquad (16.2)$$

To convert a complex number in rectangular form ($z = a + jb$) to polar form ($|z| \angle z$) one uses the relationships:

$$|z| = \sqrt{a^2 + b^2} \qquad \text{and} \qquad \angle z = \tan^{-1}\frac{b}{a}$$

Applying these relationships to Eq. (16.2) gives

$$G(j\omega) = \frac{1}{\sqrt{\omega^2\tau^2 + 1}} \angle \tan^{-1}(-\omega\tau) \qquad (16.3)$$

The quantities on the right side of Eq. (16.3) are familiar. In Chap. 5 we found that, after sufficient time had elapsed, the response of a first-order system to a sinusoidal input of frequency ω is also a sinusoid of frequency ω. Furthermore, we saw that the ratio of the amplitude of the response to that of the input is $1/\sqrt{\omega^2\tau^2 + 1}$ and the phase difference between output and input is $\tan^{-1}(-\omega\tau)$. Hence, we have shown here that for the frequency response of a first-order system,

$$\text{AR} = |G(j\omega)|$$

$$\text{Phase angle} = \angle G(j\omega)$$

That is, to obtain the amplitude ratio (AR) and phase angle, one merely substitutes $j\omega$ for s in the transfer function and then takes the magnitude and argument (or angle) of the resulting complex number, respectively.

Example 16.1. Rework Example 5.2. The pertinent transfer function is

$$G(s) = \frac{1}{0.1s + 1}$$

The frequency of the bath-temperature variation is given as $10/\pi$ cycles/min which is equivalent to 20 rad/min.
  Hence, let

$$s = 20j$$

to obtain

$$G(20j) = \frac{1}{2j + 1}$$

In polar form, this is

$$G(20j) = \frac{1}{\sqrt{5}} \angle - 63.5°$$

which agrees with the previous result.

Generalization

At this point, it is necessary to ascertain whether or not we may generalize the result of the last section to other systems. This can be done by checking the result for second-order systems, third-order systems, etc. However, it is more satisfying to prove the general validity of the result as follows. (The reader may, if desired, accept the result as general and skip to Example 16.2. We remark here that an important restriction on this rule is that it applies only to systems whose transfer functions yield stable responses.)

An nth-order linear system is characterized by an nth-order differential equation:

$$a_n \frac{d^n Y}{dt^n} + a_{n-1} \frac{d^{n-1} Y}{dt^{n-1}} + \cdots + a_1 \frac{dY}{dt} + a_0 Y = X(t) \qquad (16.4)$$

where Y is the output variable and $X(t)$ is the forcing function or input variable. For specific cases of Eq. (16.4), refer to Eq. (5.5) for a first-order system and Eq. (8.4) for a second-order system. If $X(t)$ is sinusoidal

$$X(t) = A \sin \omega t$$

the solution of Eq. (16.4) will consist of a complementary solution, and a particular solution of the form

$$Y_p(t) = C_1 \sin \omega t + C_2 \cos \omega t \qquad (16.5)$$

If the system is stable, the roots of the characteristic equation of (16.4) all lie to the left of the imaginary axis and the complementary solution will vanish exponentially in time. Then Y_p is the quantity previously defined as the sinusoidal or *frequency response*. If the system is not stable, the complementary solution grows exponentially and the term frequency response has no physical significance because $Y_p(t)$ is inconsequential.

The problem now is the evaluation of C_1 and C_2 in Eq. (16.5). Since we are interested in the amplitude and phase of $Y_P(t)$, Eq. (16.5) is rewritten as

$$Y_p = D_1 \sin (\omega t + D_2) \qquad (16.6)$$

as was done previously [compare to Eq. (5.23) and related equations].

It will be convenient to change $X(t)$ and $Y_p(t)$ from trigonometric to exponential form, using the identity

$$\sin \theta = \frac{e^{j\theta} - e^{-j\theta}}{2j}$$

Thus,

$$X(t) = \frac{A}{2j}(e^{j\omega t} - e^{-j\omega t}) \qquad (16.7)$$

and from Eq. (16.6)

$$Y_p(t) = \frac{D_1}{2j}[e^{j(\omega t + D_2)} - e^{-j(\omega t + D_2)}] \qquad (16.8)$$

Substitution of Eqs. (16.7) and (16.8) into Eq. (16.4) yields:

$$\frac{D_1 e^{j(\omega t + D_2)}}{2j}[a_n(j\omega)^n + a_{n-1}(j\omega)^{n-1} + \cdots + a_1(j\omega) + a_0]$$

$$-\frac{D_1 e^{-j(\omega t + D_2)}}{2j}[a_n(-j\omega)^n + a_{n-1}(-j\omega)^{n-1} + \cdots + a_1(-j\omega) + a_0]$$

$$= \frac{A}{2j}(e^{j\omega t} - e^{-j\omega t}) \qquad (16.9)$$

The coefficients of $e^{j\omega t}$ on both sides of Eq. (16.9) must be equal. Hence,

$$D_1 e^{jD_2}[a_n(j\omega)^n + a_{n-1}(j\omega)^{n-1} + \cdots + a_1(j\omega) + a_0] = A \qquad (16.10)$$

Equation (16.10) will be satisfied if and only if

$$\left| \frac{1}{a_n(j\omega)^n + a_{n-1}(j\omega)^{n-1} + \cdots + a_1(j\omega) + a_0} \right| = \frac{D_1}{A} \qquad (16.11)$$

$$\measuredangle \frac{1}{a_n(j\omega)^n + a_{n-1}(j\omega)^{n-1} + \cdots + a_1(j\omega) + a_0} = D_2$$

But D_1/A and D_2 are the AR and phase angle of the response, respectively, as may be seen from Eq. (16.6) and the forcing function. Furthermore, from Eq. (16.4) the transfer function relating X and Y is

$$\frac{Y(s)}{X(s)} = \frac{1}{a_n s^n + a_{n-1}s^{n-1} + \cdots + a_1 s + a_0} \qquad (16.12)$$

Equations (16.11) and (16.12)* establish the general result.

Example 16.2. Find the frequency response of the system with the general second-order transfer function and compare the results with those of Chap. 8. The transfer function is

$$\frac{1}{\tau^2 s^2 + 2\zeta\tau s + 1}$$

Putting $s = j\omega$ yields

$$\frac{1}{1 - \tau^2\omega^2 + j2\zeta\omega\tau}$$

which may be converted to the polar form

$$\frac{1}{\sqrt{(1 - \omega^2\tau^2)^2 + (2\zeta\omega\tau)^2}} \measuredangle \tan^{-1}\left(\frac{-2\zeta\omega\tau}{1 - \omega^2\tau^2}\right)$$

---

*In writing this equation, it is assumed that X and Y have been written as deviation variables, so that initial conditions are zero.

Hence,

$$AR = \frac{1}{\sqrt{(1 - \omega^2\tau^2)^2 + (2\zeta\omega\tau)^2}} \qquad (16.13)$$

$$\text{Phase angle} = \tan^{-1}\frac{-2\zeta\omega\tau}{1 - \omega^2\tau^2}$$

which agree with Eq. (8.40).

Transportation Lag

The response of a transportation lag is not described by Eq. (16.4). Rather, a transportation lag is described by the relation

$$Y(t) = X(t - \tau) \qquad (16.14)$$

which states that the output Y lags the input X by an interval of time τ. If X is sinusoidal,

$$X = A \sin \omega t$$

then from Eq. (16.14)

$$Y = A \sin \omega(t - \tau) = A \sin (\omega t - \omega\tau)$$

It is apparent that the AR is unity and the phase angle is $(-\omega\tau)$.

To check the substitution rule of the previous section, recall that the transfer function is given by

$$G(s) = \frac{Y(s)}{X(s)} = e^{-\tau s}$$

Putting $s = j\omega$,

$$G(j\omega) = e^{-j\omega\tau}$$

Then,

$$AR = |e^{-j\omega\tau}| = 1 \qquad (16.15)$$

$$\text{Phase angle} = \measuredangle e^{-j\omega\tau} = -\omega\tau$$

and the validity of the rule is verified.

Example 16.3. The stirred-tank heater of Chap. 1 has a capacity of 15 gal. Water is entering and leaving the tank at the constant rate of 600 lb/min. The heated water that leaves the tank enters a well-insulated section of 6-in.-ID pipe. Two feet from the tank, a thermocouple is placed in this line for recording the tank temperature, as shown in Fig. 16.1. The electrical heat input is held constant at 1,000 kw.

If the inlet temperature is varied according to the relation

$$T_i = 75 + 5 \sin 46t$$

where T_i is in degrees Fahrenheit and t is in minutes, find the eventual behavior of the thermocouple reading T_m. Compare this with the behavior of the tank temperature

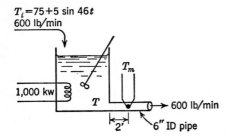

$T_i = 75 + 5 \sin 46t$
600 lb/min

T_m

1,000 kw

T

→ 600 lb/min

2′   6″ ID pipe

FIGURE 16-1
Tank-temperature system for Example 16.3.

T. It may be assumed that the thermocouple has a very small time constant and effectively measures the true fluid temperature at all times.

The problem is to find the frequency response of T_m to T_i. Deviation variables must be used. Define the deviation variable T_i' as

$$T_i' = T_i - 75 = 5 \sin 46t$$

To define a deviation variable for T_m, note that, if T_i were held at 75°F, T_m would come to the steady state satisfying

$$q_s = wC(T_{m_s} - T_{i_s})$$

This may be solved for T_{m_s}:

$$T_{m_s} = \frac{q_s}{wC} + T_{i_s} = \frac{(1,000)(1,000)(0.0569)}{(600)(1.0)} + 75 = 170°F$$

Hence, define

$$T_m' = T_m - 170$$

Now the overall system between T_i' and T_m' is made up of two components in series: the tank and the 2-ft section of pipe. The transfer function for the tank is

$$G_1(s) = \frac{1}{\tau_1 s + 1}$$

where, as we have seen before, τ_1 is given by

$$\tau_1 = \frac{\rho V}{w} = \frac{(60.3)(15)}{(600)(7.48)} = 0.202 \, \text{min}$$

The transfer function of the 2-ft section of pipe, which corresponds to a transportation lag, is

$$G_2(s) = e^{-\tau_2 s}$$

where τ_2 is the length of time required for the fluid to transverse the length of pipe. This is

$$\tau_2 = \frac{L}{v} = \frac{(2)(60.3)(0.197)}{600} = 0.0396 \, \text{min}$$

The factor 0.197 is the cross-sectional area of the pipe in square feet.

Since the two systems are in series, the overall transfer function between T_i' and T_m' is

$$\frac{T_m'}{T_i'} = \frac{e^{-\tau_2 s}}{\tau_1 s + 1} = \frac{e^{-0.0396s}}{0.202s + 1}$$

To find the AR and phase lag, we merely substitute $s = 46j$ and take the magnitude and argument of the resulting complex number. However, note that we have previously derived the individual frequency responses for the first-order system and transportation lag. The overall transfer function is the product of the individual transfer functions, hence, its magnitude will be the product of the magnitudes and its argument the sum of the arguments of the individual transfer functions. In general, if

$$G(s) = G_1(s)G_2(s)\cdots G_n(s)$$

then

$$|G(j\omega)| = |G_1(j\omega)|\,|G_2(j\omega)|\cdots|G_n(j\omega)|$$
$$\measuredangle G(j\omega) = \measuredangle G_1(j\omega) + \measuredangle G_2(j\omega) + \cdots + \measuredangle G_n(j\omega)$$

This rule makes it very convenient to find the frequency response of a number of systems in series.

Using Eq. (16.3) for the tank,

$$\text{AR} = \frac{1}{\sqrt{(46) \times 0.202)^2 + 1}} = \frac{1}{9.35} = 0.107$$

$$\text{Phase angle} = \tan^{-1}[(-46)(0.202)] = -84°$$

For the section of pipe, the AR is unity, so that the overall AR is just 0.107. The phase lag due to the pipe may be obtained from Eq. (16.15) as

$$\text{Phase angle} = -\omega\tau_2 = -(46)(0.0396) = -1.82 \text{ rad} = -104°$$

The overall phase lag from T_i' to T_m' is the sum of the individual lags,

$$\measuredangle\frac{T_m'}{T_i'} = -84 - 104 = -188°$$

Hence

$$T_m = 170 + 0.535\sin(46t - 188°)$$

For comparison, a plot of T_i', T_m', and T' is given in Fig. 16.2, where

$$T' = \text{tank temperature} - 170°F$$

It should be emphasized that this plot applies only after sufficient time has elapsed for the complementary solution to become negligible. This restriction applies to all the forthcoming work on frequency response. Also, note that, for convenience of scale, the tank and thermocouple temperatures have been plotted as $2T'$ and $2T_m'$, respectively.

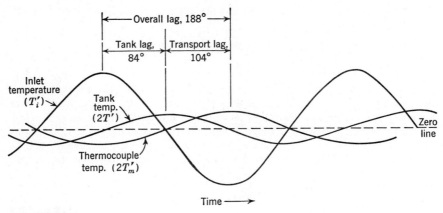

FIGURE 16-2
Temperature variation in Example 16.3.

A Control Problem

An interesting conclusion may be reached from a study of Fig. 16.2. Suppose that we are trying to control the tank temperature, using the deviation between the thermocouple reading and the set point as the error. A block diagram for proportional control might appear as in Fig. 16.3, where T_i' is replaced by U, T' by C, and T_m' by B to conform with our standard block-diagram nomenclature. The variable R denotes the deviation of the set point from 170°F and is the desired value of the deviation C. The value of R is assumed to be zero in the following analysis (control at 170°F). The following arguments, while not rigorous, serve to give some insight regarding application of frequency response to control system analysis.

The heat being added to the tank is given in deviation variables as $-K_cB$. With reference to Fig. 16.2, which shows the response of the *uncontrolled* tank to a sinusoidal variation in U, it can be seen that the peaks of U and B are almost exactly opposite because the phase difference is 188°. This means that, if the loop

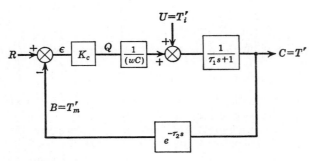

FIGURE 16-3
Proportional control of heated, stirred tank.

were closed, the control system would have a tendency to add *more* heat when the inlet temperature T_i is at its high peak, because B is then negative and $-K_c B$ becomes positive. (Recall that the set point R is held constant at zero.) Conversely, when the inlet temperature is at a low point, the *tendency* will be for the control system to add less heat because B is positive. This is precisely opposite to the way the heat input should be controlled.

Therefore, the possibility of an unstable control system exists for this particular sinusoidal variation in frequency. Indeed, we shall demonstrate in Chap. 17 that, *if K_c is taken too large,* the tank temperature will oscillate with increasing amplitude for *all* variations in U and hence we have an unstable control system. The fact that such information may be obtained by study of the *frequency response* (i.e., the particular solution for a sinusoidal forcing function) justifies further study of this subject.

BODE DIAGRAMS

Thus far, it has been necessary to calculate AR and phase lag by direct substitution of $s = j\omega$ into the transfer function for the particular frequency of interest. It can be seen from Eqs. (16.3), (16.13), and (16.15) that the AR and phase lag are functions of frequency. There is a convenient graphical representation of their dependence on the frequency that largely eliminates direct calculation. *This is called a Bode diagram and consists of two graphs: logarithm of AR versus logarithm of frequency, and phase angle versus logarithm of frequency.* The Bode diagram will be shown in Chap. 17 to be a convenient tool for analyzing control problems such as the one discussed in the preceding section. The remainder of the present chapter is devoted to developing this tool and presenting Bode diagrams for the basic components of control loops.

First-Order System

The AR and phase angle for the sinusoidal response of a first-order system are

$$AR = \frac{1}{\sqrt{\tau^2 \omega^2 + 1}} \qquad (16.16)$$

$$\text{Phase angle} = \tan^{-1}(-\omega\tau) \qquad (16.17)$$

It is convenient to regard these as functions of $\omega\tau$ for the purpose of generality. From Eq. (16.16)

$$\log AR = -\tfrac{1}{2} \log[(\omega\tau)^2 + 1] \qquad (16.18)$$

The first part of the Bode diagram is a plot of Eq. (16.18). The true curve is shown as the solid line on the upper part of Fig. 16.4. Some asymptotic considerations can simplify this plot. As $(\omega\tau) \to 0$, Eq. (16.16) shows that AR $\to 1$. This is indicated by the low-frequency asymptote on Fig. 16.4. As $(\omega\tau) \to \infty$, Eq. (16.18) becomes asymptotic to

$$\log AR = -\log(\omega\tau)$$

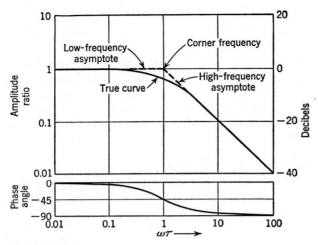

FIGURE 16-4
Bode diagram for first-order system.

which is a line of slope -1, passing through the point

$$\omega\tau = 1 \qquad AR = 1$$

This line is indicated as the high-frequency asymptote in Fig. 16.4. The frequency $\omega_c = 1/\tau$, where the two asymptotes intersect, is known as the *corner frequency;* it may be shown that the deviation of the true AR curve from the asymptotes is a maximum at the corner frequency. Using $\omega_c = 1/\tau$ in Eq. (16.16) gives

$$AR = \frac{1}{\sqrt{2}} = 0.707$$

as the true value, whereas the intersection of the asymptotes occurs at $AR = 1$. Since this is the maximum deviation and is an error of less than 30 percent, for engineering purposes it is often sufficient to represent the curve entirely by the asymptotes. Alternately, the asymptotes and the value of 0.707 may be used to sketch the curve if more accuracy is required.

In the lower half of Fig. 16.4, we have shown the phase curve as given by Eq. (16.17). Since

$$\phi = \tan^{-1}(-\omega\tau) = -\tan^{-1}(\omega\tau)$$

it is evident that ϕ approaches $0°$ at low frequencies and $-90°$ at high frequencies. This verifies the low- and high-frequency portions of the phase curve. At the corner frequency, $\omega_c = 1/\tau$,

$$\phi_c = -\tan^{-1}(\omega_c\tau) = -\tan^{-1}(1) = -45°$$

There are asymptotic approximations available for the phase curve, but they are not so accurate or so widely used as those for the AR. Instead, it is convenient to note that the curve is symmetric about $-45°$.

It should be stated that, in a great deal of the literature on control theory, amplitude ratios (or gains) are reported in decibels. The decibel is defined by

$$\text{Decibels} = 20 \log_{10}(\text{AR})$$

Thus, an AR of unity corresponds to zero decibels and an amplitude ratio of 0.1 corresponds to -20 decibels. The abbreviation for the decibel is db. The value of the AR in decibels is given on the right-hand ordinate of Fig. 16.4.

First-Order Systems in Series

The advantages of the Bode plot become evident when we wish to plot the frequency response of systems in series. As shown in Example 16.3, the rules for multiplication of complex numbers indicate that the AR for two first-order systems in series is the product of the individual ARs:

$$\text{AR} = \frac{1}{\sqrt{\omega^2 \tau_1^2 + 1} \sqrt{\omega^2 \tau_2^2 + 1}} \tag{16.19}$$

Similarly, the phase angle is the sum of the individual phase angles

$$\phi = \tan^{-1}(-\omega\tau_1) + \tan^{-1}(-\omega\tau_2) \tag{16.20}$$

Since the AR is plotted on a logarithmic basis, multiplication of the ARs is accomplished by addition of logarithms on the Bode diagram. The phase angles are added directly. The procedure is best illustrated by an example.

Example 16.4. Plot the Bode diagram for the system whose overall transfer function is

$$\frac{1}{(s + 1)(s + 5)}$$

To put this in the form of two first-order systems in series, it is rewritten as

$$\frac{\tfrac{1}{5}}{(s + 1)(\tfrac{1}{5}s + 1)} \tag{16.21}$$

The time constants are $\tau_1 = 1$ and $\tau_2 = \frac{1}{5}$. The factor $\frac{1}{5}$ in the numerator corresponds to the steady-state gain.

From Eqs. (16.21) and (16.19)

$$\text{AR} = \frac{\tfrac{1}{5}}{\sqrt{\omega^2 + 1}\sqrt{(\omega/5)^2 + 1}}$$

Hence,

$$\log \text{AR} = \log \frac{1}{5} - \frac{1}{2}\log(\omega^2 + 1) - \frac{1}{2}\log\left[\left(\frac{\omega}{5}\right)^2 + 1\right]$$

or

$$\log \text{AR} = \log \tfrac{1}{5} + \log(\text{AR})_1 + \log(\text{AR})_2 \tag{16.22}$$

where $(AR)_1$ $(AR)_2$ are the ARs of the individual first-order systems, each with unity gain. Equation (16.22) shows that the overall AR is obtained, on logarithmic coordinates, by adding the individual ARs and a constant corresponding to the steady-state gain.

The individual ARs must be plotted as functions of log ω rather than log $(\omega\tau)$ because of the different time constants. This is easily done by shifting the curves of Fig. 16.4 to the right or left so that the corner frequency falls at $\omega = 1/\tau$. Thus, the individual curves of Fig. 16.5 are placed so that the corner frequencies fall at $\omega_{c_1} = 1$ and $\omega_{c_2} = 5$. These curves are added to obtain the overall curve shown. Note that in this case the logarithms are negative and the addition is downward. To complete the AR curve, the factor log $\frac{1}{5}$ should be added to the overall curve. This would have the effect of shifting the entire curve down by a constant amount. Instead of doing this, the factor $\frac{1}{5}$ is incorporated by plotting the overall curve as AR/$\frac{1}{5}$ instead of AR. This procedure is usually more convenient.

Asymptotes have also been indicated on Fig. 16.5. The sum of the individual asymptotes gives the overall asymptote, which is seen to be a good approximation to the overall curve. The overall asymptote has a slope of zero below $\omega = 1$, -1 for ω between 1 and 5, and -2 above $\omega = 5$. Its slope is obtained by simply adding the slopes of the individual asymptotes.

To obtain the phase angle, the individual phase angles are plotted and added according to Eq. (16.20). The factor $\frac{1}{5}$ has no effect on the phase angle, which approaches $-180°$ at high frequency.

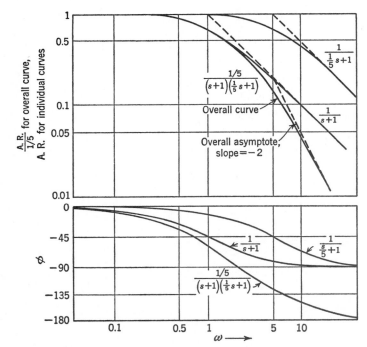

FIGURE 16-5
Bode diagram for $0.2/[(s + 1)(0.2s + 1)]$.

Graphical Rules for Bode Diagrams

Before proceeding to a development of the Bode diagram for other systems, it is desirable to summarize the graphical rules that were utilized in Example 16.4.

Consider a number of systems in series. As shown in Example 16.3, the overall AR is the product of the individual ARs, and the overall phase angle is the sum of the individual phase angles. Therefore,

$$\log (AR) = \log (AR)_1 + \log (AR)_2 + \cdots + \log (AR)_n \qquad (16.23)$$

and

$$\phi = \phi_1 + \phi_2 + \cdots + \phi_n$$

where n is the total number of systems. Therefore, the following rules apply to the true curves or to the asymptotes on the Bode diagram:

1. The overall AR is obtained by adding the individual ARs. For this graphical addition, an individual AR that is above unity on the frequency response diagram is taken as positive; an AR that is below unity is taken as negative. To understand this, recall that the logarithm of a number greater than one is positive and the logarithm of a number less than one is negative. A convenient way to combine two or more individual AR curves is to use a pair of dividers to transfer distances at a selected value of ω.
2. The overall phase angle is obtained by addition of the individual phase angles.
3. The presence of a constant in the overall transfer function shifts the entire AR curve vertically by a constant amount and has no effect on the phase angle. It is usually more convenient to include a constant factor in the definition of the ordinate.

These rules will be of considerable value in later examples. Let us now proceed to develop Bode diagrams for other control system components.

The Second-Order System

As shown in Example 16.2, the frequency response of a system with a second-order transfer function

$$G(s) = \frac{1}{\tau^2 s^2 + 2\zeta\tau s + 1}$$

is given by Eq. (16.13), repeated here for convenience,

$$AR = \frac{1}{\sqrt{(1 - \omega^2\tau^2)^2 + (2\zeta\omega\tau)^2}} \qquad (16.13)$$

$$\text{Phase angle} = \tan^{-1}\frac{-2\zeta\omega\tau}{1 - (\omega\tau)^2}$$

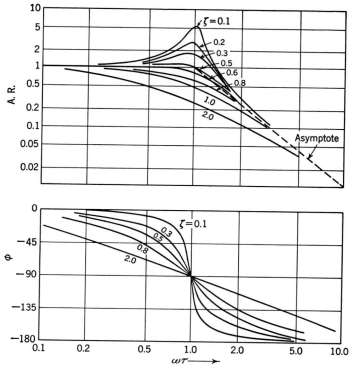

FIGURE 16-6
Bode diagram for second-order system $1/(\tau^2 s^2 + 2\zeta\tau s + 1)$.

If $\omega\tau$ is used as the abscissa for the general Bode diagram, it is clear that ζ will be a parameter. That is, there is a different curve for each value of ζ. These curves appear as in Fig. 16.6.

The calculation of phase angle as a function of ω from Eq. (16.13) requires careful attention. The calculation can be done most clearly with the aid of a plot of $\tan^{-1} x$ (or arctan x) as shown in Fig. 16.7. As $\omega\tau$ goes from zero to unity, we see from Eq. (16.13) that the argument of the arctan function goes from 0 to $-\infty$ and the phase angle goes from $0°$ to $-90°$ as shown by the branch from A to B in Fig. 16.7. As $\omega\tau$ crosses unity from a value less than unity to a value greater than unity, the sign of the argument of the arctan function in Eq. (16.13) shifts from negative to positive. To preserve continuity in angle as $\omega\tau$ crosses unity, the phase angle must go from $-90°$ to $-180°$ as $\omega\tau$ goes from unity to $+\infty$ and the branch of the arctan function goes from C to D (in Fig. 16.7).

The arctan function available in calculators and digital computers normally covers the principal branches of the arctan function, shown as BAE in Fig. 16.7. For this reason, one must be very careful in calculating the phase angle with Eq. (16.13). If a calculator programmed for the principal branches of the arctan

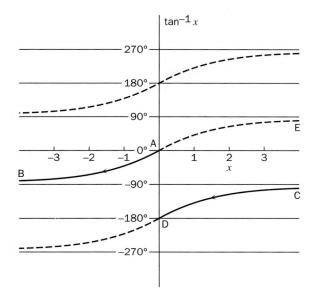

FIGURE 16-7
Use of plot of $\tan^{-1}x$ for computing phase angle of second-order system.

function is used and the argument is positive, one obtains the correct phase angle by subtracting 180° from the answer given by the calculator. Notice that for $\omega\tau = 1$, the phase angle is $-90°$, independently of ζ. This verifies that all phase curves intersect at $-90°$ as shown in Fig. 16.6.

We may now examine the amplitude curves obtained from Eq. (16.13). For $\omega\tau \ll 1$, the AR, or gain, approaches unity. For $\omega\tau \gg 1$, the AR becomes asymptotic to the line

$$AR = \frac{1}{(\omega\tau)^2}$$

This asymptote has slope -2 and intersects the line $AR = 1$ at $\omega\tau = 1$. The asymptotic lines are indicated on Fig. 16.6. For $\zeta \geq 1$, we have shown that the second-order system is equivalent to two first-order systems in series. The fact that the AR for $\zeta \geq 1$ (as well as for $\zeta < 1$) attains a slope of -2 and phase of -180 is, therefore, consistent.

Figure 16.6 also shows that, for $\zeta < 0.707$, the AR curves attain maxima in the vicinity of $\omega\tau = 1$. This can be checked by differentiating the expression for the AR with respect to $\omega\tau$ and setting the derivative to zero. The result is

$$(\omega\tau)_{\max} = \sqrt{1 - 2\zeta^2} \qquad \zeta < 0.707 \qquad (16.24)$$

for the value of $\omega\tau$ at which the maximum AR occurs. The value of the maximum AR, obtained by substituting $(\omega\tau)_{\max}$ into Eq. (16.13) is

$$(AR)_{\max} = \frac{1}{2\zeta\sqrt{1 - \zeta^2}} \qquad \zeta < 0.707$$

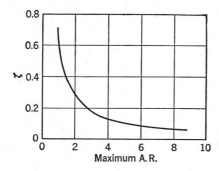

FIGURE 16-8
Maximum AR versus damping for second-order system.

A plot of the maximum AR against ζ is given in Fig. 16.8. The frequency at which the maximum AR is attained is called the resonant frequency and is obtained from Eq. (16.24),

$$\omega_r = \frac{1}{\tau}\sqrt{1 - 2\zeta^2} \tag{16.25}$$

The phenomenon of resonance is frequently observed in our everyday experience. A vase may vibrate when the stereo is playing a particular note. As a car decelerates, perceptible vibrations may occur at particular speeds. A suspension bridge oscillates violently when scouts march across stepping at a certain cadence.

It may be seen that AR values exceeding unity are attained by systems for which $\zeta < 0.707$. This is in sharp contrast to the first-order system, for which the AR is always less than unity.

The curves of Fig. 16.6 for $\zeta < 1$ are not simple to construct, particularly in the vicinity of the resonant frequency. Fortunately, almost all second-order control system components for which we shall want to construct Bode diagrams have $\zeta > 1$. That is, they are composed of two first-order systems in series. Actually, the curves of Fig. 16.6 are presented primarily because they are useful in analyzing the *closed-loop frequency response* of many control systems.

Transportation Lag

As shown by Eq. (16.15), the frequency response for $G(s) = e^{-\tau s}$ is

$$AR = 1$$

$$\phi = -\omega\tau \text{ radians} \quad \text{or} \quad \phi = -57.2958 \, \omega\tau \text{ degrees}$$

In this expression, ω is in radians and 57.2958 is the number of degrees in one radian. There is no need to plot the AR since it is constant at 1.0. On logarithmic coordinates, the phase angle appears as in Fig. 16.9, where $\omega\tau$ is used as the abscissa to make the figure general. The transportation lag contributes a phase lag, which increases without bound as ω increases. Note that it is necessary to convert $\omega\tau$ from radians to degrees to prepare Fig. 16.9.

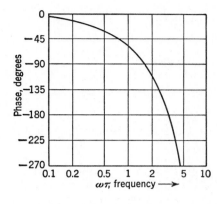

FIGURE 16-9
Phase characteristic of transportation lag.

Proportional Controller

A proportional controller with transfer function K_c has amplitude ratio K_c and phase angle zero at all frequencies. No Bode diagram is necessary for this component.

Proportional-Integral Controller

This component has the ideal transfer function

$$G(s) = K_c\left(1 + \frac{1}{\tau_I s}\right)$$

Accordingly, the frequency response is given by

$$\text{AR} = |G(j\omega)| = K_c\left|1 + \frac{1}{\tau_I j\omega}\right| = K_c\sqrt{1 + \frac{1}{(\omega\tau_I)^2}}$$

$$\text{Phase} = \measuredangle G(j\omega) = \measuredangle\left(1 + \frac{1}{\tau_I j\omega}\right) = \tan^{-1}\left(-\frac{1}{\omega\tau_I}\right)$$

The Bode plot of Fig. 16.10 uses $(\omega\tau_I)$ as the abscissa. The constant factor K_c is included in the ordinate for convenience. Asymptotes with a corner

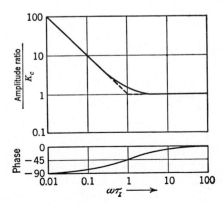

FIGURE 16-10
Bode diagram for PI controller.

frequency of $\omega_c = 1/\tau_I$ are indicated. The verification of Fig. 16.10 is recommended as an exercise for the reader.

Proportional-Derivative Controller

The transfer function is

$$G(s) = K_c(1 + \tau_D s)$$

The reader should show that this has amplitude and phase behavior that is just the inverse of the first-order system

$$\frac{1}{\tau s + 1}$$

Hence, the Bode plot is as shown in Fig. 16.11. The corner frequency is $\omega_c = 1/\tau_D$.

This system is important because it introduces phase lead. Thus, it can be seen that using PD control for the tank temperature-control system of Example 16.3 would decrease the phase lag at all frequencies. In particular, 180° of phase lag would not occur until a higher frequency. This may exert a stabilizing influence on the control system. In the next chapter, we shall look in detail at designing stabilizing controllers using Bode diagram analysis. It is appropriate to conclude this chapter with a summarizing example.

Example 16.5. Plot the Bode diagram for the open-loop transfer function of the control system of Fig. 16.12. This system might represent PD control of three tanks in series, with a transportation lag in the measuring element.

The open-loop transfer function is

$$G(s) = \frac{10(0.5s + 1)e^{-s/10}}{(s + 1)^2(0.1s + 1)}$$

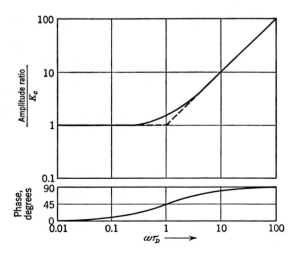

FIGURE 16-11
Bode diagram for PD controller.

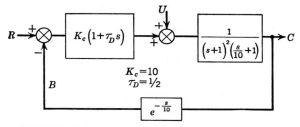

FIGURE 16-12
Block diagram of control system for Example 16.5.

The individual components are plotted as dashed lines in Fig. 16.13. Only the asymptotes are used on the AR portion of the graph. Here it is easiest to plot the factor $(s + 1)^{-2}$ as a line of slope -2 through the corner frequency of 1. For the phase-angle graph, the factor $(s + 1)^{-1}$ is plotted and added in twice to form the overall curve. The overall curves are obtained by the graphical rules previously presented. For comparison, the overall curves obtained without derivative action [i.e., by not adding in the curves corresponding to $(0.5s + 1)$] are also shown. It should be noted, that, on the asymptotic AR diagram, the slopes of the individual curves are added to obtain the slope of the overall curve.

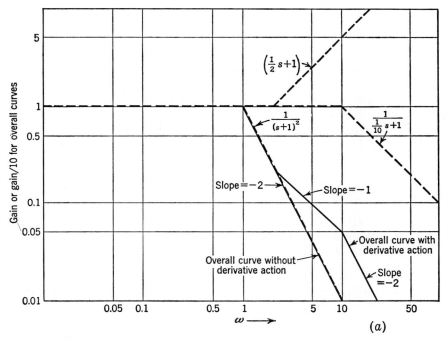

FIGURE 16-13
Bode diagram for Example 16.5: (*a*) amplitude ratio; (*b*) phase angle.

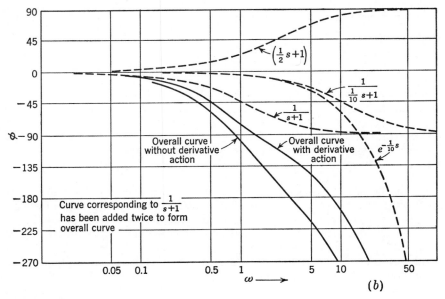

FIGURE 16-13 *(Continued)*
Bode diagram for Example 16.5: (*a*) amplitude ratio; (*b*) phase angle.

PROBLEMS

16.1. For each of the following transfer functions, sketch the gain versus frequency, asymptotic Bode diagram. For each case, find the actual gain and phase angle at $\omega = 10$. *Note:* It is not necessary to use log-log paper; simply rule off decades on rectangular paper.

(*a*) $\dfrac{100}{(10s + 1)(s + 1)}$

(*b*) $\dfrac{10s}{(s + 1)(0.1s + 1)^2}$

(*c*) $\dfrac{s + 1}{(0.1s + 1)(10s + 1)}$

(*d*) $\dfrac{s - 1}{(0.1s + 1)(10s + 1)}$

(*e*) $(10s + 1)^2$

(*f*) $(10 + s)^2$

16.2. A temperature bath in which the temperature varies sinusoidally at various frequencies is used to measure the frequency response of a temperature-measuring element *B*. The apparatus is shown in Fig. P16.2. A standard thermocouple *A*, for which the time constant is 0.1 min for the arrangement shown in the sketch, is placed near the element to be measured. The response of each temperature-measuring element is recorded simultaneously on a two-channel recorder. The phase lag between

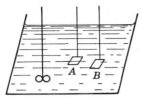

FIGURE P16-2

the two chart records at different frequencies is shown in the table. From these data, show that it is reasonable to consider element B as a first-order process and calculate the time constant. Describe your method clearly.

Frequency, cycles/min	Phase lag of B behind A, deg
0.1	7.1
0.2	12.9
0.4	21.8
0.8	28.2
1.0	29.8
1.5	26.0
2.0	23.6
3.0	18.0
4.0	14.2

16.3. Plot the asymptotic Bode diagram for the PID controller:

$$G(s) = K_c\left(1 + \tau_D s + \frac{1}{\tau_I s}\right)$$

where $K_c = 10$, $\tau_I = 1$, $\tau_D = 100$. Label corner frequencies and give slopes of asymptotes.

16.4. One way of experimentally measuring frequency response is to plot the output sine wave versus the input sine wave. The results of such a plot look like the figure shown in Fig. P16.4. This is the sinusoidal *deviation* in output versus sinusoidal *deviation* in input and appears as an ellipse centered at the origin. Show how to obtain the AR and phase lag from this plot.

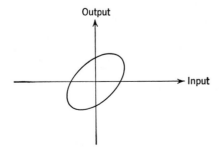

FIGURE P16-4

16.5. For the transfer function shown below, sketch carefully the gain versus frequency portion of the asymptotic plot of the Bode diagram. Determine the actual (exact) value of gain and phase angle at $\omega = 1$. Determine the phase angle as $\omega \to \infty$.

$$G(s) = \frac{2(0.1s + 1)}{s^2(10s + 1)}$$

Indicate very clearly the slopes of the asymptotic bode diagram of $G(s)$.

16.6. (*a*) Plot accurately and neatly the Bode diagram for the process shown in Fig. P16.6 using log-log paper for gain vs. frequency and semi-log paper for phase vs. frequency. Plot the frequency as rad/min.

(*b*) Find the amplitude ratio and phase angle for Y/X at $\omega = 1$ rad/min and $\omega = 4$ rad/min.

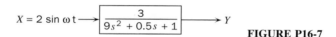

$$X \longrightarrow \boxed{\frac{1}{2s + 1}} \longrightarrow \boxed{\frac{7}{5s + 1}} \longrightarrow \boxed{e^{-0.5s}} \longrightarrow Y$$

FIGURE P16-6

16.7. For the system shown in Fig. P16.7, determine accurately the phase angle in degrees between $Y(t)$ and $X(t)$ for $\omega = 0.5$. Determine the lag between the input wave and the output wave.

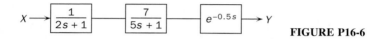

$$X = 2 \sin \omega t \longrightarrow \boxed{\frac{3}{9s^2 + 0.5s + 1}} \longrightarrow Y$$

FIGURE P16-7

16.8. (*a*) For the transfer function given below, sketch carefully the asymptotic approximation of gain vs. frequency. Show detail such as slopes of asymptotes.

$$G(s) = \frac{(1 + s)^2}{s}$$

(*b*) Find the actual (exact) value of gain and phase angle for $\omega = 1$ and for $\omega = 2$.

16.9. Derive expressions for amplitude ratio and phase angle as functions of ω for the transfer function $G(s) = 1/(s^2 - 1)$.

16.10. The data given in the following table represent experimental, frequency response data for a process consisting of a first-order process and a transportation lag. Determine the time constant and the transportation lag parameter. Write the transfer function for the process, giving numerical values of the parameters.

Frequency, cpm	Gain	Phase angle, deg
0.01	1.0	0.0
0.02	1.0	−2.0
0.04	1.0	−6.0
0.06	1.0	−7.0
0.08	1.0	−8.5
0.10	1.0	−11.0
0.15	1.0	−17.0
0.20	1.0	−23.0
0.30	1.0	−36.0
0.40	0.98	−48.0
0.60	0.94	−73.0
0.80	0.88	−96.0
1.00	0.83	−122.0
1.50	0.71	−180.0
2.00	0.61	−239.0
4.00	0.37	—
6.00	0.26	—
8.00	0.20	—
10.00	0.16	—
20.00	0.080	—
40.00	0.041	—

CHAPTER
17

CONTROL SYSTEM DESIGN BY FREQUENCY RESPONSE

The purpose of this chapter is twofold. First, it will be indicated that the stability of a control system can usually be determined from the Bode diagram of its open-loop transfer function. Then methods will be presented for rational selection of controller parameters based on this Bode diagram. The material to be presented here is one of the more useful design aspects of the subject of frequency response.

Tank-Temperature Control System

It was indicated in the discussion following Example 16.3 that the control system of Fig. 17.1 might offer stability problems because of excessive phase lag. To review, this system represents proportional control of tank temperature with a delay in the feedback loop. The factor $\frac{1}{600}$ is the process sensitivity $1/wC$, which gives the ultimate change in tank temperature per unit change in heat input Q. The proportional sensitivity K_c, in Btu per hour per degree of temperature error, is to be specified by the designer.

The open-loop transfer function for this system is

$$G(s) = \frac{(K_c/600)e^{-0.0396s}}{0.202s + 1} \tag{17.1}$$

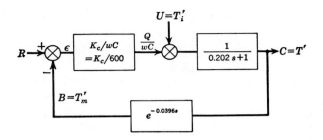

FIGURE 17-1
Control system for stirred-tank heater of Example 16.3.

The Bode diagram for $G(s)$ is plotted in Fig. 17.2. As usual, the constant factor $K_c/600$ is included in the definition of the ordinate for AR. At the frequency of 43 rad/min, the phase lag is exactly $180°$ and

$$\frac{AR}{K_c/600} = 0.12$$

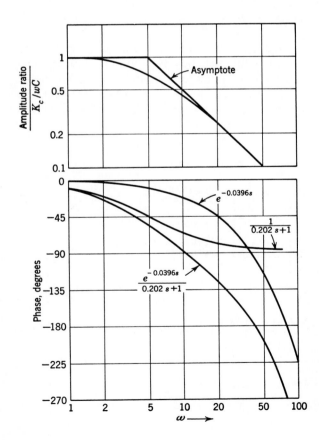

FIGURE 17-2
Bode diagram for open-loop transfer function of control system for stirred-tank heater: $(K_c/wC)e^{-\tau_2 s}$ $[1/(\tau_1 s + 1)]$. (Block diagram shown in Fig. 17.1.)

Therefore, if a proportional gain of 5000 Btu/(hr)(°F) is used,

$$AR = 0.12\frac{5000}{600} = 1$$

This is the AR between the signals ϵ and B. Note that it is dimensionless as it must be, since ϵ and B both have the units of temperature.

The control system is redrawn for $K_c = 5000$ in Fig. 17.3a, with the loop opened. That is, the feedback signal B is disconnected from the comparator. It is imagined that a set point disturbance

$$R = \sin 43t$$

is applied to the opened loop. Then, since the open-loop AR and phase lag are unity and 180°

$$B = \sin(43t - 180°) = -\sin 43t$$

Now imagine that, at some instant of time, R is set to zero and simultaneously the loop is closed. Figure 17.3b indicates that the closed loop continues to oscillate indefinitely. *This oscillation is theoretically sustained even though both R and U are zero.*

Now suppose K_c is set to slightly higher value and the same experiment repeated. This time, the signal ϵ is amplified slightly each time it passes around the loop. Thus, if K_c is set to 5001, after the first time around the loop the signal ϵ becomes $(5001/5000) \sin 43t$. After the second time, it is $(5001/5000)^2 \sin 43t$, etc. The phase-angle relations are not affected by changing K_c. We thus conclude

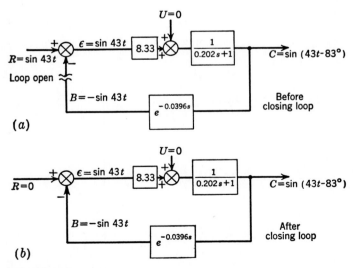

(a)

(b)

FIGURE 17-3
Sustained closed-loop oscillation.

that, for $K_c > 5000$, the response is unbounded, since it oscillates with *increasing amplitude*.

Using the definition of stability presented in Chap. 14, it is concluded that the control system is unstable for $K_c > 5000$ because it exhibits an unbounded response to the bounded input described above. (The bounded input is zero in this case, for $U = R = 0$.) The condition $K_c > 5000$ corresponds to

$$AR > 1$$

for the open-loop transfer function, at the frequency 43 rad/min, where the open-loop phase lag is 180°.

This argument is not rigorous. We know the response B only if ϵ remains *constant in amplitude* because of the definition of frequency response. If, however, the change in K_c is very small, so that ϵ is amplified infinitesimally, then B will closely approximate the frequency response. While this does not *prove* anything, it shows that we are justified in suspecting instability and that closer investigation is warranted. A rigorous proof of stability requires application of the Nyquist stability criterion [See Coughanowr and Koppel (1965) or Kuo (1987)], which uses the theory of complex variables. For our purposes, it is sufficient to proceed with heuristic arguments.

The Bode Stability Criterion

It is tempting to generalize the results of the analysis of the tank-temperature control system to the following rule. *A control system is unstable if the open-loop frequency response exhibits an AR exceeding unity at the frequency for which the phase lag is 180°.* This frequency is called the *crossover frequency*. The rule is called the *Bode stability criterion*.

Actually, since the discussion of the previous section was based on heuristic arguments, this rule is not quite general. It applies readily to systems for which the gain and phase curves decrease continuously with frequency. However, if the phase curve appears as in Fig. 17.4, the more general Nyquist criterion must usually be used to determine stability. Other exceptions may occur. Fortunately, most process control systems can be analyzed with the simple Bode criterion, and it therefore finds wide application.

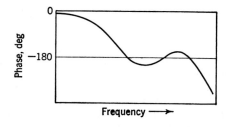

FIGURE 17-4
Phase behavior of complex system for which Bode criterion is not applicable.

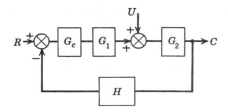

FIGURE 17-5
Block diagram for general control system.

Application of the criterion requires nothing more than plotting the open-loop frequency response. This may be based on the theoretical transfer function, if it is available, as we have done for the tank-temperature system. If the theoretical system dynamics are not known, the frequency response may be obtained experimentally. To do this, the open-loop system is disturbed with a sine-wave input at several frequencies. At each frequency, records of the input and output waves are compared to establish AR and phase lag. The results are plotted as a Bode diagram. This experimental technique will be illustrated in more detail in Chap. 19.

For the remainder of this chapter, we accept the Bode stability criterion as valid and use it to establish control system design procedure.

Gain and Phase Margins

Let us consider the general problem of selecting $G_c(s)$ for the system of Fig. 17.5. Suppose the open-loop frequency response, when a particular controller $G_c(s)$ is tried, is as shown in the Bode diagram of Fig. 17.6. The crossover frequency, at which the phase lag is 180°, is noted as ω_{co} on the Bode diagram. At this frequency, the AR is A. If A exceeds unity, we know from the Bode criterion that the system is unstable and that we have made a poor selection of $G_c(s)$. In Fig. 17.6 it is assumed that A is less than unity and therefore the system is stable.

It is necessary to ascertain to what degree the system is stable. Intuitively, if A is only slightly less than unity the system is "almost unstable" and may be expected to behave in a highly oscillatory manner even though it is theoretically stable.* Furthermore, the constant A is determined by physical parameters of the system, such as time constants. These can be only estimated and may actually change slowly with time because of wear or corrosion. Hence, a design for which A is close to unity does not have an adequate safety factor.

To assign some quantitative measure to these considerations, the concept of gain margin is introduced. Using the nomenclature of Fig. 17.6,

$$\text{Gain margin} = \frac{1}{A}$$

---

* Again, heuristic arguments are used. This statement is self-evident to the reader who has studied Chap. 15, where it is shown that the roots of the characteristic equation vary continuously with system parameters. Proof of the statement requires the Nyquist stability criterion.

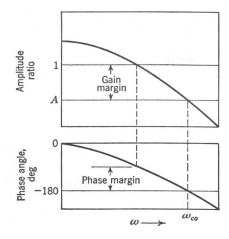

FIGURE 17-6
Open-loop Bode diagram for typical control system.

Typical specifications for design are that the gain margin should be greater than 1.7. This means that the AR at crossover could increase by a factor of 1.7 over the design value before the system became unstable. The design value of the gain margin is really a safety factor. *As such, its value varies considerably with the application and designer.* A gain margin of unity or less indicates an unstable system.

Another margin frequently used for design is the phase margin. As indicated in Fig. 17.6, it is the difference between 180° and the phase lag at the frequency for which the gain is unity. The phase margin therefore represents the additional amount of phase lag required to destabilize the system, just as the gain margin represents the additional gain for destabilization. *Typical design specifications are that the phase margin must be greater than* 30°. A negative phase margin indicates an unstable system.

Example 17.1. Find a relation between relative stability and the phase margin for the control system of Fig. 17.7. A proportional controller is to be used.

This block diagram corresponds to the stirred-tank heater system, for which the block diagram has been given in Fig. 13.6. The particular set of constants is

$$\tau = \tau_m = 1 \qquad \frac{1}{wC} = 1$$

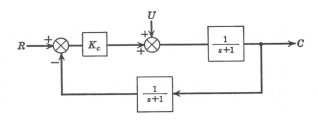

FIGURE 17-7
Block diagram for Example 17.1.

These are to be regarded as fixed, while the proportional gain K_c is to be varied to give satisfactory phase margin. The *closed-loop* transfer function for this system is given by Eq. (13.17), rewritten for our particular case as

$$\frac{C}{R} = \frac{K_c}{1 + K_c} \frac{s + 1}{\tau_2^2 s^2 + 2\tau_2 \zeta_2 s + 1} \tag{17.2}$$

where $\tau_2 = \sqrt{\dfrac{1}{1 + K_c}}$

$\zeta_2 = \sqrt{\dfrac{1}{1 + K_c}}$

Since the closed-loop system is second-order, it can never be *unstable*. The shape of the response of the closed-loop system to a unit step in R must resemble the curves of Fig. 8.2. The meaning of *relative stability* is illustrated by Fig. 8.2. The lower ζ_2 is made, the more oscillatory and hence the "less stable" will be the response. Therefore, a relationship between phase margin and ζ_2 will give the relation between phase margin and relative stability.

To find this relation the open-loop Bode diagram is prepared and is shown in Fig. 17.8. The simplest way to proceed from this diagram is as follows: consider a typical frequency $\omega = 4$. If the open-loop gain were 1 at this frequency, then since the phase angle is $-152°$, the phase margin would be $28°$. To make the open-loop gain 1 at $\omega = 4$, it is required that

$$K_c = \frac{1}{0.062} = 16.1$$

Then

$$\zeta_2 = \sqrt{\frac{1}{1 + K_c}} = 0.24$$

Hence, a point on the curve of ζ_2 versus phase margin is

$$\zeta_2 = 0.24 \qquad \text{phase margin} = 28°$$

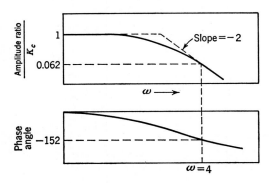

FIGURE 17-8
Bode diagram for system of Example 17.1.

Other points are calculated similarly at different frequencies, and the resulting curve is shown in Fig. 17.9. From this figure it is seen that ζ_2 decreases with decreasing phase margin and that, if the phase margin is less than $30°$, ζ_2 is less than 0.26. From Fig. 8.2, it can be seen that the response of this system for $\zeta_2 < 0.26$ is highly oscillatory, hence relatively unstable, compared with a response for the system with phase margin $50°$ and $\zeta_2 = 0.4$.

For the particular system of Example 17.1, it was shown that the response became more oscillatory as the phase margin was decreased. This result generalizes to more complex systems. Thus, the phase margin is a useful design tool for application to systems of higher complexity, where the transient response cannot be easily determined and a plot such as Fig. 17.9 cannot be made. To repeat, the rule of thumb is that the phase margin must be greater than $30°$.

A similar statement can be made about the gain margin. As the gain margin is increased, the system response generally becomes less oscillatory, hence more stable. A control system designer will often try to make *both* the gain and phase margins equal to or greater than specified minimum values, typically 1.7 and $30°$. Note that, for the case of Example 17.1, the gain margin is always infinite because the phase lag never quite reaches $180°$. However, the phase margin requirement of $30°$ necessitates that $\zeta_2 > 0.26$, hence $K_c < 14$, which means that an offset of $\frac{1}{15}$ [see Eq. (17.2)] must be accepted. This illustrates the importance of considering both margins. The reader should refer to Fig. 17.6 to see that both margins exist simultaneously.

Example 17.2. Specify the proportional gain K_c for the control system of Fig. 16.12. The Bode diagram for the particular case $K_c = 10$ is presented in Fig. 16.13. The gain is to be specified for the two cases:

1. $\tau_D = 0.5$ min
2. $\tau_D = 0$ (no derivative action)

1. Consider first the gain margin. The crossover frequency for the curve with derivative action is 8.0 rad/min. At this frequency, the open-loop gain is 0.062 if the value of K_c is unity. (Including the factor of 1/10 in the ordinate is actually

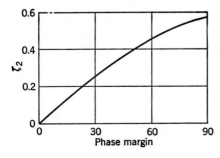

FIGURE 17-9
Damping versus phase margin for system of Fig. 17.7.

equivalent to plotting the case $K_c = 1$.) Therefore, according to the Bode criterion, the value of K_c necessary to destabilize the loop is 1/0.062 or 16. To achieve a gain margin of 1.7, K_c must be taken as 16/1.7 or 9.4. To achieve proper phase margin, note that the frequency for which the phase lag is 150° (phase margin is 30°) is 5.3 rad/min. At this frequency, a value for K_c of 1/0.094 or 10.6 will cause the open-loop gain to be unity. Since this is higher than 9.4, we use 9.4 as the design value of K_c. The resulting phase margin is then 38°.

2. Proceeding exactly as in case 1 but using the curve in Fig. 16.13 for no derivative action, it is found that $K_c = 5.3$ is needed for satisfactory gain margin and $K_c = 3.7$ for satisfactory phase margin. Hence K_c is taken as 3.7 and the resulting gain margin is 2.4.

To see the advantage of adding derivative control in this case, note from Fig. 16.12 that the final value of C for a unit-step change in U is $1/(1 + K_c)$ for any value of τ_D. The addition of the derivative action allows increase of the value of K_c from 3.7 to 9.4 while maintaining approximately the same relative stability in terms of gain and phase margins. This reduces the offset from 21 percent of the change in U to 9.6 percent of the change in U.

The reader is cautioned that the values of K_c selected in this way should be regarded as initial approximations to the actual values, which give "optimal" control of the system of Fig. 16.12. More will be said about this matter later in this chapter in conjunction with the two-tank chemical-reactor control system of Chap. 11.

Thus far, nothing has been said about upper limits on the gain and phase margins. Referring to Example 17.1 and Fig. 8.2, it is seen that, if ζ_2 is too large, the response is sluggish. In fact, Fig. 8.2 suggests that for the system of Fig. 17.7 one should choose a value of ζ_2 low enough to give a short rise time without causing excessive response time and overshoot. In other words, one wants the most rapid response that has sufficient relative stability. The results of Example 17.1 generalize to many systems of higher complexity, in terms of margin. Hence, the designer frequently chooses the controller so that either the gain or phase margin is equal to its lowest acceptable value and the other margin is (probably) above its lowest acceptable value. This was the procedure followed in Example 17.2. In almost every situation, the designer faces this conflict between speed of response and degree of oscillation. In addition, if integral action is not used, the amount of the offset must be considered.

The concepts of gain and phase margin are useful in selecting K_c for proportional action. However, for additional modes of control such as PD, these concepts are difficult to apply in practice. Consider the selection of K_c and τ_D in Example 17.2. For a different value of τ_D the derivative contribution is shifted to the right or left on the Bode diagram of Fig. 16.13. This means that a different value of K_c will provide the proper margins. A typical design procedure is to select the value of τ_D for which the value of K_c resulting in a 30° phase margin is maximized. The motivation for this choice is that the offset will be minimized. However, the procedure is clearly trial and error. In the case of three-mode control, there are two parameters, τ_I and τ_D, which must be varied by trial to meet various design criteria. Fortunately, for this case and others there are simple rules for directly establishing values of the control parameters that usually give satisfactory gain and phase margins. These are the Ziegler-Nichols rules, which we develop in the next section.

Ziegler-Nichols Controller Settings

Consider selection of a controller G_c for the general control system of Fig. 17.5. We first plot the Bode diagram for the final control element, the process, and the measuring element in series, $G_1G_2H(j\omega)$. It should be emphasized that the controller is omitted from this plot. Suppose the diagram appears as in Fig. 17.6. As noted on the figure, the crossover frequency for these three components in series is ω_{co}. At the crossover frequency, the overall gain is A, as indicated. According to the Bode criterion, then, the gain of a proportional controller which would cause the system of Fig. 17.5 to be on the verge of instability is $1/A$. We define this quantity to be the ultimate gain K_u. Thus

$$K_u = \frac{1}{A} \tag{17.3}$$

The ultimate period P_u is defined as the period of the sustained cycling that would occur if a proportional controller with gain K_u were used. From the discussion of Fig. 17.3, we know this to be

$$P_u = \frac{2\pi}{\omega_{co}} \qquad \text{time/cycle} \tag{17.3a}$$

The factor of 2π appears, so that P_u will be in units of time per cycle rather than time per radian. It should be emphasized that K_u and P_u are easily determined from the Bode diagram of Fig. 17.6.

The Ziegler-Nichols settings for controllers are determined directly from K_u and P_u according to the rules summarized in Table 17.1. Unfortunately, specifications of K_c and τ_D for PD control cannot be made using only K_u and P_u. In general, the values $0.6K_u$ and $P_u/8$, which correspond to the limiting case of no integral action in a three-mode controller, are too conservative. That is, the

TABLE 17.1
Ziegler-Nichols Controller Settings

Type of control	$G_c(s)$	K_c	τ_I	τ_D
Proportional	K_c	$0.5K_u$		
Proportional-integral (PI)	$K_c\left(1 + \dfrac{1}{\tau_I s}\right)$	$0.45K_u$	$\dfrac{P_u}{1.2}$	
Proportional-integral-derivative (PID)	$K_c\left(1 + \dfrac{1}{\tau_I s} + \tau_D s\right)$	$0.6K_u$	$\dfrac{P_u}{2}$	$\dfrac{P_u}{8}$

resulting system will be too stable. There exist methods for this case which are in principle no more difficult to use than the Ziegler-Nichols rules. One of these is selection of τ_D for maximum K_c at 30° phase margin, which was discussed above. Another method, which utilizes the step response and avoids trial and error, is presented in Chap. 19.

The reasoning behind the Ziegler-Nichols selection of values of K_c is relatively clear. In the case of proportional control only, a gain margin of 2 is established. The addition of integral action introduces more phase lag at all frequencies (see Fig. 16.10); hence a lower value of K_c is required to maintain roughly the same gain margin. Adding derivative action introduces phase lead. Hence, more gain may be tolerated. This was demonstrated in Example 17.2. However, by and large the Ziegler-Nichols settings are based on experience with typical processes and should be regarded as first estimates.

Example 17.3. Using the Ziegler-Nichols rules, determine K_c and τ_I for the control system shown in Fig. 17.10.

For this problem, the computation will be done without plotting a Bode diagram; however, the reader may wish to do the problem with such a diagram. We first obtain the crossover frequency by applying the Bode stability criterion:

$$-180° = -\tan^{-1}(\omega) - 57.3(1.02)(\omega)$$

The value 57.3 converts radians to degrees. Solving this equation by trial and error gives for the crossover frequency, $\omega_{co} = 2$ rad/min. The amplitude ratio (AR) at the crossover frequency for the open loop can be written

$$\text{AR} = K_c \frac{1}{\sqrt{1 + \omega^2}}(1) = \frac{K_c}{2.24}$$

where we have used Eq. (16.16) for the first-order system and the fact that the amplitude ratio for a transport lag is one. According to the Bode criterion, the AR is 1.0 at the crossover frequency when the system is on the verge of instability. Inserting AR = 1 into the above equation and solving for K_c gives $K_{cu} = 2.24$. From the Ziegler-Nichols rules of Table 17.1, we obtain

$$K_c = 0.45 K_{cu} = (0.45)(2.24) = 1.01$$

and

$$\tau_I = P_u/1.2 = [2\pi/\omega_{co}]/1.2 = [2\pi/2]/1.2 = 2.62 \text{ min.}$$

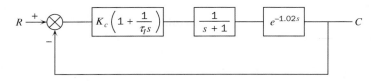

FIGURE 17-10
Block diagram for Example 17.3.

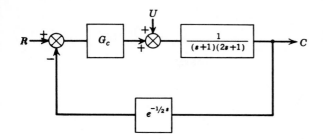

FIGURE 17-11
Block diagram for two-tank chemical-reactor system.

Example 17.4. Using the Ziegler-Nichols rules, determine controller settings for various modes of control of the two-tank chemical-reactor system of Chap. 11. The block diagram is reproduced in Fig. 17.11.

For convenience, the process gain K and the controller gain K_c are combined into an overall gain K_1. The equivalent controller transfer function is regarded as

$$G_c = K_1\left(1 + \frac{1}{\tau_I s} + \tau_D s\right)$$

where K_1 (as well as τ_I and τ_D) is to be selected by the Ziegler-Nichols rules. The required value of K_c is then easily determined as

$$K_c = \frac{K_1}{K}$$

where $K = 0.09$ for the present case (see Chap. 11.)

The Bode diagram for the transfer function *without the controller*

$$\frac{e^{-(1/2)s}}{(s + 1)(2s + 1)}$$

is prepared by the usual procedures and is shown in Fig. 17.12. From this figure, it is found that

$$\omega_{co} = 1.56 \text{ rad/min}$$

$$K_{1_u} = \frac{1}{0.145} = 6.9 \tag{17.4}$$

$$P_u = \frac{2\pi}{1.56} = 4.0 \text{ min/cycle}$$

Hence, the Ziegler-Nichols control constants determined from Table 17.1 and Eq. (17.4) are given in Table 17.2.

A plot comparing the open-loop frequency responses *including the controller* for the three cases, using the controller constants of Table 17.2, is given in Fig. 17.13. This figure shows quite clearly the effect of the phase lead due to the derivative action. The resulting gain and phase margins are listed in Table 17.3. From this table it may be seen that the margins are adequate and generally conservative.

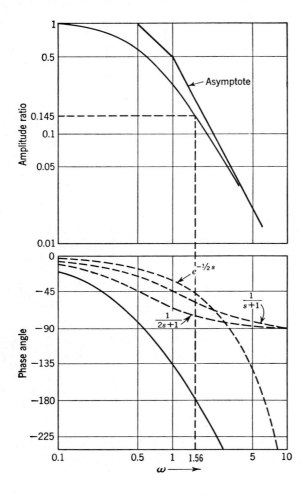

FIGURE 17-12
Bode diagram for
$e^{-0.5s}/[(s + 1)(2s + 1)]$.

Note that to obtain the Bode diagram for systems including the PID controller, the controller transfer function is rewritten as

$$K_c\left(1 + \frac{1}{\tau_I s} + \tau_D s\right) = K_c \frac{\tau_D \tau_I s^2 + \tau_I s + 1}{\tau_I s} \qquad (17.5)$$

This is second-order in the numerator and has integral action in the denominator. In general, the numerator factors into first-order factors; hence it contributes two

TABLE 17.2

Control	K_I	τ_I	τ_D
P	3.5		
PI	3.1	3.3	
PID	4.2	2.0	0.50

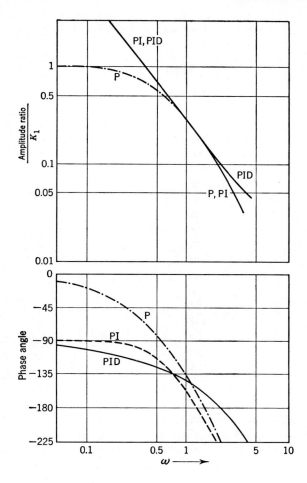

FIGURE 17-13
Open-loop Bode diagrams for various controllers with system of Fig. 17.11.

curves similar to that of Fig. 16.11 to the overall diagram. For the Ziegler-Nichols settings it is seen from Table 17.1 that $\tau_I = 4\tau_D$. Making this substitution into Eq. (17.5)

$$G_c = K_c \frac{4\tau_D^2 s^2 + 4\tau_D s + 1}{4\tau_D s} = \frac{K_c(2\tau_D s + 1)^2}{4\tau_D s} \qquad (17.6)$$

shows that the numerator is equivalent to two PD components in series. This AR is represented by a high-frequency asymptote of slope $+2$ passing through the fre-

TABLE 17.3

Control	Gain margin	Phase margin
P	2.0	45°
PI	1.9	33°
PID	2.6	34°

quency $\omega = 1/2\tau_D$ and a low-frequency asymptote on the line AR = 1. It should be emphasized that these special considerations apply only to the Ziegler-Nichols settings. In the general case, the two times constants obtained by factoring the numerator of Eq. (17.5) will be different. The Bode plot of the denominator follows from

$$\frac{1}{\tau_I \, j\omega} = |\frac{1}{\omega\tau_I}|\angle - 90°$$

The gain is a straight line of slope -1 passing through the point (AR = 1, ω = $1/\tau_I$). The phase lag is 90° at all frequencies. Plotting of the overall Bode diagram for the PID case to check the results of Fig. 17.13 is recommended as an exercise for the reader.

Transient Responses

For instructive purposes, the two-tank reactor system of Fig. 17.11 was simulated on a computer. Responses of $C(t)$ to a unit-step change in $R(t)$ are shown in Fig. 17.14. These responses were obtained using the Ziegler-Nichols controller settings determined in Example 17.4.

The responses to a step load change were also obtained on a computer. These are the curves of Fig. 10.7 that were discussed in Chap. 10 to illustrate the function of the various modes of control. A load change for this system corresponds to a change in the inlet concentration of reactant to tank 1 (refer to Fig. 11.1). As process control engineers, we would be more interested in controlling against this kind of disturbance than against a set-point change because the set point or desired product concentration is likely to remain relatively fixed. In other words, this is a regulator problem and the curves of Fig. 10.7 are those we would use to determine the quality of control.

However, the step change in set point is frequently used to test control systems despite the fact that the system will be primarily subject to load changes during actual operation. The reason for this is the existence of well-established terminology used to describe the step response of the underdamped second-order system. This terminology, which was presented in Chap. 8, is used to assign quantitative measure to responses that are not truly second-order, such as those

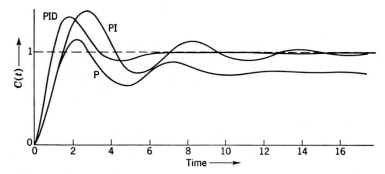

FIGURE 17-14
Closed-loop response to step change in $R(t)$ for control system of Fig. 17.11, using various control modes (obtained by computer).

of Fig. 17.14. Of course, the terminology can be applied only to responses that *resemble* damped sinusoids. Values of the various parameters determined for the responses of Fig 17.14 are summarized in Table 17.4. Offset, realized only with proportional control, is included for completeness.

It can be seen from Fig. 17.14 and Table 17.4 that addition of integral action eliminates offset at the expense of a more oscillatory response. When derivative action is also included, the response is much faster (lower rise time) and much less oscillatory (lower response time). The large overshoots realized in all three cases are characteristic of systems with relatively large time delays. In this case the controller is receiving information about the concentration in the second reactor that was true $\frac{1}{2}$ min ago. This is to be compared with the reactor time constants of 1 and 2 min. Hence, it is not surprising that the system overshoots before the controller can take sufficient action.

Figure 17.15 is presented for two purposes: (1) to illustrate that the Ziegler-Nichols controller settings should be regarded as first guesses rather than fixed values and (2) to show the effects of changing the various controller settings. These figures, which were obtained on a computer, are transient responses to step changes in set point for the three-mode PID control. They show the effects of individually varying the three control parameters K_c, τ_I, and τ_D.

As an example of the use of these figures, suppose that it is decided that the maximum overshoot that can be tolerated is 25 percent. Figure 17.15*a* shows that overshoot may be reduced by decreasing K_c at the expense of a considerably more sluggish response. From Fig. 17.15*b*, we see that overshoot may be reduced by increasing τ_I (decreasing integral action) at a lesser expense in speed of response. Thus, for $\tau_I = 5$ min, the overshoot is reduced to 20 percent without a serious sacrifice in speed. The overshoot cannot be significantly reduced by changing τ_D, as can be seen from Fig. 17.15*c*. However, the speed of response may be significantly increased by increasing the derivative action, at the expense of more oscillation before the response has settled (higher decay ratio, lower period). From this brief study of these figures, it may be concluded that, to decrease overshoot without seriously slowing the response, a combination of changes should be made. A possible combination, which should be tried, is to reduce K_c slightly and to increase τ_I and τ_D moderately. These changes would probably be tried on the actual reactor system when it is put into operation. Such adjustments from the preliminary settings are usually made by experienced control engineers, using trial procedures that are more art than science. For this reason, we leave the problem of adjustment at this point.

TABLE 17.4

Control	Overshoot	Decay ratio	Rise time, min	Response time, min	Period of oscillation, min	Offset
P	0.49	0.26	1.3	10.4	5.0	0.21
PI	0.46	0.29	1.5	11.8	5.5	0
PID	0.42	0.05	0.9	4.9	5.0	0

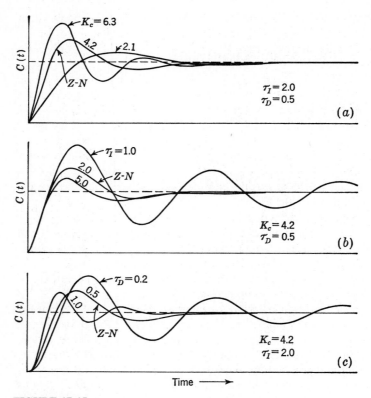

FIGURE 17-15
Effects of varying controller settings on system response. (*Z-N* indicates response using Ziegler-Nichols settings.)

PROBLEMS

17.1. Calculate the value of gain K_c needed to produce continuous oscillations in the control system shown in Fig. P17.1 when
(a) n is 2.
(b) n is 3.
Do not use a graph for this calculation.

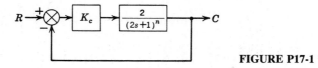

FIGURE P17-1

17.2. (a) Plot the asymptotic Bode diagram $|B/\epsilon|$ versus ω for the control system shown in Fig. P17.2.

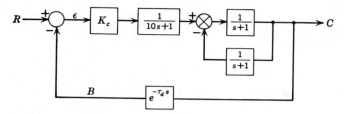

FIGURE P17-2

(b) The gain K_c is increased until the system oscillates continuously at a frequency of 3 rad/min. From this information, calculate the transportation lag parameter τ_d.

17.3. The frequency response for the block G_p in Fig. P17.3 is given in the following table:

f, cycles/min	Gain	Phase angle, degrees
0.06	1.60	− 68
0.08	1.40	− 88
0.10	1.20	− 105
0.15	0.84	− 145
0.20	0.61	− 177
0.30	0.35	− 235
0.40	0.22	
0.60	0.11	
0.80	0.066	

G_p contains a distance velocity lag $e^{-\tau s}$ with $\tau = 1$ (this transfer function is included in the data given in the table).

(a) Find the value of K_c needed to produce a phase margin of 30° for the system if $\tau_I = 0.2$.

(b) Using the value of K_c found in part (a) and using $\tau_I = 0.2$, find the percentage change in the parameter τ to cause the system to oscillate continuously with constant amplitude.

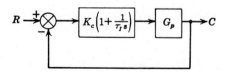

FIGURE P17-3

17.4. The system shown in Fig. P17.4 is controlled by a proportional controller. The concentration of salt in the solution leaving the tank is controlled by adding a concentrated solution through a control valve. The following data apply:

1. Concentration of concentrated salt solution $C_1 = 25$ lb salt/ft^3 solution.

2. Controlled concentration $C = 0.1$ lb salt/ft^3 solution.

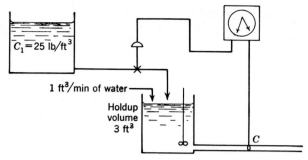

FIGURE P17-4

3. Transducer: The pen on the controller moves full scale when the concentration varies from 0.08 to 0.12 lb/ft^3. This relationship is linear. The pen moves 4.25 inches during full-scale travel.

4. Control valve: The flow through the control valve varies from 0.002 to 0.0006 cfm with a change of valve-top pressure from 3 to 15 psi. This relationship is linear.

5. Distance velocity lag: It takes 1 min for the solution leaving the tank to reach the concentration-measuring element at the end of the pipe.

6. Neglect lags in the valve and transducer.

    (a) Draw a block diagram of the control system. Place in each block the appropriate transfer function. Calculate all the constants and give the units.

    (b) Using a frequency-response diagram and the Ziegler-Nichols rules, determine the settings of the controller.

    (c) Using the controller settings of part (b) calculate the offset when the set point is changed by 0.02 unit of concentration.

17.5. The stirred-tank heater system shown in Fig. P17.5 is controlled by a PI controller. The following data apply:

> w, flow rate of liquid through the tanks: 250 lb/min
> Holdup volume of each tank: 10 ft^3
> Density of liquid: 50 lb/ft^3
> Transducer: A change of 1°F causes the controller pen to move 0.25 inch.
> Final control element: A change of 1 psi from the controller changes the heat input q by 400 Btu/min. The final control element is linear. q

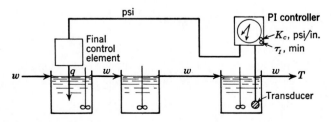

FIGURE P17-5

(a) Draw a block diagram of the control system. Show in detail such things as units and numerical values of the parameters.
(b) Determine the controller settings by the Ziegler-Nichols rules.
(c) If the control system is operated with *proportional mode only*, using the value of K_c found in part (b), determine the flow rate w at which the system will be on the verge of instability and oscillate continuously. What is the frequency of this oscillation?

17.6. The transfer function of a process and measurement element connected in series is given by

$$\frac{e^{-0.4s}}{(2s + 1)^2}$$

(a) Sketch the open-loop Bode diagram (gain and phase) for a control system involving this process and measurement lag.
(b) Specify the gain of a proportional controller to be used in this control system.

17.7. (a) For the control system shown in Fig. P17.7, determine the transfer function C/U.
(b) For $K_c = 2$ and $\tau_D = 1$, find $C(1.25)$ and the offset if $U(t) = u(t)$, a unit-step.
(c) Sketch the open-loop Bode diagram for $K_c = 2$ and $\tau_D = 1$. For the upper part of the diagram (AR versus ω), show the asymptotic approximation. Include in the open-loop Bode diagram the transfer function for the controller.
(d) From the Bode diagram, what do you conclude about stability of the closed-loop system?

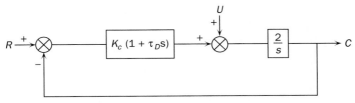

FIGURE P17-7

17.8. The proportional controller of the temperature-control system shown in Fig. P17.8 is properly tuned to give a good transient response for a standard set of operating

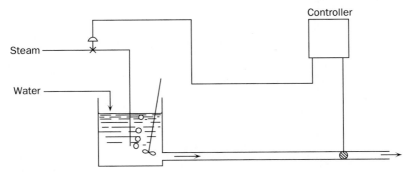

FIGURE P17-8

conditions. If changes are made in the operating conditions, the control system may become more or less stable. If the changes listed below are made *separately*, determine whether the system becomes more stable, less stable, or remains the same. Try to use the Bode stability criterion and sketches of frequency response graphs to solve this problem.

1. Controller gain increases.
2. Length of pipe between measuring element and tank increases.
3. Measuring element is inserted in tank.
4 Integral action is provided in controller.
5. A larger valve is used (i.e., one with a higher C_v value).

17.9. For each control system shown in Fig. P17.9, determine the characteristic equation of the closed-loop response and determine the value of K_c that will cause the system to be on the verge of instability (i.e., find the ultimate gain, K_{cu}). If *possible*, use the Routh test. Note that the feedback element for System B is an approximation to e^{-2s}.

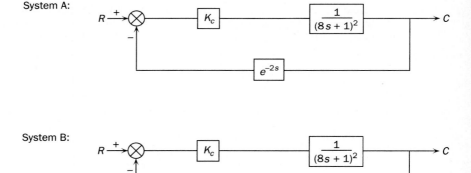

System A:

System B:

FIGURE P17-9

17.10. (*a*) For the system shown in Fig. P17.10 determine the value of K_c that will give 30° of phase margin.
   (*b*) If a PI controller with $\tau_I = 2$ is used in place of the proportional controller, determine the value of K_c for 30° of phase margin.

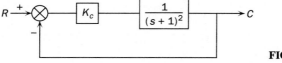

FIGURE P17-10

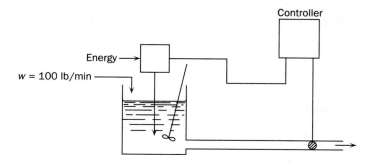

Controller

Energy

$w = 100$ lb/min

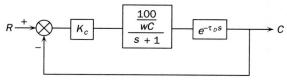

$$R \xrightarrow{+} \bigotimes \xrightarrow{} \boxed{K_c} \xrightarrow{} \boxed{\dfrac{\dfrac{100}{wC}}{s+1}} \xrightarrow{} \boxed{e^{-\tau_D s}} \xrightarrow{} C$$

FIGURE P17-11

17.11. A stirred-tank heating process and its block diagram are shown in Fig. P17.11. The control system is tuned by the Ziegler-Nichols method, and the ultimate frequency, ω_u is 2 rad/min.

  (a) Determine the value of K_c by the Ziegler-Nichols method of tuning.
  (b) What is the length of the pipe between the tank and the measuring element?
  (c) What are the gain margin and the phase margin for the control system when K_c is set to the Ziegler-Nichols value found in part (a).

Data on process:

$$\rho, \text{ density of fluid } = 62 \text{ lb/ft}^3$$
$$C, \text{ heat capacity of fluid } = 1.0 \text{ Btu/(lb)(°F)}$$
$$\text{inside diameter of pipe } = 2.0 \text{ in.}$$

PART
V

PROCESS APPLICATIONS

CHAPTER
18

ADVANCED CONTROL STRATEGIES

Up to this point, the control systems considered have been single-loop systems involving one controller and one measuring element. In this chapter, several multiloop systems will be described; these include cascade control, feedforward control, ratio control, Smith predictor control, and internal model control. The first three have found wide acceptance in industry. Smith predictor control has been known for about thirty years, but it was considered impractical until the modern microprocessor-based controllers provided the simulation of transport lag. Internal model control, which is new and is based on a rigorous mathematical foundation and an accurate model of the process, has been the subject of intense research for the past ten years. The controller hardware and instrumentation for all of these systems are readily available from manufacturers. Since this chapter is quite long, the reader may wish to select the type of advanced control strategy that is of particular interest. The descriptions of the five strategies are independent and need not be read in the order presented.

CASCADE CONTROL

To provide motivation for the study of cascade control, consider the single-loop control of a jacketed kettle as shown in Fig. 18.1a. The system consists of a kettle through which water, entering at temperature T_i, is heated to T_o by the flow of hot oil through a jacket surrounding the kettle. The temperature of the water in the kettle is measured and transmitted to the controller, which in turn adjusts the flow of hot oil through the jacket. This control system is satisfactory for controlling the kettle temperature; however, if the temperature of the oil-supply should drop, the kettle temperature can undergo a large prolonged excursion from the set point before control is again established. The reason for this is that the controller does not take corrective action until the effect of the drop in oil-supply temperature

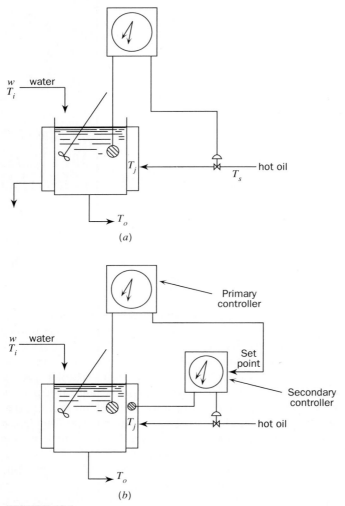

FIGURE 18-1
(a) Single-loop control of a jacketed kettle (b) cascade control of a jacketed kettle.

has worked itself through the system of several resistances to reach the measuring element.

To prevent the sluggish response of kettle temperature to a disturbance in oil-supply temperature, the control system shown in Fig. 18.1b is proposed. In this system, which includes two controllers and two measuring elements, the output of the primary controller is used to adjust the set point of a secondary controller, which is used to control the jacket temperature. Under these conditions, the primary controller adjusts indirectly the jacket temperature. If the oil temperature should drop, the secondary control loop will act quickly to maintain the jacket temperature close to the value determined by the set point that is adjusted by the

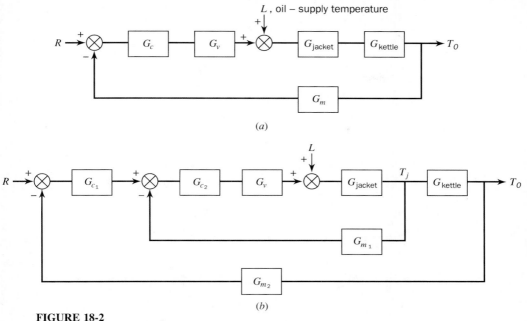

FIGURE 18-2
Block diagram: (*a*) single-loop conventional control (*b*) cascade control.

primary controller. This system shown in Fig. 18.1*b* is called a *cascade* control system. The primary controller is also referred to as the master controller and the secondary controller is referred to as the slave controller.

A simplified block diagram of the single-loop system is shown in Fig. 18.2*a*. Figure 18.2*b*, which is a block diagram representation of the cascade control system, shows clearly that an inner loop has been added to the conventional control system.

Analysis of Cascade Control

To develop the closed-loop transfer functions for a cascade control system, consider the general block diagram shown in Fig. 18.3. In this diagram, the load disturbance U enters between two blocks of the plant and the inner loop encloses this load disturbance.

To determine the transfer function C/R, the inner loop is reduced to one block by the method shown in Chapter 12. The result is shown in Fig. 18.3*b*, and the block diagram of Fig. 18.3*b* can be used to give the result

$$\frac{C}{R} = \frac{G_{c_1} G_a G_3}{1 + G_{c_1} G_a G_3 H_1} \tag{18.1}$$

where $G_a = \dfrac{G_{c_2} G_1 G_2}{1 + G_{c_2} G_1 G_2 H_2}$

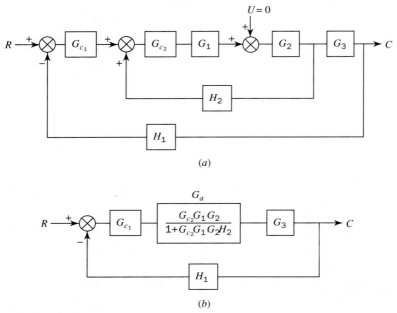

FIGURE 18-3
Block diagram for cascade control for set-point change.

To obtain the transfer function relating output to load, C/U, the block diagram of Fig. 18.3a is rearranged by placing the transfer function $G_{c_2}G_1$ in the feedback paths of the primary and secondary loops; the new arrangement is shown in Fig. 18.4a. Since R = 0 for the case under consideration, the block diagram can be redrawn as shown in Fig. 18.4b. This diagram, which has the same form as the one in Fig. 18.3a, can now be reduced to the form shown in Fig. 18.4c. Application of the rules of Chapter 12 to Fig. 18.4c finally gives

$$\frac{C}{U} = \frac{G_3}{G_1 G_{c_2}} \frac{G_a}{1 + G_a G_{c_1} H_1 G_3} \tag{18.2}$$

where G_a is the same as given in Eq. (18.1).

Example 18.1. To compare conventional control with cascade control, consider the conventional control system of Fig. 18.5a in which a third-order process is under PI control. A cascade version of this single-loop control system is shown in Fig. 18.5b in which an inner-loop having proportional control encloses the load disturbance U.

To obtain a response of the conventional control system for use in comparison with the response of the cascade system, the block diagram of Fig. 18.5a was simulated on a computer. The values of K_c and τ_I were chosen by trial and error to give the response to a step change in set point shown as Curve I of Fig. 18.6; this response, which has a decay ratio of about $\frac{1}{4}$, was obtained with $K_c = 2.84$ and $\tau_I = 5$. The Ziegler-Nichols settings ($K_c = 3.65$ and $\tau_I = 3.0$) gave a set-point response that was too oscillatory. Having obtained satisfactory controller settings

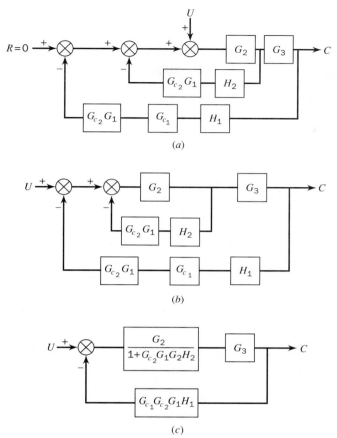

FIGURE 18-4
Block diagram for cascade control for load change.

(K_c = 2.84 and τ_I = 5.0), the response of the system to a step change in U of 4 units is shown as Curve II of Fig. 18.7. The load response for no control (i.e., K_c = 0) is also shown as Curve I for comparison.

The cascade control system of Fig. 18.5b was also simulated to obtain a load response. The controller gain K_{c_2} of the inner loop was chosen arbitrarily to be 10.0. This value was chosen to be high in order to obtain a fast-responding inner loop, a desirable situation for cascade control. Because of the introduction of the inner loop, the dynamics of the control system have changed and it is necessary to tune the primary controller parameters for a good response to a step change in set point. By trial and error, primary controller settings of K_{c_1} = 1.0 and τ_I = 0.63 were found that produced the response to a unit step in set point, shown as Curve II in Fig. 18.6. The use of Ziegler-Nichols settings produced a less desirable response.

Using the controller parameters found from the step change in set point (K_{c_1} = 1.0 , τ_I = 0.63), the response of the cascade system to a step change in load of 4 units was obtained and is shown as Curve III of Fig. 18.7. As shown

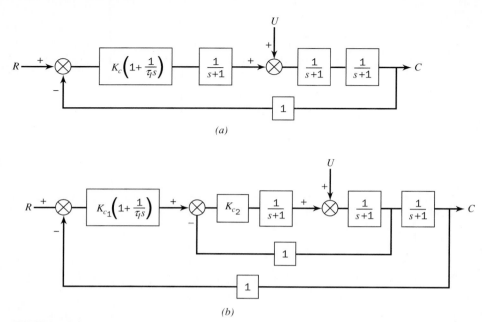

FIGURE 18-5
Block diagrams for Example 18.1:(*a*) Single-loop conventional control (*b*) cascade control.

in Fig. 18.7, the load response for the cascade control system is far superior to the load response of the conventional control system. The maximum deviation of the cascade response has been reduced by a factor of about four and the frequency of oscillation has nearly doubled.

Generalizations

Cascade control is especially useful in reducing the effect of a load disturbance that moves through the control system slowly. The inner loop has the effect of reducing the lag in the outer loop, with the result that the cascade system responds more quickly with a higher frequency of oscillation. Example 18.2 will illustrate this effect of cascade control.

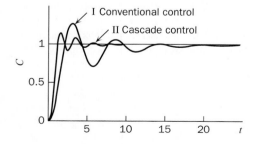

FIGURE 18-6
Responses to step change in set point for single-loop control and cascade control for Example 18.1. I Conventional control with $K_c = 2.84$, $\tau_I = 5$; II Cascade control with $K_{c_i} = 1.0$, $\tau_I = 0.63$, $K_{c_2} = 10$.

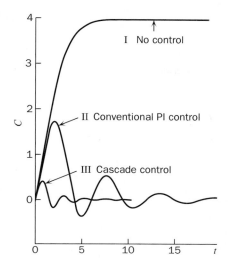

FIGURE 18-7

Responses to step change in load for Example 18.1. I no control; II conventional control with $K_c = 2.84$, $\tau_I = 5$; III cascade control with $K_{c_1} = 1.0$, $\tau_I = 0.63$, $K_{c_2} = 10$.

The choice of control action and tuning of the primary and secondary controllers for a cascade control system must be given careful consideration. The control action for the inner loop is often proportional with the gain set to a high value. The rationale for the use of proportional control rather than two- or three-mode control is that tuning is simplified and any offset associated with proportional control of the inner loop can be handled by the presence of integral action in the primary controller. The gain of the secondary controller should be set to a high value to give a tight inner loop that responds quickly to load disturbance; however, the gain should not be so high that the inner loop is unstable. Although the primary control loop can provide stable control even when the inner loop is unstable, it is considered unwise to have an unstable inner loop because the system will go unstable if the primary controller is placed in manual operation or if there is a break in the outer loop.

The action for the primary controller is generally PI or PID. The integral action is needed to reduce offset when sustained changes in load or set point occur. The problem of adjusting a primary controller is essentially the same as for a single-loop control system. Since the addition of the inner loop can change the dynamics of the outer loop significantly, the primary controller must be re-tuned when the inner loop is closed or when the secondary controller settings are changed.

The microprocessor-based controllers available today can implement cascade control very easily. A discussion of such controllers will be given in a later chapter.

Example 18.2. The claim is often made that cascade control gives a better response than conventional control because the lags in the outer loop are reduced. To illustrate this benefit, consider the conventional control and the cascade control of a third-order plant in Figs. 18.8a and b. The inner loop of the cascade system surrounds two of

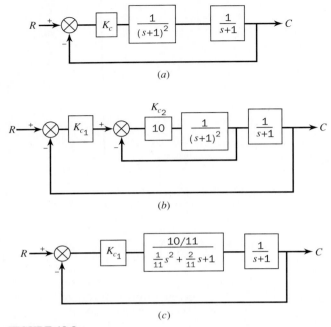

FIGURE 18-8
Block diagram for Example 18.2.

the first-order blocks in the plant. To simplify the discussion, the load disturbance is not shown since we are interested only in the closed-loop dynamics. The equivalent single-loop control system of the cascade system, shown in Fig. 18.8c, was obtained by the usual method for reducing a loop to a single block.

Comparing Fig. 18.8a with Fig. 18.8c shows that the use of cascade control has replaced a second-order critically damped system represented by the first two blocks of the plant $[1/(s + 1)^2]$ with the following underdamped second-order system:

$$\frac{K}{\tau^2 s^2 + 2\zeta\tau s + 1}$$

where $K = 10/11$

$\tau = \sqrt{1/11}$

$\zeta = \sqrt{1/11}$

This second-order underdamped system, for which τ and ζ are small, responds much faster than the critically damped second-order transfer function of the first two blocks of the open-loop system. Consequently, the cascade system will respond faster with a higher frequency of oscillation as we have already seen in the simulated response of Fig. 18.6.

FEEDFORWARD CONTROL

If a particular load disturbance occurs frequently in a control process, the quality of control can often be improved by the addition of feedforward control. Consider the composition control system shown in Fig. 18.9a in which a concentrated stream of control reagent containing water and solute is used to control the concentration of the stream leaving a three-tank system. The stream to be processed passes through a preconditioning stirred tank where composition fluctuations are smoothed out before the outlet stream is mixed with control reagent. A three-tank system has been chosen for ease of computation in a numerical example that follows.

In the conventional feedback control system shown in Fig. 18.9a, the measurement of composition in the third tank is sent to a controller, which generates

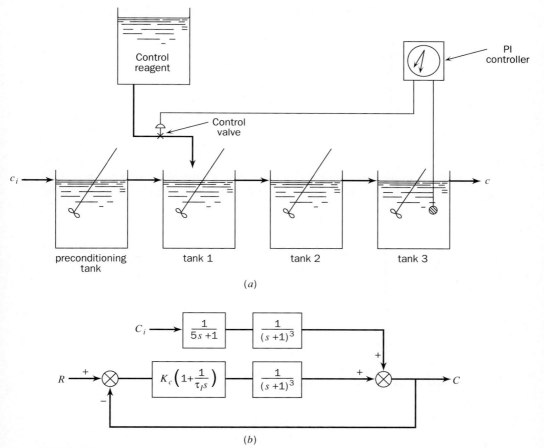

(a)

(b)

FIGURE 18-9
Composition control system: (a) physical process; (b) block diagram.

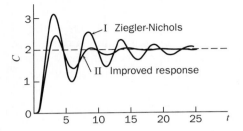

FIGURE 18-10

Responses to a step change in set point for PI control.

Curve I: Ziegler-Nichols settings: $K_c = 3.65$, $\tau_I = 3.0$; Curve II: Settings for improved response: $K_c = 2.84$, $\tau_I = 5.0$.

a signal that opens or closes the control valve, which in turn supplies concentrated reagent to the first tank. The block diagram corresponding to the control system of Fig. 18.9a is shown in Fig. 18.9b.* To obtain some specific control system responses, numerical values of the time constants of the tanks have been chosen as shown in Fig. 18.9b. To study the response of this control system, the block diagram shown in Fig. 18.9b was simulated on a computer. The values of K_c and τ_I were chosen by trial and error to give the response to a step change in set point shown in Curve II of Fig. 18.10; this response, which has a decay ratio of about $\frac{1}{4}$, was obtained with $K_c = 2.84$ and $\tau_I = 5.0$. The Ziegler-Nichols settings ($K_c = 3.65$ and $\tau_I = 3.0$) give a set-point response shown as Curve I of Fig. 18.10, which is too oscillatory. Having obtained satisfactory settings for the controller ($K_c = 2.84$, $\tau_I = 5.0$), the response of the system to a step change in C_i of 10 units was obtained and is shown as Curve I in Fig. 18.11. Note that the response is oscillatory and has a long tail. This response illustrates the fact that the feedback control system does not begin to respond until the load disturbance has worked its way through the forward loop and reaches the measuring element, with the result that the composition can move far from the set point during the transient.

---

*In Figure 18.9a, concentration is denoted by c (lower-case letter). In the block diagram of the process in Fig. 18.9b, the symbol for concentration is denoted by C (capital letter) to denote a deviation variable. This use of symbols follows the procedure established in Chap. 5.

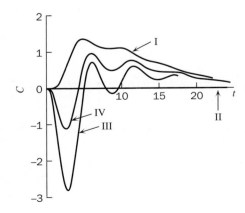

FIGURE 18-11

Responses to a step change in load for feedforward-feedback control.

Curve I: PI control with $K_c = 2.84$, $\tau_I = 5.0$

Curve II: FF control with $K_c = 2.84$, $\tau_I = 5.0$, $G_f = -1/(5s + 1)$

Curve III: FF control with $K_c = 2.84$, $\tau_I = 5.0$, $G_f = -1$

Curve IV: FF control with $K_c = 2.84$, $\tau_I = 5.0$, $G_f = -0.5$

If the change in load disturbance (C_i) can be detected as soon as it occurs in the inlet stream, this information can be fed forward to a second controller that adjusts the control valve in such a way as to prevent any change in the outlet composition from the set point. A controller that uses information fed forward from the source of the load disturbance is called a *feedforward* controller. The block diagram that includes the feedforward controller (G_f) as well as the feedback controller (G_c) is shown in Fig. 18.12.

Analysis of Feedforward Control

The response of C to changes in C_i and R can be written from Fig. 18.12 as follows:

$$C(s) = G_1(s)G_p(s)C_i(s) + G_f(s)G_p(s)C_i(s) + G_c(s)G_p(s)E(s) \qquad (18.3)$$

where $E(s) = R(s) - C(s)$

In order to determine the transfer function of $G_f(s)$ that will prevent any change in the control variable C from its set point R, which is 0, we solve Eq. (18.3) for $G_f(s)$ with $C = 0$, $R = 0$. The result is

$$G_f(s) = -G_1(s) \qquad (18.4)$$

For the example under consideration in Fig. 18.12.

$$G_f(s) = -1/(5s + 1) \qquad (18.5)$$

This transfer function can be implemented easily with control hardware now available.

If the load response of the control system in Fig. 18.12, with $G_f(s)$ given by Eq. (18.5), were obtained for a step change in C_i, there would be no deviation of C from the set point (i.e., perfect control). This response is shown as Curve II in Fig. 18.11, which, of course is a horizontal line at $C = 0$.

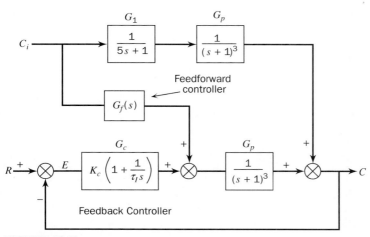

FIGURE 18-12
Control system with feedback and feedforward controllers.

Rather than use the $G_f(s)$ of Eq. (18.5) in the feedforward controller, one can try using only the constant term of $G_f(s)$, that is,

$$G_f(s) = -1$$

The response for $G_f = -1$ gives Curve III in Fig. 18.11; this response has a very large undershoot before the feedback controller returns C to the set point. If we try using $G_f(s) = -0.5$, we obtain Curve IV of Fig. 18.11; the undershoot is less in this case, but the response is still unsatisfactory. As shown by Curves III and IV, omitting the dynamic part of $G_f(s)$ can give very poor results. The success of using a feedforward controller depends on accurate knowledge of the process model, a luxury that may not be available in many applications.

Implementing Feedforward Transfer Functions

In applications of feedforward control, $G_f(s)$ may take the form of a lead expression, such as $G_f(s) = 1 + \tau_f s$. When this occurs, it is necessary to approximate $1 + \tau_f s$ by a lead-lag expression, such as

$$G_f(s) = (1 + \tau_f s)/(1 + \beta \tau_f s)$$

where $\beta \ll 1$. To see how $G_f(s)$ takes the form of a lead expression, consider the load disturbance, c_i, of Fig. 18.9 to enter tank 2. Since no change in concentration occurs in the stream entering the preconditioning tank, we may eliminate it from the diagram for the case under consideration to obtain the diagram in Fig. 18.13.

Adding feedforward control and feedback control to the system in Fig. 18.13 gives the block diagram of Fig. 18.14. The diagram shown in Fig. 18.14 is the same as that in Fig. 18.12 with the exception that the disturbance C_i enters tank 2 instead of the preconditioning tank. As shown previously, the response of C to a change in C_i and R can be written directly from Fig. 18.14 as follows:

$$C(s) = G_1(s)C_i(s) + G_f(s)G_p(s)C_i(s) + G_c(s)G_p(s)E(s) \qquad (18.6)$$

where $E(s) = R(s) - C(s)$

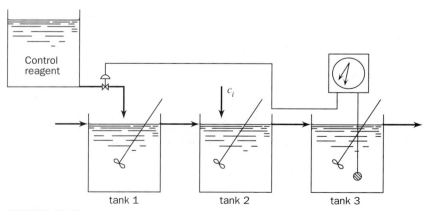

FIGURE 18-13
Composition control with disturbance to second tank.

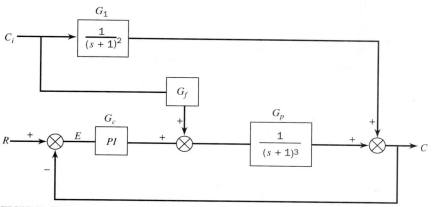

FIGURE 18-14
Feedforward-feedback control for system in Fig. 18.13.

In order for C not to change from the set point R, which is 0, we solve Eq.(18.6) for $G_f(s)$ with $C = 0$ and $R = 0$ to obtain:

$$G_f(s) = -\frac{G_1(s)}{G_p(s)} \tag{18.7}$$

Introducing the expressions for $G_1(s)$ and $G_p(s)$ from Fig. 18.14 into Eq. (18.7) gives

$$G_f(s) = -(s + 1) \tag{18.8}$$

It is not practical to implement $-(s + 1)$. To see this, consider the response of $-(s + 1)$ to a step change as shown in Fig. 18.15. There is no hardware that will produce an impulse as shown in Fig. 18.15; however, one can approximate $-(s + 1)$ by means of a lead-lag transfer function of the form.

$$\frac{Y(s)}{X(s)} = -\frac{\tau_f s + 1}{\beta \tau_f s + 1} \tag{18.9}$$

where $\beta \ll 1$

If we let $\beta = 0.1$ and $\tau_f = 1$ for the control system under consideration, we obtain as an approximation to Eq. (18.8)

$$G_f(s) = -\frac{s + 1}{0.1s + 1} \tag{18.10}$$

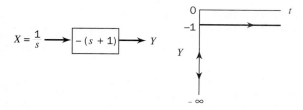

FIGURE 18-15
Step response for $-(s + 1)$.

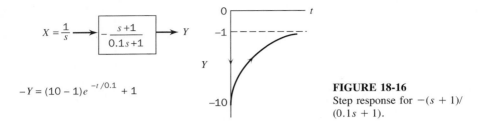

FIGURE 18-16
Step response for $-(s + 1)/(0.1s + 1)$.

The response of this transfer function to a step input is shown in Fig. 18.16. The effect of this transfer function, $-(s + 1)/(0.1s + 1)$, on the output of the feedforward controller for a step change in load is to give a sudden drop in flow followed by a fast exponential increase in the flow to a steady-state flow of -1. Note that for the parameters chosen for the transfer functions in Fig. 18.14, a unit increase in C_i must eventually be compensated by a unit decrease in the signal from the feedforward controller if there is to be no change in the process output. The sudden, initial drop in flow may be too abrupt for the control hardware, in which case the output would saturate. In practice, β can be increased (perhaps to 0.5) in order to reduce the magnitude of the initial drop.

The effect of using $G_f(s) = -(s + 1)/(0.1s + 1)$ with feedback control is shown in Fig. 18.17. The responses shown, which were obtained by simulation, are for a unit-step change in C_i. Curve I is for the case of feedback control only with $K_c = 2.84$ and $\tau_I = 5.0$. Curve II is for feedforward-feedback control using Eq. (18.10) for $G_f(s)$ and $K_c = 2.84$ and $\tau_I = 5.0$. One can see that the overshoot for the feedforward-feedback response has been reduced significantly.

Tuning Rules for Feedforward-Feedback Control

In the practical application of feedforward control, one does not have a block diagram with transfer functions as shown in Figs. 18.12 and 18.14. For such a practical situation, one can still tune the feedforward controller by introducing a step change in the disturbance that enters the feedforward controller (C_i in Fig. 18.14) and then applying some tuning rules. The rules to be discussed here are from a training film on feedforward control produced by the Foxboro Co. (1978).

Feedforward Rules

In describing these rules, reference will be made to the general block diagram for a feedforward-feedback system shown in Fig. 18.18. It is assumed that $G_f(s)$ will be a lead-lag transfer function of the form

$$G_f(s) = K_f(T_1 s + 1)/(T_2 s + 1) \tag{18.11}$$

where K_f = steady-state gain of the feedforward controller

T_1, T_2 = time constants of dynamic part of the feedforward controller

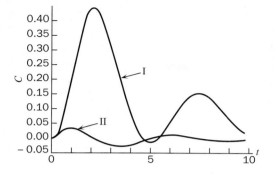

FIGURE 18-17

Comparison of conventional feedback control with feedforward-feedback control for system shown in Fig. 18.14.

Curve I: PI control with $K_c = 2.84$ and $\tau_I = 5$

Curve II: Feedforward-feedback control with $K_c = 2.84$, $\tau_I = 5$, and $G_f = -(s + 1)/(0.1s + 1)$

Commercial microprocessor-based controllers provide this lead-lag transfer function.

The tuning rules listed below are explained with the help of Fig. 18.19. In that figure, a unit step is selected for the distubrance C_i and K_f has been taken as -1. In practice, K_f will, of course, depend on the particular process being controlled.

1. Remove the control action in $G_c(s)$ by setting the controller to manual.
2. Set the feedback controller to the computed steady-state gain (K_f) necessary to compensate ultimately for a step change in C_i. This means that the dynamic portion of $G_f(s)$ will be removed and only the constant term (K_f) will remain.
3. Make a step change in C_i and observe the open-loop transient of C. The general shapes of the response to be expected are shown in Fig. 18.19.
4. If the response shown in Fig. 18.19a occurs, lead must predominate in $G_f(s)$ of Eq. (18.11) (i.e., $T_1 > T_2$). If the response of Fig. 18.19b occurs, lag must predominate in $G_f(s)$ (i.e., $T_1 < T_2$). The values of T_1 and T_2 in Eq. (18.11) are found by use of the information in Table 18.1. The value of K_f in Eq. (18.11) has been obtained in step 2.

The next example will help clarify the use of these tuning rules.

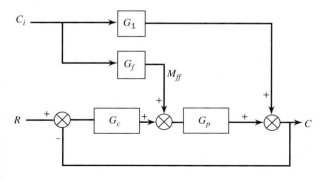

FIGURE 18-18

Feedforward-feedback control system.

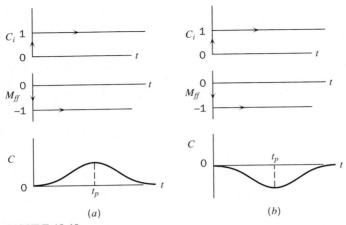

FIGURE 18-19
Open-loop response to determine lead-lag time constants in feedforward tuning rules: (*a*) lead must predominate in G_f; (*b*) lag must predominate in G_f.

Example 18.3. *Use of feedforward tuning rules.* Apply the feedforward tuning rules to the system in Fig. 18.14. Since this example is concerned with the application of the tuning rules to a system for which a mathematical model is not generally available, the reader should assume that the transfer functions for $G_1(s)$ and $G_p(s)$ in Fig. 18.14 are unknown. The determination of $G_f(s)$ is to be obtained solely by information from open-loop transients.

We must first determine the steady-state gain (K_f) for the system of Fig. 18.14. If a step change in C_i is made, C will undergo a transient and eventually level out at a steady-state value. If the controller parameters are properly selected, the value of C at the end of the transient will be the same as it was before the transient occured. By computation or experiment, one can determine the value of K_f needed to obtain no change in C. For the system in Fig. 18.14, one can see that K_f of Eq. (18.11) must be equal to -1.

We must now apply the feedforward tuning rules to obtain T_1 and T_2 in Eq. (18.11). After removing the feedback controller action $[G_c(s)]$ we have the equivalent diagram shown in Fig. 18.20. A unit-step change in C_i produces the

TABLE 18.1
Tuning parameters for feedforward control

Predominant mode	T_1	T_2
Lead	$1.5t_p$	$0.7t_p$
Lag	$0.7t_p$	$1.5t_p$

$$G_f(s) = K_f \frac{T_1 s + 1}{T_2 s + 1}$$

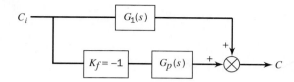

FIGURE 18-20
Open-loop feedforward test to determine parameters for G_f.

transient for C shown as Curve I in Fig. 18.21. Comparing the shape of the transient with those of Fig. 18.19, we see that lead must predominate in $G_f(s)$. The peak value occurs at $t_p = 2$. Applying the rules in Table 18.1 gives

$$T_1 = 1.5t_p = 3$$
$$T_2 = 0.7t_p = 1.4$$

The feedforward controller transfer function is therefore

$$G_f(s) = -(3s + 1)/(1.4s + 1) \tag{18.12}$$

It is of interest to show the response of C for feedforward only when the feedforward transfer function of Eq. (18.12) is used; the result for a unit-step change in C_i is shown as Curve II in Fig. 18.21.

When the $G_f(s)$ of Eq. (18.12) is used and the controller parameters for $G_c(s)$ are $K_c = 2.84$ and $\tau_I = 5.0$, the feedforward-feedback response to a unit-step change in C_i is shown as Curve II in Fig. 18.22. For comparison, the response for feedback control only is also shown in Fig. 18.22.

RATIO CONTROL

An important control problem in chemical industry is the combining of two or more streams to provide a mixture having a desired ratio of components. Examples of this mixing operation include the blending of reactants entering a chemical reactor or for the injection of a fuel-air mixture into a furnace.

In Fig. 18.23a is shown a control system for blending two liquid streams A and B to produce a mixed stream having the ratio K_r in units of mass B/mass A. Stream A, which is uncontrolled, is used to adjust the flow of stream B so that the desired ratio is maintained. The measured signal for stream A is multiplied by

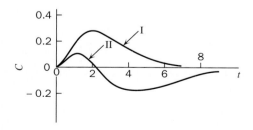

FIGURE 18-21
Open-loop response for step change in C_i for Example 18.3.
Curve I: $G_f = -1$
Curve II: $G_f = -(3s + 1)/(1.4s + 1)$

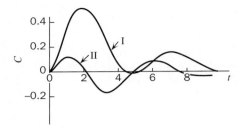

FIGURE 18-22
Comparison of conventional feedback control with feedforward-feedback control for Example 18.3.
Curve I: PI control with
$K_c = 2.84$, $\tau_I = 5.0$
Curve II: Feedforward-feedback control with
$K_c = 2.84$, $\tau_I = 5.0$, and $G_f = -(3s + 1)/(1.4s + 1)$

the desired ratio K_r to provide a signal that is the set point for the flow-control loop for stream B. The parameter K_r can be adjusted to the desired value. Control hardware is available to perform the multiplication of two control signals.

A block diagram of the ratio control system is shown in Fig. 18.23b. In a flow-control loop, the dynamic elements consist of the controller, the flow-

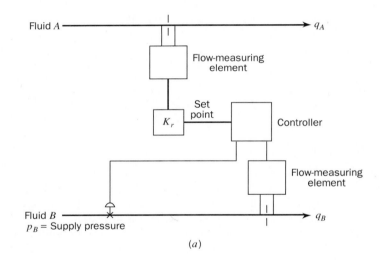

(a)

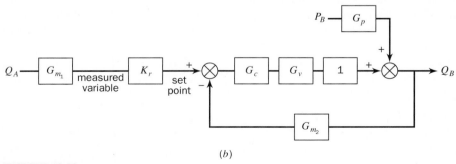

(b)

FIGURE 18-23
(a) Ratio control system; (b) block diagram for ratio control (set point $= G_{m_1} K_r Q_A$).

measuring element, and the control valve. For incompressible fluids, there is no lag between the change in valve position and the corresponding flow rate. For this reason, the transfer function between the valve and the measurement of flow rate is simply unity. The block diagram also shows a transfer function G_p that relates the flow rate of B to the supply pressure of B. A transfer function G_{m_1} is also shown that represents the dynamic lag of the flow measuring element for stream A.

From the block diagram, the flow of B may be written:

$$Q_B = \frac{G_{m_1} K_r G_c G_v}{1 + G_c G_v G_{m_2}} Q_A + \frac{G_p}{1 + G_c G_v G_{m_2}} P_B$$

The control action for a flow-control system is usually PI. The integral action is needed to eliminate offset and thereby establish a precise ratio of the mixed streams of A and B. Derivative action is usually avoided in flow control because the signal from a flow-measuring element is inherently noisy. The presence of derivative action would amplify the noise and give poor control.

DEAD-TIME COMPENSATION (SMITH PREDICTOR)

Processes that contain a large transport lag [exp $(-\tau_D s)$] can be difficult to control because a disturbance in set point or load does not reach the output of the process until τ_D units of time have elapsed. The control strategy to be described here, which is also known as dead-time compensation, attempts to reduce the deleterious effect of transport lag. Dead-time compensation, which is also referred to as a Smith predictor, was first described by O. J. M. Smith (1957).

Consider the single-loop control system of Fig. 18.24 in which the process transfer function $G_p(s)$ is to be modeled by

$$G_p(s) = G(s)e^{-\tau_D s} \tag{18.13}$$

The right side of Eq. (18.13) is the product of a transport lag [exp$(-\tau_D s)$] and a transfer function $G(s)$, which has minimum phase characteristics, such as $1/(\tau s + 1)$. For convenience in developing the dead-time compensation method, only a change in set point R will be considered. If a step change is made in R,

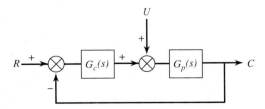

FIGURE 18-24
Control system.

the disturbance will not break through and appear at C until τ_D units of time elapse. Up to time τ_D, no control action occurs, with the result that the overall closed-loop response will be sluggish and generally unsatisfactory. To overcome this difficulty, Smith suggested that $G_p(s)$ be modeled according to Eq. (18.13) and that additional feedback paths be inserted into Fig. 18.24 as shown in Fig. 18.25a. If $G_p(s)$ is modeled exactly by Eq. (18.13), a close study of Fig. 18.25a shows that the signals entering comparator A will be identical; as a result, the signals cancel and cause the output of comparator A to be zero. The net effect is to completely eliminate the outer feedback path; this simplification is shown in Fig. 18.25b.

The system of Fig. 18.25b is now much easier to control because the transport lag is not present in the loop. Of course, in the real system the transport lag is still present; we have eliminated it in a mathematical sense from the feedback path by the additional feedback paths of Fig. 18.25a and the assumption that the process transfer function, $G_p(s)$ can be modeled exactly as shown in Eq. (18.13). To achieve the simplification suggested by Fig. 18.25b we must now face reality and realize that the signal C_1 in Fig. 18.25b is not available to feed back. Only the signal C can be measured and fed back to the controller. In terms of controller hardware implementation, the diagram of Fig. 18.25a is redrawn in Fig. 18.26a to show which portion of the diagram will be implemented with controller hardware. Figure 18.26b, which is another way to represent Fig. 18.26a, is a form sometimes presented in the literature for dead-time compensation. The reader may legitimately ask whether or not hardware exists to actually implement what is shown within the dotted lines in Fig. 18.26. Until the appearance

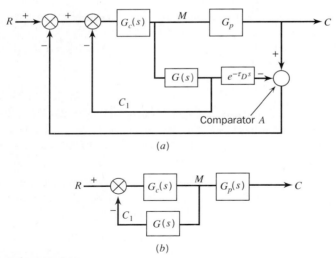

(a)

(b)

FIGURE 18-25
(a) Dead-time compensation (Smith predictor) block diagram; (b) Equivalent diagram for part (a) when $G_p = G(s)e^{-\tau_D s}$.

of microprocessor-based controllers, the answer was no. However, today many commercially available controllers provide dead-time $[\exp(-\tau_D s)]$ and $G(s)$ in the form of a first-order lag $[1/(\tau s + 1)]$. Features such as these will be discussed in Chap. 35.

The recommended procedure for applying dead-time compensation is as follows:

1. By means of an open-loop test of the process, model $G_p(s)$ by the transfer function

$$\frac{1}{\tau s + 1}e^{-\tau_D s}$$

In this step, we have chosen $G(s)$ of Fig. 18.26a to be first-order. Many processes in chemical engineering can be modeled by a first-order lag with dead-time.

2. By means of appropriate hardware, implement the controller portion of Fig.

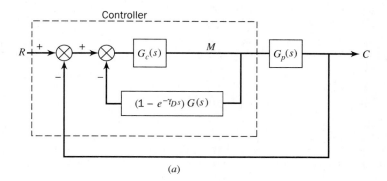

(a)

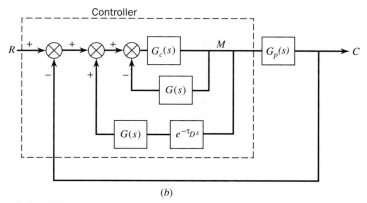

(b)

FIGURE 18-26
Hardware implementation of dead-time compensation.

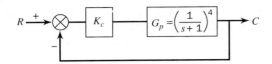

FIGURE 18-27
Control system for Example 18.4.

18.26a or Fig. 18.26b. If $G_p(s)$ can be exactly modeled by a first-order process with dead-time, the response of the control system in Fig. 18.26 will be equivalent to the response obtained for the system in Fig. 18.25b in which the loop involves the control of a first-order process. In most practical situations, there will be some mismatch between $G_p(s)$ and its model of first-order with dead-time. The greater the mismatch, the greater the deterioration in control response from the ideal situation of Fig. 18.25b. The application of the dead-time compensation technique and the effect of mismatch between $G_p(s)$ and its model will be illustrated in the next example.

Example 18.4. *Dead-time compensation.* Consider the control system shown in Fig. 18.27 in which the process is fourth-order; thus

$$G_p(s) = \left(\frac{1}{s+1}\right)^4$$

In a practical situation, we would not know the transfer function of the process. In this example, we have taken the process model to be fourth-order to provide a system sufficiently complex to show considerable transfer lag.

One can show for the system in Fig. 18.27 that the ultimate gain and the corresponding period are: $K_{c_u} = 4.0$ and $P_u = 2\pi$. Using the Ziegler-Nichols rules, one gets $K_c = 2.0$. The response for a unit-step change in set point for $K_c = 2$ is shown in Curve I of Fig. 18.29. Notice that the decay ratio is about $\frac{1}{4}$.

We shall now use the dead-time compensation method to control the process

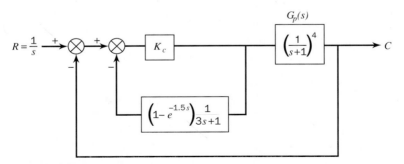

FIGURE 18-28
Dead-time compensation for Example 18.4.

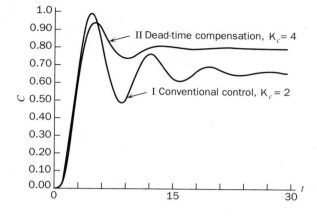

FIGURE 18-29
Comparison of response for conventional control with response for dead-time compensation for Example 18.4.

in Fig. 18.27. If one fits the step response of $(s+1)^{-4}$ to a first-order with dead-time model, one obtains

$$\frac{1}{3s+1}e^{-1.5s}$$

This model was obtained from a unit-step response using a least squares fit procedure. We can now draw the diagram for the dead-time compensation system as shown in Fig. 18.28. The system shown in Fig. 18.28 was simulated by computer in order to compare the responses of the two control systems as shown in Fig. 18.29. Using a K_c of 2.0 (the Ziegler-Nichols value) for the conventional control we see from Curve I that the response is quite oscillatory and has an offset of 0.333 as required for this value of gain. Using a K_c of 4.0 for the dead-time compensation, we see that the response is less oscillatory and the offset is 0.20. It should be noted that if a K_c of 4.0 were applied to the conventional control system, the system would be on the verge of instability since a K_c of 4.0 is the ultimate gain.

In conclusion, the dead-time compensation has permitted the use of a higher value of K_c, reduced the offset, and produced a less oscillatory response. The dead-time compensation response shown in Fig. 18.29 can be improved by adding integral action to the controller and tuning the controller parameters.

To successfully apply dead-time compensation to the control of a process, one must have an accurate model of the process, such as a first-order with dead-time model. The parameters in this model (τ and τ_D) can be considered as controller parameters along with the controller parameters of $G_c(s)$. For the case of dead-time compensation with proportional control in Example 18.4, we actually have three controller parameters: K_c, τ_D, and τ. If the process dynamics $[G_p(s)]$ changes, all three parameters may need adjustment in order to achieve good control.

INTERNAL MODEL CONTROL

Internal model control (IMC) has been the subject of intense research since about 1980. This method of control, which is based on an accurate model of the process, leads to the design of a control system that is stable and robust. A robust control system is one that maintains satisfactory control in spite of changes in the dynamics of the process. In applying the IMC method of control system design, the following information must be specified:

Process model
Model uncertainty
Type of input (step, ramp, etc.)
Performance objective (integral square error, overshoot, etc.)

In many industrial applications for control systems, none of the above items is available, with the result that the system usually performs in a less than optimum manner. Determining the mathematical model and its uncertainty can be a difficult task. When the process is not sufficiently understood to obtain a mathematical model by applying fundamental principles, one must obtain a model experimentally. A discussion on the modeling of a process is presented in the next chapter. The choice of a performance objective is subjective and often arbitrary. In the IMC method, the integral square error is implied.

A simple description of the IMC method will be presented here. The interested reader is advised to consult the book by Morari and Zafiriou (1989) for a full treatment of internal model control. The literature on IMC is difficult to understand without a good foundation in control theory and mathematics. A full treatment of IMC is beyond the scope of this text. It is hoped that the simple treatment given here will stimulate interest in this important new area of process control.

Internal Model Control Structure

A block diagram of an IMC system is shown in Fig. 18.30a. Notice that the diagram is similar to the diagram for the Smith predictor method shown in Fig. 18.25a. In this diagram, G is the transfer function of the process and G_m is the model of the process. Although G and G_m are called the transfer functions of the process, they actually include the valve and the process. The transfer function of the measuring element is taken as 1.0. The portion of the diagram that is implemented by the computer includes the IMC controller and the model; this portion is surrounded by the dotted boundary.

In order to compare the IMC structure of Fig. 18.30a with the conventional control structure, the diagram of Fig. 18.30a has been rearranged as shown in Fig. 18.30b. For convenience, the transfer function through which the load U passes has been omitted. We show only the output from the load block (U_1). We may use the structure in Fig. 18.30b to relate the IMC controller to the conventional controller. Replacing the inner loop of Fig. 18.30b with a single block gives the structure shown in Fig. 18.30c. Since this structure is the conventional single-

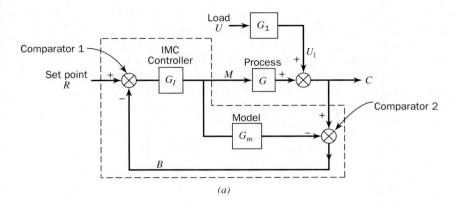

(a)

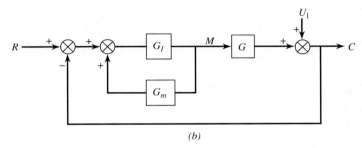

(b)

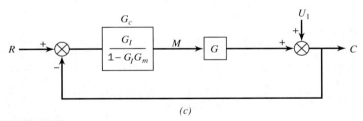

(c)

FIGURE 18-30
Internal model control structures: (a) basic structure, (b) alternate structure, (c) structure equivalent
to conventional control.

loop control structure, we can identify the single controller block as G_c. After one
designs the IMC controller (G_I) by the method to be described, one can determine
the equivalent conventional controller G_c by the relation

$$G_c = G_I/(1 - G_I G_m) \tag{18.14}$$

For the structure shown in Fig. 18.30a, one can show that

$$C = U_1 + \frac{GG_I}{1 + G_I(G - G_m)}[R - U_1] \tag{18.15}$$

If the model exactly matches the process (i.e., $G_m = G$), the only signal entering comparator 1 in Fig. 18.30a is U_1. (The signals from G and G_m are equal and cancel each other in going through comparator 2.) Since U_1 is not the result of any processing by the transfer functions in the forward loop, U_1 is not a feedback signal but an independent signal that is equivalent to R in its effect on the output C. In fact, there is no feedback when $G = G_m$ and we have an open-loop system as shown in Fig. 18.31. In this case the stability of the control system depends only on G_I and G_m. If G_I and G_m are stable, the control system is stable.

Ideally, we should like to have C track R without lag when only a set-point change occurs (i.e., $U_1 = 0$). In order for this to occur, we see from Fig. 18.31 or Eq. (18.15) that $G_IG = 1$ or since $G = G_m$, we may write $G_IG_m = 1$. Solving for G_I gives

$$G_I = 1/G_m \qquad (18.16)$$

Equation (18.16) simply states that the IMC controller should be the inverse of the transfer function of the process model. Keep in mind that Eq. (18.16) is based on the assumption that the model exactly matches the process.

For the case of only a change in load U_1 (i.e., $R = 0$), we should like to have the output C remain unchanged (i.e., $C = 0$). In order for this to occur, we see again from Fig. 18.31 or Eq. (18.15) that $G_IG_m = 1$; this leads to the same result as given by Eq. (18.16).

Even if there is no mismatch between the model and the process, the application of Eq. (18.16) will usually lead to a transfer function that cannot be implemented because it will be unstable, requires prediction, or requires pure differentiation. For example, if $G_m = 1/(\tau s + 1)$, the application of Eq. (18.16) gives

$$G_I = \tau s + 1$$

This result is equivalent to an ideal PD controller, which cannot be implemented because of the derivative term. If $G_m = e^{-\tau s}/(\tau_1 s + 1)$, we obtain

$$G_I = (\tau_1 s + 1)e^{\tau s}$$

The term $e^{\tau s}$, which represents prediction, cannot be implemented. If $G_m = (1 - s)/[(1 + s)(\tau s + 1)]$

$$G_I = [(1 + s)(\tau s + 1)]/(1 - s)$$

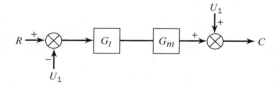

FIGURE 18-31
IMC structure when model matches process ($G_m = G$).

The term $1 - s$ in the denominator means that a pole is in the right half plane, which leads to an unstable controller. With such difficulties of implementation of the internal model controller, one might ask if any practical result can be obtained. These difficulties can be overcome by application of the following simplified procedure.

Design of IMC Controllers

In using these rules, only a step change in disturbance is considered. The procedure for disturbances other than a step response is more complicated and beyond the scope of the limited discussion presented here.

1. Separate the process model G_m into two terms

$$G_m = G_{m_a} G_{m_m} \qquad (18.17)$$

where G_{m_a} is a transfer function of an all-pass filter. An all-pass filter is one for which $|G_{m_a}(j\omega)| = 1$ for all ω. Examples are $e^{-\tau_d s}$ and $(1 - s)/(1 + s)$. G_{m_m} is a transfer function that has minimum phase characteristics. A system has non-minimum phase characteristics if its transfer function contains zeros in the right half plane or transport lags, or both. Otherwise, a system has minimum phase characteristics. For a step change in disturbance $(R = 1/s$ or $U_1 = 1/s)$, G_I is determined by

$$G_I = 1/G_{m_m} \qquad (18.18)$$

For a disturbance other than a step change, obtaining G_I is more complicated and the reader is referred to Morari and Zafiriou (1989).

The results of applying Eq. (18.18) will yield a transfer function that is stable and does not require prediction; however, it will have terms that cannot be implemented because they require pure differentiation (e.g., $\tau s + 1$).

2. To obtain a practical IMC controller, one multiplies G_I in step 1 by a transfer function of a filter, $f(s)$. The simplest form recommended by Morari and Zafiriou is given by

$$f(s) = 1/(\lambda s + 1)^n \qquad (18.19)$$

where λ is a filter parameter and n is an integer. The practical IMC controller (G_I) can now be expressed as

$$G_I = f/G_{m_m} \qquad (18.20)$$

The value of n is selected large enough to give a result for G_I that does not require pure differentiation. For the simple treatment of IMC design presented here, λ will be considered as a tunable parameter. In the full treatment of IMC given by Morari and Zafiriou, λ can be related to the model uncertainty. In practice, model uncertainty may not be available, in which case one is forced to treat λ as a tunable parameter.

3. If one wants to obtain the conventional controller transfer function G_c, use is made of Eq. (18.14), with G_I obtained from Eq. (18.20). For many simple process models, G_c turns out to be equivalent to a PID controller multiplied by a first-order transfer function; thus

$$G_c = K_c \left(1 + \tau_D s + \frac{1}{\tau_I s} \right) \left(\frac{1}{\tau_1 s + 1} \right) \qquad (18.21)$$

where K_c, τ_D, τ_I and τ_1 are functions of λ and the parameters in G_I and G_m. The examples that follow will illustrate the application of this simplified procedure for designing an IMC controller.

Example 18.5. *Internal model control.* Design an IMC controller for the process which, is first-order:

$$G_m = K/(\tau s + 1)$$

For this case $G_{m_a} = 1$ and $G_{m_m} = K/(\tau s + 1)$. Applying Eq. (18.18) gives

$$G_I = 1/G_{m_m} = (\tau s + 1)/K$$

In order to be able to implement this transfer function let $f(s) = 1/(\lambda s + 1)$. The IMC controller becomes

$$G_I = \frac{1}{K} \frac{\tau s + 1}{\lambda s + 1}$$

This result is a lead-lag transfer function that can be implemented with modern microprocessor-based controllers. We may now obtain G_c from Eq. (18.14)

$$G_c = \frac{G_I}{1 - G_I G_m}$$

Introducing the expressions for G_I and G_m into this equation gives

$$G_c = \frac{\dfrac{\tau s + 1}{K(\lambda s + 1)}}{1 - \dfrac{\tau s + 1}{K(\lambda s + 1)} \dfrac{K}{\tau s + 1}} = \frac{\tau s + 1}{K \lambda s} = \frac{\tau}{\lambda K} \left(1 + \frac{1}{\tau s} \right)$$

This result is in the form of a PI controller:

$$G_c = K_c \left(1 + \frac{1}{\tau_I s} \right) \qquad K_c = \tau/\lambda K \qquad \tau_I = \tau$$

Although this design procedure results in the equivalence of a PI controller, only one parameter (λ) must be used to tune the controller. This is a distinct advantage over the use of a conventional controller in which both K_c and τ_I must be tuned.

Example 18.6. *Internal model control.* Design an IMC controller for a process which is first-order with transport lag:

$$G = K \frac{e^{-\tau_d s}}{\tau s + 1}$$

In the model of this process, use as an approximation to the transport lag a first-order Padé approximation [See Eq. (8.47)], thus

$$e^{-\tau_d s} = \frac{1 - (\tau_d/2)s}{1 + (\tau_d/2)s}$$

The model becomes

$$G_m = K\frac{1 - (\tau_d/2)s}{1 + (\tau_d/2)s}\frac{1}{\tau s + 1}$$

For this model,

$$G_{m_a} = \frac{1 - (\tau_d/2)s}{1 + (\tau_d/2)s} \qquad \text{(an all-pass filter)}$$

and

$$G_{m_m} = \frac{K}{\tau s + 1}$$

Following the same steps as used in Example 18.5, we obtain for the IMC controller

$$G_I = \frac{1}{K}\frac{\tau s + 1}{\lambda s + 1}$$

It is instructive to see the form G_c takes for this example. Applying Eq. (18.14) gives

$$G_c = \frac{G_I}{1 - G_I G_m} = \frac{\dfrac{\tau s + 1}{K(\lambda s + 1)}}{1 - \dfrac{\tau s + 1}{K(\lambda s + 1)}\dfrac{K[1 - (\tau_d/2)s]}{[1 + (\tau_d/2)s](\tau s + 1)}}$$

This may be reduced algebraically to the form given by Eq. (18.21) with

$$K_c = \frac{2\tau + \tau_d}{2(\lambda + \tau_d)}$$

$$\tau_I = \tau + (\tau_d/2)$$

$$\tau_D = \frac{\tau\tau_d}{2\tau + \tau_d}$$

$$\tau_1 = \frac{\lambda\tau_d}{2(\lambda + \tau_d)}$$

The response of this first-order with transport lag system for several values of λ and for $K = 1$, $\tau = 1$, $\tau_d = 1$ is given in Fig. 18.32. The values of K_c, τ_I, τ_d, and τ_1 obtained from the above relations are shown in Table 18.2. Notice that once a model is accepted, the tuning of the modified IMC controller [Eq. (18.21)] depends only on the choice of λ. For the range of λ used, Fig. 18.32 shows that the step response is only slightly oscillatory for all values of λ and the fastest response is for $\lambda = 0.5$. Also notice that λ affects only K_c and τ_1. This example shows that the design of a controller by the IMC method is a straightforward procedure and leads to a controller that requires the adjustment of only one parameter, λ.

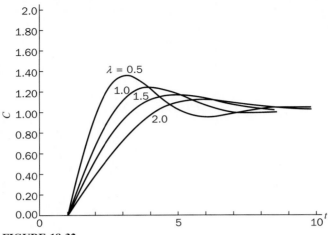

FIGURE 18-32
Response for IMC-designed controller of Example 18.6.

It is instructive to compare the response for the IMC-derived controller with the response for a PI controller using Ziegler-Nichols settings. The responses, which are given in Fig. 18.33, show that for this particular example the controller using Z-N settings produces a response with less overshoot and a higher frequency of oscillation than the controller designed by the IMC method.

These two examples show clearly how the parameters of the conventional controller G_c are related to the parameters of the model and the filter.

The treatment of internal model control presented here has been limited to single input/single output continuous systems for which the disturbance is a step change. Furthermore, we have not discussed the use of model uncertainty in selecting the filter parameters. Internal model control has been extended to sampled-data control systems and to multiple input/multiple output systems. IMC is a new approach to the design of control systems that considers the process model as an essential part of the control system design. As the method becomes better understood it will most likely affect the design of industrial control systems. Microcomputer-based controllers now have the capability of implementing many of the control algorithms designed by the IMC method. There is no longer a need to be tied to the classical control algorithms.\

TABLE 18.2
IMC derived controller settings for Example 18.6

λ	0.5	1.0	1.5	2.0
K_c	1.0	0.75	0.60	0.5
τ_I	1.5	1.5	1.5	1.5
τ_D	0.33	0.33	0.33	0.33
τ_1	0.167	0.25	0.30	0.33

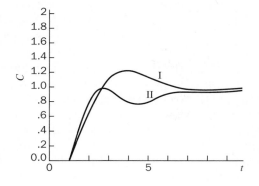

FIGURE 18-33
Comparison of response for IMC controller and conventional controller for Example 18.6:
I IMC-derived controller with $\lambda = 1.0$,
II PI controller with Ziegler-Nichols settings ($K_c = 1.02$, $\tau_I = 2.84$).

SUMMARY

In this chapter, we have examined five advanced control strategies. The first three on cascade control, feedforward-feedback control, and ratio control are advanced only in the sense that each strategy is more complex than the single-loop systems we have encountered up to this chapter. These three strategies are used extensively in industry and modern microprocessor-based controllers can implement them easily. The other strategies, on Smith prediction and internal model control (IMC), are less likely to be used in industry and are closely related in their block diagram structure. Of the five control strategies, the IMC method has the most rigorous mathematical foundation and is presently the focus of intense academic research. Three of the strategies, feedforward-feedback, Smith prediction, and IMC, are dependent on accurate models of the processes for their application.

Cascade control is especially useful in reducing the effect of a load disturbance that is located far from the control variable and which moves through the system slowly. The presence of the inner control loop reduces the lag in the outer loop with the result that the cascade system responds more quickly to a load disturbance.

If a particular load disturbance occurs frequently, the quality of control can often be improved by applying feedforward control. Ideally the transfer function of the feedforward controller is obtained from knowledge of the model of the process. In cases where the feedforward controller transfer function requires prediction (for example $\tau_f s + 1$), one must be satisfied with an approximation of the feedforward controller, which takes the form of a lead-lag transfer function. When a model of the process does not exist, the feedforward controller can be tuned after doing some open-loop step tests that relate the control variable to the load disturbance. To provide for load disturbances that cannot be measured or anticipated, feedforward control is always combined with feedback control in a practical situation.

Ratio control is widely used in industry in the blending of two component streams (*A* and *B*) to produce a mixed stream of desired composition (i.e., ratio of components). Ratio control is essentially a flow-control problem in which the flow measurement of stream A (the wild stream) is used to compute the set point for the flow of stream B so that the desired ratio of components will be obtained.

The Smith predictor control scheme (dead-time compensation) was developed to improve the control of a system having a large transport lag. The method is based on a model of the process that is first-order with dead time. By introducing inner loops that contain elements of the transfer function of the model, the control system is transformed ideally to one without transport lag, a system that is much easier to control. This ideal situation occurs when the process and the model are in exact agreement. In reality, the success of the Smith predictor strategy depends on the degree of agreement between process and model.

Internal model control resembles the Smith predictor strategy in terms of the structure of the block diagram. To apply the IMC method, one must have an accurate model of the process, the model uncertainty, the type of disturbance (step, ramp, etc.) and the performance objective (integral of square error). The method, which is based on a rigorous mathematical foundation, leads to an IMC controller that is the best that can be designed in terms of the performance objective. The IMC structure can be reduced to a conventional control structure in which the conventional controller is related to the IMC controller and the parameters of the model. For many simple processes with simple disturbance (impulse, step, etc.), the equivalent conventional controller based on the IMC design method turns out to be the equivalent of a PID controller.

PROBLEMS

18.1. (*a*) Obtain G_f for the feedforward-feedback system shown in Fig. P18.1 so that C does not change when a disturbance in C_i occurs. Would there be any problem in implementing this G_f?

(*b*) If G_f is to be a lead-lag transfer function

$$\frac{T_1 s + 1}{T_2 s + 1}$$

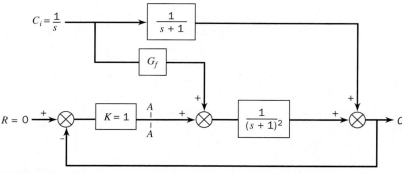

FIGURE P18-1

determine T_1 and T_2 by the Foxboro rule. How do you determine whether lead or lag is to predominate? Use $t_p = 1.0$ in the Foxboro rule.

(c) When feedforward-feedback control is present, sketch the response $C(t)$ when $C_i = 1/s$ and when G_f from part (a) is used.

(d) Repeat (c) when G_f from part (b) is used. Only a rough sketch that suggests the transient response is expected in this case.

(e) Determine $C(t)$ when $C_i = 1/s$, and $G_f = -1$, and the feedback loop is broken at AA. Obtain the numerical value of $C(t)$ at $t = 0.5, 1.0,$ and 1.5.

CHAPTER
19

CONTROLLER TUNING AND PROCESS IDENTIFICATION

The selection of a controller type (P, PI, PID) and its parameters (K_c, τ_I, τ_D) is intimately related to the model of the process to be controlled. The adjustment of the controller parameters to achieve satisfactory control is called *tuning*. The selection of the controller parameters is essentially an optimization problem in which the designer of the control system attempts to satisfy some criterion of optimality, the result of which is often referred to as "good" control. The process of tuning can vary from a trial-and-error attempt to find suitable control parameters for "good" control to an elaborate optimization calculation based on a model of the process and a specific criterion of optimal control. In many applications, there is no model of the process and the criterion for good control is only vaguely defined. A typical criterion for good control is that the response of the system to a step change in set point or load should have minimum overshoot and one-quarter decay ratio. Other criteria may include minimum rise time and minimum settling time.

In the first part of this chapter, some of the widely used tuning rules for continuous controllers will be presented. In the second part of the chapter, methods for determining the model of a process from experimental tests will be described. Determining the model of a process experimentally is referred to as *process identification*.

282

CONTROLLER TUNING

Before presenting tuning rules, some discussion of the effect of each mode in a PID controller on the transient response of a controlled process will be instructive.

Selection of Controller Modes

Consider a typical loop as shown in Fig. 19.1 in which the process is second-order and the measuring element is a transport lag. (The transfer function of the valve is taken as 1.) Load responses for this process for four types of controllers (P, PD, PI, PID) are shown in Fig. 19.2. For each response curve, the process was subjected to a unit-step change in load ($U = 1/s$) and the controller parameters were selected by tuning rules to be presented later. Regardless of the specific tuning rules used, the responses shown in Fig. 19.2 are typical of well-tuned controllers for systems found in industry. The nature of the response for each type controller will now be described. (The reader should also refer to Figs. 10.7 and 17.14 to reinforce this discussion.)

PROPORTIONAL CONTROL. As shown in Fig. 19.2, proportional control produces an overshoot followed by an oscillatory response, which levels out at a value that does not equal the set point; this ultimate displacement from the set point is the offset.

PROPORTIONAL-DERIVATIVE CONTROL. For this case the response exhibits a smaller overshoot and a smaller period of oscillation compared to the response for proportional control. The offset that still remains is less than that for proportional control.

PROPORTIONAL-INTEGRAL CONTROL. In this case, the response has about the same overshoot as proportional control, but the period is larger; however, the response returns to the set point (offset $= 0$) after a relatively long settling time. The most beneficial influence of the integral action in the controller is the elimination of offset.

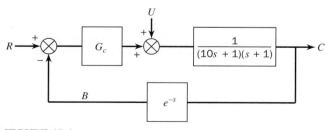

FIGURE 19-1
Typical control system used to study the effect of controller modes on load responses shown in Fig. 19.2.

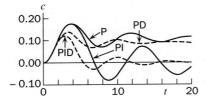

FIGURE 19-2
Load response of a typical control system using various modes of control (process shown in Fig. 19.1).

PROPORTIONAL-INTEGRAL-DERIVATIVE CONTROL. As one might expect, the use of PID control combines the beneficial features of PD and PI control. The response has lower overshoot and returns to the set point more quickly than the responses for the other types of controllers.

From the nature of the responses just described, we can make the following generalizations. Integral action, which is present in PI and PID controllers, eliminates offset. The addition of derivative action speeds up the response by contributing to the controller output a component of the signal that is proportional to the rate of change of the process variable.

For simple, low-order (first or second-order) processes that can tolerate some offset, P or PD control is satisfactory. For processes that cannot tolerate offset and are of low order, PI control is required. For processes that are of high-order (those with transport lag or many first-order lags in series), PID control is needed to prevent large overshoot and long settling time.

Before the availability of microprocessor-based controllers, it was customary to select a controller based on price. Pneumatic and electronic controllers with proportional action were the least expensive and those with PID action were the most expensive. It was considered uneconomical to purchase a controller with more control actions than needed by the process. Today this price incentive no longer exists in the selection of the type of controller, for the modern microprocessor-based controller comes with all three actions, as well as other functions such as lead-lag and transport lag. A discussion of the features of modern controllers will be given in Chap. 35. There is probably little justification to select a P or PD controller for most processes. The PI controller is often the choice because it eliminates offset and requires only two parameter adjustments. Tuning a PID controller is more difficult because three parameters must be adjusted. The presence of derivative action can also cause the controller output to be very jittery if there is much noise in the signals. We now turn our attention to some of the criteria for good control that are used to judge whether or not a control system is well tuned.

Criteria for Good Control

Before we can be satisfied with the response of a control system for a choice of control parameters, we must have some concept of what we want as an ideal response. Most operators of processes know what they want in the form of a response to a change in set point or load. For example, a response that gives minimum overshoot and $\frac{1}{4}$ decay ratio is often considered as a satisfactory response. In many cases, tuning is done by trial and error until such a response is obtained. In order to compare different responses that use different sets of controller parameters, a criterion that reduces the entire response to a single number, or a *figure of merit,* is desirable.

One criterion that is often used to evaluate a response of a control system is the integral of the square of the error with respect to time (ISE). The definition of ISE is as follows:

Integral of the square of the error (ISE)

$$\text{ISE} = \int_0^\infty e^2 dt \tag{19.1}$$

where e is the usual error (i.e., set point $-$ control variable). For a stable system for which there is no offset (i.e., $e(\infty) = 0$), Eq. (19.1) produces a single number as a figure of merit. The objective of the designer is to obtain the minimum value of ISE by proper choice of control parameters. A response that has large errors and persists for a long time will produce a large ISE. For the cases of P and PD control, where offset occurs, the integral given by Eq. (19.1) does not converge. In these cases, one can use a modified integrand, which replaces the error $r(t) - c(t)$, by $c(\infty) - c(t)$. Since $c(\infty) - c(t)$ does approach zero as t goes to infinity, the integral will converge and serve as a figure of merit.

Two other criteria often used in process control are defined as follows:

Integral of the absolute value of error (IAE)

$$\text{IAE} = \int_0^\infty |e| \, dt \tag{19.2}$$

Integral of time-weighted absolute error (ITAE)

$$\text{ITAE} = \int_0^\infty |e| t \, dt \tag{19.3}$$

Each of the three figures of merit given by Eqs. (19.1), (19.2), and (19.3) have different purposes. The ISE will penalize (i.e., increase the value of ISE) the response that has large errors, which usually occur at the beginning of a response, because the error is squared. The ITAE will penalize a response which has errors that persist for a long time. The IAE will be less severe in penalizing a response for large errors and treat all errors (large and small) in a uniform manner. The ISE figure of merit is often used in optimal control theory because it can be used more easily in mathematical operations (for example differentiation) than the figures of merit, which use the absolute value of error. In applying the tuning rules to be discussed in the next section, these figures of merit can be used in comparing responses that are obtained with different tuning rules.

TUNING RULES

Ziegler-Nichols Rules (Z-N)

These rules were first proposed by Ziegler and Nichols (1942), who were engineers for a major control hardware company in the United States (Taylor Instrument Co.). Based on their experience with the transients from many types of processes, they developed a *closed-loop tuning method* still used today in one form or another. The method is described as a closed-loop method because the controller remains in the loop as an active controller in automatic mode. This closed-loop method will be contrasted with an open-loop tuning method to be discussed later. We have already discussed the Ziegler-Nichols rules in Chap. 17 as a natural consequence of our study of frequency response. Ziegler and Nichols did not suggest that the ultimate gain (K_{cu}) and ultimate period (P_u) be computed from frequency response calculations based on the model of the process. They intended that K_{cu} and P_u be obtained from a closed-loop test of the actual process. When the rules were first proposed, frequency response methods and process models were not generally available to the control engineers. The rules are presented below, and are in the form that one would use for actual application to a real process.

1. After the process reaches steady state at the normal level of operation, remove the integral and derivative modes of the controller, leaving only proportional control. On some PID controllers, this requires that the integral time (τ_I) be set to its maximum value and the derivative time (τ_D) to its minimum value. On modern controllers (microprocessor-based), the integral and derivative modes can be removed completely from the controller.

2. Select a value of proportional gain (K_c), disturb the system, and observe the transient response. If the response decays, select a higher value of K_c and again observe the response of the system. Continue increasing the gain in small steps until the response first exhibits a sustained oscillation. The value of gain and the period of oscillation that correspond to the sustained oscillation are the ultimate gain (K_{cu}) and the ultimate period (P_u).

   Some very important precautions to take in applying this step of the tuning method are given in the next section.

3. From the values of K_{cu} and P_u found in the previous step, use the Ziegler-Nichols rules given in Table 19.1 to determine controller settings (K_c, τ_I, τ_D). This table is the same as Table 17.1 in Chap. 17.

   Although variations in the tuning rules given in Table 19.1 are used by industry, the same approach of using K_{cu} and P_u to obtain controller parameters is used. The Ziegler-Nichols rules generally provide conservative (and safe) controller settings. The Z-N settings should be considered as only approximate settings for satisfactory control. Fine tuning of the controller settings is usually required to get an improved control response.

   The experimental determination of K_{cu} and P_u described in step 2 can be replaced by a computation using frequency response methods if an accurate model of the process, valve, and measuring element is known. This type of calculation was done in Chap. 17.

TABLE 19.1
Ziegler-Nichols controller settings

Type of control	$G_c(s)$	K_c	τ_I	τ_D
Proportional (P)	K_c	$0.5K_u$		
Proportional-integral (PI)	$K_c\left(1 + \dfrac{1}{\tau_I s}\right)$	$0.45K_u$	$\dfrac{P_u}{1.2}$	
Proportional-integral-derivative (PID)	$K_c\left(1 + \dfrac{1}{\tau_I s} + \tau_D s\right)$	$0.6K_u$	$\dfrac{P_u}{2}$	$\dfrac{P_u}{8}$

PRECAUTIONS TO TAKE IN APPLYING THE Z-N METHOD. Some discussion is needed to avoid some pitfalls in applying step 2 of the Z-N method to obtain K_{cu} and P_u. These precautions are concerned with the type and size of the disturbance that induces the response and with the avoidance of using a limit cycle as the indication that the system is on the threshold of instability.

The simplest way to introduce a disturbance is to move the set point away from the control variable for a short time and then return the set point to its original value. This procedure, which is equivalent to introducing a pulse function in the error, causes the system to respond and yet stay within a narrow band surrounding the normal operating point of the process.

An alternate type of disturbance would be to introduce a small step change in set point. If step changes in set point are used to induce transients, the successive step changes should alternate around the normal operating point of the process. It is also important to make the disturbance as small as possible, especially as the gain of the controller is increased, so that the valve and other components do not exceed their physical limits.

When the valve moves to its limits during a closed-loop transient, we say that the valve saturates. Under these conditions, a sustained oscillation occurs, which is called a limit cycle. The limit cycle that is caused by saturation is a nonlinear phenomenon, which will be covered in Chap. 33 on nonlinear control. If a limit cycle occurs, the gain that produces it and the period of the cycle should not be used in the Ziegler-Nichols rules. Since the limit cycle will appear to the observer to be the same as a sustained oscillation when the system is on the verge of instability, the novice will often mistakenly use the information derived from the limit cycle (controller gain and period) to obtain controller settings. A simple way to know if one has a limit cycle is to observe the swing in pressure to the valve. If the limits of the valve (e.g., 3 psi to 15 psi) are reached repeatedly during the oscillatory response, one has a limit cycle and the controller gain and period should not be used to determine controller settings. It is for this reason step 2 states that K_c should be increased in small steps until the response first exhibits a sustained oscillation.

To appreciate the use of step 2 of the tuning method, one should have some laboratory experience in tuning a real process, or at least a computer simulation of a process. The experienced operator can develop some short cuts to finding the ultimate gain and ultimate period.

Cohen and Coon Rules (C-C)

The next method of tuning to be discussed is an *open-loop method*, in which the control action is removed from the controller by placing it in manual mode and an open-loop transient is induced by a step change in the signal to the valve. This method was proposed by Cohen and Coon (1953) and is often used as an alternative to the Z-N method. Figure 19.3 shows a typical control loop in which the control action is removed and the loop opened for the purpose of introducing a step change (M/s) to the valve. The step response is recorded at the output of the measuring element. The step change to the valve is conveniently provided by the output from the controller, which is in manual mode. The response of the system (including the valve, process, and measuring element) is called the *process reaction curve;* a typical process reaction curve exhibits an S-shape as shown in Fig. 19.4. After presenting the Cohen and Coon method of tuning, the basis for their recommendations will be discussed. The C-C method is summarized in the following steps:

1. After the process reaches steady state at the normal level of operation, switch the controller to manual. In a modern controller, the controller output will remain at the same value after switching as it had before switching. (This is called "bumpless" transfer.)
2. With the controller in manual, introduce a small step change in the controller output that goes to the valve and record the transient, which is the process reaction curve (Fig. 19.4).
3. Draw a straight line tangent to the curve at the point of inflection, as shown in Fig. 19.4. The intersection of the tangent line with the time axis is the apparent transport lag (T_d); the apparent first-order time constant (T) is obtained from

$$T = B_u/S \qquad (19.4)$$

where B_u is the ultimate value of B at large t and S is the slope of the tangent line. The steady-state gain that relates B to M in Fig. 19.3 is given by

$$K_p = B_u/M \qquad (19.5)$$

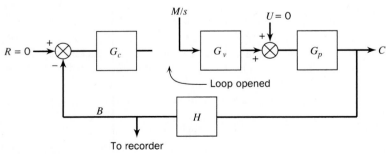

FIGURE 19-3
Block diagram of a control loop for measurement of the process reaction curve.

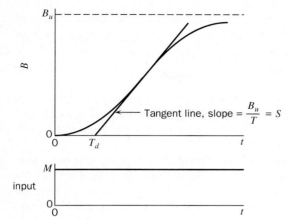

Tangent line, slope $= \dfrac{B_u}{T} = S$

FIGURE 19-4
Typical process reaction curve showing graphical construction to determine first-order with transport lag model.

4. Using the values of K_p, T, and T_d from step 3, the controller settings are found from the relations given in Table 19.2.

Notice in Table 19.2 that all of the controller settings are a function of the dimensionless group T_d/T, the ratio of the apparent transport lag to the apparent time constant. Also K_c is inversely proportional to K_p.

TABLE 19.2
Cohen-Coon controller settings

Type of control	Parameter setting
Proportional (P)	$K_c = \dfrac{1}{K_p}\dfrac{T}{T_d}\left(1 + \dfrac{T_d}{3T}\right)$
Proportional-integral (PI)	$K_c = \dfrac{1}{K_p}\dfrac{T}{T_d}\left(\dfrac{9}{10} + \dfrac{T_d}{12T}\right)$
	$\tau_I = T_d\,\dfrac{30 + 3T_d/T}{9 + 20T_d/T}$
Proportional-derivative (PD)	$K_c = \dfrac{1}{K_p}\dfrac{T}{T_d}\left(\dfrac{5}{4} + \dfrac{T_d}{6T}\right)$
	$\tau_D = T_d\,\dfrac{6 - 2T_d/T}{22 + 3T_d/T}$
Proportional-integral-derivative (PID)	$K_c = \dfrac{1}{K_p}\dfrac{T}{T_d}\left(\dfrac{4}{3} + \dfrac{T_d}{4T}\right)$
	$\tau_I = T_d\,\dfrac{32 + 6T_d/T}{13 + 8T_d/T}$
	$\tau_D = T_d\,\dfrac{4}{11 + 2T_d/T}$

The rationale for the C-C tuning method begins with the representation of the S-shaped process reaction curve by a first-order with transport lag model; thus

$$G_p(s) = \frac{K_{\dot{p}} e^{-T_d s}}{Ts + 1} \qquad (19.6)$$

Using the system expressed by Eq. (19.6), Cohen and Coon obtained by theoretical means the controller settings given in Table 19.2. Their computations required that the response have $\frac{1}{4}$ decay ratio, minimum offset, minimum area under the load-response curve, and other favorable properties.

In applying the C-C tuning method, an important task is the graphical construction, shown in Fig. 19.4, which reduces the process reaction curve to the first-order with the transport lag model given by Eq. (19.6). To understand the basis for the graphical procedure, consider the response of the transfer function of Eq. (19.6) to a step change in input; the resulting transient is shown in Fig. 19.5. After $t = T_d$, the response is a first-order response. The point of inflection of the curve in Fig. 19.5 occurs at $t = T_d$ and the slope of the tangent line at this point is related to the time constant by the relation:

$$S = B_u/T$$

Solving for T gives the expression in Eq. (19.4). The response after $t = T_d$, shown in Fig. 19.5, was also presented in Fig. 5.6.

The attempt to model the process reaction curve by the method shown in Fig. 19.4 is crude and does not give a very good fit. Finding the point of inflection and drawing a tangent line at this point is quite difficult, especially if the data for the process reaction curve are not accurate and if they scatter. A better method for fitting the process reaction curve to a first-order with transport lag model is to perform a least-square fit of the data. The disadvantage to this fitting procedure is the time and effort required. An example to be presented later will study the effect of the type of model fitting procedure on the selection of controller parameters.

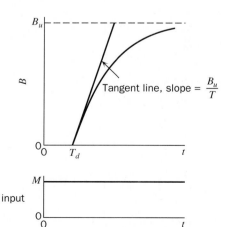

FIGURE 19-5
Step response for a first-order with transport lag model.

More recently, Lopez et al. (1967) studied the tuning of controllers with error-integral criteria for the first-order with transport lag model of Eq. (19.6). The error-integral criteria that they considered were ISE, IAE, and ITAE. In their work, a search procedure was used to find the controller parameters that minimized each particular figure of merit. Their results, developed for T_d/T varying from 0 to 1.0, were presented in graphical form and as empirical equations that were fitted to their graphical results. Their results, which can be considered as a variation of the C-C tuning method, were not compared with the C-C method. The interested reader may wish to compare the method of Cohen and Coon and the method of Lopez et al. as a project.

To illustrate the two methods of controller tuning just presented, the system shown in Fig. 19.1 was simulated by use of a computer program called TUTSIM. (This simulation software is described in Chap. 35.) Table 19.3 gives the values of the controller parameters obtained by applying each tuning method; Figure 19.6 shows the resulting transients. Since the Z-N method does not give a rule for a PD controller, the settings listed for a PD controller under the Z-N heading of Table 19.3 were obtained by using a theoretical frequency response calculation in which the design was based on 30° phase margin and a maximum K_c. No general conclusions can be made about the relative merits of the two tuning methods from the results shown in Fig. 19.6, since these results apply to one specific example. About all that can be said is that for this specific example, both methods give reasonable first guesses of the control parameters.

Example 19.1. For the control system shown in Fig. 19.7, determine controller settings for a PI controller using the Z-N method and the C-C method. This problem will be instructive because the transfer function of the model is already in the form of first-order with transport lag, which is the form used by Cohen and Coon to derive their tuning rules.

C-C method. Since the transfer function of the plant is in the form of Eq. (19.6), we obtain T and T_d immediately without having to draw a tangent line through the point of inflection, i.e., $T = 1$ and $T_d = 1$. We also observe from the block

TABLE 19.3
Controller settings for the system of Fig. 19.1

Control Type	Parameter	Closed-loop method (Z-N method)	Open-loop method (C-C method)
P	K_c	6.4	8.1
PI	K_c	5.8	7.0
	τ_I	5.6	4.4
PD	K_c	11.4*	9.8
	τ_D	1.0*	0.43
PID	K_c	7.7	10.5
	τ_I	3.4	3.9
	τ_D	1.6	0.59

* Obtained by design for 30° phase margin and maximum K_c.

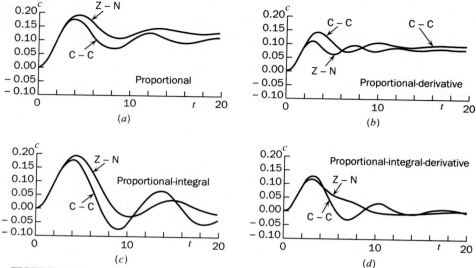

FIGURE 19-6

Comparison of load responses for the system of Fig. 19.1 using controller settings obtained by the Ziegler-Nichols method (Z-N) and the Cohen and Coon method (C-C).

diagram that $K_p = 1$. Substituting these values into the appropriate equations of Table 19.2 gives

$$K_c = \frac{T}{K_p T_d}\left(0.9 + \frac{T_d}{12T}\right) = \frac{1}{1}\left(0.9 + \frac{1}{12}\right) = 0.983$$

and

$$\tau_I = T_d \frac{30 + 3T_d/T}{9 + 20T_d/T} = \frac{30 + 3}{9 + 20} = 1.14$$

Using these values for K_c and τ_I, the step response shown in Fig. 19.8 was obtained by simulation.

Z-N Method. Application of the Bode criterion from Chap. 17 gives the following results

$$\omega_{co} = 2.03 \quad \text{or} \quad P_u = 2\pi/\omega_{co} = 3.09$$
$$K_{cu} = 2.26$$

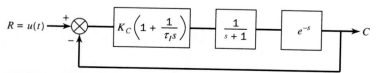

FIGURE 19-7

Process for Example 19.1.

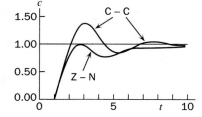

FIGURE 19-8
Response to unit step in set point for the system in Fig. 19.7 (Example 19.1).

The details for obtaining these results will not be given here since this type calcula-tion was covered in depth in Chap. 17. Applying the Z-N rules for PI control from Table 19.1 gives:

$$K_c = 0.45K_{cu} = (0.45)(2.26) = 1.02$$

and

$$\tau_I = P_u/1.2 = 3.09/1.2 = 2.58$$

The step response for these controller settings is shown in Fig. 19.8. The ISE value for each response was calculated out to a sufficiently long time (10 units of time) for the integral to converge; the results are as follows:

$$\text{C-C response: ISE} = 1.54 \quad \text{at } t = 10$$
$$\text{Z-N response: ISE} = 1.49 \quad \text{at } t = 10$$

Although the ISE values are nearly the same, the transient for the Z-N settings is better than the transient for the C-C settings. The Z-N transient has much less overshoot. The lesson to be learned from this example is that the comparison of two transients based on only one criterion (in this case, the ISE) may be mislead-ing in the selection of the best transient. It is also important to judge the quality of a transient by its actual appearance. It should be noted that for this example, in which there is a relatively large transport lag ($T_d = 1$), much of the con-tribution to the ISE occurs from $t = 0$ to $t = 1$, during which time the ISE reaches 1.0. This value of the ISE at $t = 1$ is the same, regardless of the tuning method used because the transport lag causes error to be constant from $t = 0$ to $t = 1$.

Example 19.2. For the control system shown in Fig. 19.9, determine the controller settings for a PI controller using the Z-N method and the C-C method. In this problem, the process reaction curve must be modeled by the method shown in Fig. 19.4.

C-C method. Since the transfer function of the plant is given as $1/(s + 1)^4$, we can obtain the value of T_d and T for use in the C-C method analytically. A unit-step response for the plant transfer function is

$$c(t) = 1 - \left(\frac{1}{6}t^3 + \frac{1}{2}t^2 + t + 1\right)e^{-t}$$

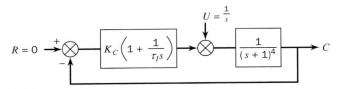

FIGURE 19-9
Process for Example 19.2.

From this result one can readily obtain the first and second derivatives; thus

$$\dot{c}(t) = \frac{1}{6}t^3 e^{-t}$$

$$\ddot{c}(t) = \frac{1}{6}e^{-t}(3t^2 - t^3)$$

The location of the inflection point on the transient $c(t)$ is obtained by setting the second derivative to zero:

$$0 = \frac{1}{6}e^{-t}(3t^2 - t^3)$$

Solving for t gives as the root of interest in this problem $t = 3$. Knowing that the point of inflection occurs at $t = 3$, we can compute the slope of the tangent line through this point to be

$$S = \dot{c}(3) = \frac{1}{6}(3)^3 e^{-3} = 0.224$$

We can now determine T_d as shown in Fig. 19.10. From the expression for $c(t)$, we obtain the value of c at the inflection point to be $c(3) = 0.353$. The value of t where the tangent line intersects the t-axis is obtained from the slope S; thus

$$\frac{0.353 - 0}{3 - T_d} = S = 0.224$$

solving for T_d gives

$$T_d = 1.42$$

Solving for T from Eq. (19.4) gives

$$T = B_u/S = 1/0.224 = 4.46$$

Having found T_d and T, we can apply the appropriate equations from Table 19.2 to get

$$K_c = 2.91 \qquad \tau_I = 2.86$$

The transient for these settings that was obtained by simulation is shown as curve C-C1 in Fig. 19.11. To our surprise, it is unstable.

Z-N method. When we apply the Z-N method for a PI controller, we obtain the following results: $K_{cu} = 4$, $P_u = 2\pi$, $K_c = 1.8$, and $\tau_I = 5.23$.

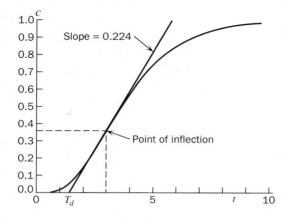

FIGURE 19-10
Process reaction curve for Example 19.2.

The transient for this set of controller parameters is also shown in Fig. 19.11. We see that the response is stable and well damped.

The lesson learned in this example is that the application of a tuning method may not produce a satisfactory transient. Fine tuning of these first guesses is usually needed.

Before abandoning the C-C method for this example, the process reaction curve was fitted to a first-order with transport lag model by means of a least square fitting procedure. Applying the least square fit procedure out to $t = 5$ produced the following results

$$T_d = 1.5 \quad \text{and} \quad T = 3.0$$

Applying the C-C method for these values of T_d and T gives

$$K_c = 2.05 \quad \text{and} \quad \tau_I = 2.49$$

Notice that the value of K_c is now considerably less than the value obtained from the fitting procedure shown in Fig. 19.10. This leads to the expectation that the response will now be stable. This expectation is fulfilled as shown by the transient labeled C-C2 in Fig. 19.11.

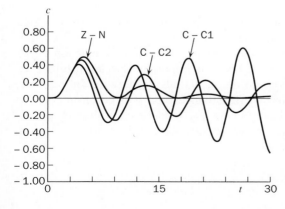

FIGURE 19-11
Comparison of transients produced by different tuning methods for Example 19.2 (process shown in Fig. 19.9). Z-N: Ziegler-Nichols method; C-C1: Cohen-Coon method based on tangent line through point of inflection; C-C2: Cohen-Coon method using model based on least square fit.

PROCESS IDENTIFICATION

Up to this point, the processes used in our control systems have been described by transfer functions that were derived by applying fundamental principles of physics and chemical engineering (e.g., Newton's law, material balance, heat transfer, fluid mechanics, reaction kinetics, etc.) to well defined processes. In practice, many of the industrial processes to be controlled are too complex to be described by the application of fundamental principles. Either the task requires too much time and effort or the fundamentals of the process are not understood. By means of experimental tests, one can identify the dynamical nature of such processes and from the results obtain a process model which is at least satisfactory for use in designing control systems. The experimental determination of the dynamic behavior of a process is called *process identification*.

The need for process models arises in many control applications, as we have seen in the use of tuning methods. Process models are also needed in developing feedforward control algorithms, self-tuning algorithms, and internal model control algorithms. Some of these advanced control strategies were discussed in the previous chapter.

Process identification provides several forms that are useful in process control; some of these forms are

Process reaction curve (obtained by step input)

Frequency response diagram (obtained by sinusoidal input)

Pulse response (obtained by pulse input)

We have already encountered the need for process identification in applying the tuning methods presented earlier in this chapter. In the case of the Z-N method, the procedure obtained one point on the open-loop frequency response diagram when the ultimate gain was found. (This point corresponds to a phase angle of $-180°$ and a process gain of $1/K_{cu}$ at the cross-over frequency ω_{co}.) In the case of the C-C method, the process identification took the form of the process reaction curve.

Step Testing

As already described in the application of the Z-N tuning method, a step change in the input to a process produces a response, which is called the process reaction curve. For many processes in the chemical industry, the process reaction curve is an S-shaped curve as shown in Fig. 19.4. It is important that no disturbances other than the test step enter the system during the test, otherwise the transient will be corrupted by these uncontrolled disturbances and will be unsuitable for use in deriving a process model. For systems that produce an S-shaped process reaction curve, a general model that can be fitted to the transient is the following second-order with transport lag model:

$$G_p(s) = \frac{K_p e^{-T_d s}}{(T_1 s + 1)(T_2 s + 1)} = \frac{Y(s)}{X(s)} \tag{19.7}$$

This model is an extension of the one used in the C-C tuning method, in which there was only one first-order term.

We shall now describe a graphical procedure for obtaining the transfer function of Eq. (19.7) from a process reaction curve.

SEMI-LOG PLOT FOR MODELING. The transfer function given by Eq. (19.7) can be obtained from a process reaction curve by a graphical method in which the logarithm of the incomplete response is plotted against time. In principle, this method can extract from the process reaction curve the two time constants in Eq. (19.7). The method, referred to as the semi-log plot method, is outlined below. The method applies for $T_1 > T_2$.

1. Determine (if transport lag is present) the time at which the process reaction curve of Fig. 19.12 first departs from the time axis; this time is taken as the transport lag T_d.
2. From the process reaction curve of Fig. 19.12, plot I versus t_1 on semi-log paper as shown in Fig. 19.13 where I is the fractional incomplete response and t_1 is the shifted time starting at T_d (i.e., $t_1 = t - T_d$). I is defined by

$$I = \frac{B_u - Y}{B_u}$$

where B_u is the ultimate value of Y.
3. Extend a tangent line through the data points at large values of t_1 (see Fig. 19.13). Refer to this tangent line as I_a and let the intersection of the tangent line with the vertical axis at $t_1 = 0$ be called P.
4. To find the time constant T_1, read from the graph in Fig. 19.13 the time at which $I_a = 0.368P$. This time is T_1.
5. Plot Δ versus t_1 where $\Delta = I_a - I$. If the data points (Δ, t_1) fall on a straight line, the system can be modeled as a second-order transfer function

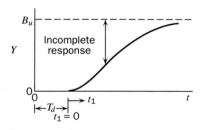

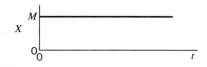

FIGURE 19-12
Process reaction curve used in the semi-log plot method of modeling.

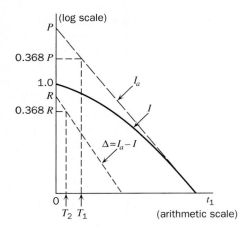

FIGURE 19-13
Graphical construction for use in modeling by semi-log plot method.

with transport lag as given by Eq. (19.7) with time constants T_1 and T_2. The value of T_2 is the time at which $\Delta = 0.368R$ where R is the intersection of the line Δ with the vertical axis at $t_1 = 0$.

If one does not get a straight line when Δ is plotted against t_1, the procedure can be extended to get more first-order time constants, T_3, T_4, and so on; however, the data must be very accurate for this method to be successful in identifying more than two time constants. Usually the data scatter, especially at large values of time, and one must be satisfied in drawing straight lines through the scattered points.

6. The process gain is simply

$$K_p = B_u/M$$

PROOF OF SEMI-LOG METHOD. By shifting the time axis from t to t_1 in Fig. 19.12 we have accounted for the effect of T_d in Eq. (19.7) and the transient to be considered (Y vs. t_1) is described by the transfer function

$$G_p(s) = \frac{K_p}{(T_1s + 1)(T_2s + 1)} = \frac{Y(s)}{X(s)} \qquad (19.8)$$

Introducing $X = M/s$ and $K_p = B_u/M$ into Eq. (19.8) gives

$$\frac{Y(s)}{B_u} = \frac{1}{s(T_1s + 1)(T_2s + 1)} \qquad (19.9)$$

The time response of this expression is given by

$$\frac{Y}{B_u} = 1 - \frac{T_1T_2}{T_1 - T_2}\left(\frac{1}{T_2}e^{-t_1/T_1} - \frac{1}{T_1}e^{-t_1/T_2}\right) \qquad (19.10)$$

This result was also given in Eq. (7.10). Letting $I = (B_u - Y)/B_u$ as was done in step 2, we obtain from Eq. (19.10)

$$I = \frac{T_1}{T_1 - T_2}e^{-t_1/T_1} - \frac{T_2}{T_1 - T_2}e^{-t_1/T_2} \qquad (19.11)$$

Assume that $T_1 > T_2$. As t_1 approaches ∞, the second term on the right side of Eq. (19.11) becomes much smaller than the first term and we can write as an approximation to Eq. (19.11) for large t_1

$$I_a = \frac{T_1}{T_1 - T_2} e^{-t_1/T_1} = P e^{-t_1/T_1} \tag{19.12}$$

where the term I_a is the approximation of I at large values of time and P is the value of I_a at $t_1 = 0$.

When $t_1/T_1 = 1$, or $t_1 = T_1$, we obtain from Eq. (19.12)

$$I_a = P e^{-1} = 0.368P \tag{19.13}$$

This proves step 4 of the graphical procedure. Now let $\Delta = I_a - I$. From Eqs. (19.11) and (19.12) we obtain

$$\Delta = \frac{T_2}{T_1 - T_2} e^{-t_1/T_2} = R e^{-t_1/T_2} \tag{19.14}$$

This relation plots as a straight line on semi-log paper.

When $t_1/T_2 = 1$, or $t_1 = T_2$, we obtain from Eq. (19.14)

$$\Delta = R e^{-1} = 0.368R$$

This proves Step 5.

To appreciate the nature of this graphical construction, the reader is encouraged to solve the problems requiring its use at the end of the chapter.

Frequency Testing

We have shown in the section on frequency response that a process having a transfer function $G(s)$ can be represented by a frequency response diagram (or Bode plot) by taking the magnitude and phase angle of $G(j\omega)$. This procedure can be reversed to obtain $G(s)$ from an experimentally determined frequency response diagram. The procedure requires that a device be available to produce a sinusoidal signal over a range of frequencies. We describe such a device as a *sine wave generator*. In frequency testing of an industrial process, a sinusoidal variation in pressure is applied to the top of the control valve so that the manipulated variable can be varied sinusoidally over a range of frequencies. The block diagram that applies during frequency testing is the same as the one of Fig. 19.3 with the step input (M/s) replaced by a sinusoidal signal. The sine wave generator used to test electronic devices operates at frequencies that are too high for many slow moving chemical processes. For frequency testing of chemical processes, special low-frequency generators must be built that can produce a sinusoidal variation in pressure to a control valve. To preserve the sinusoidal signal in the flow of manipulated variable through the valve, the valve must be linear.

In the 1960s when frequency response methods were first introduced to chemical engineers as a means for process identification, several chemical and petroleum companies constructed mobile units containing low-frequency sine wave generators and recorders that could be moved to processing units in a plant for the purpose of frequency testing.

The great disadvantage of frequency testing is that it takes a long time to collect frequency response data over a range of frequencies that can be used to construct frequency response plots. The time is especially long for chemical processes, often having long time constants measured in minutes or even hours. The frequency test at a given frequency must last long enough to make sure that the transients have disappeared and only the ultimate periodic response is represented by the data. Frequency testing usually ties up plant equipment too long to be recommended as a means of process identification. Step testing and pulse testing take much less time and can usually provide satisfactory process identification.

Pulse Testing

Pulse testing is similar to step testing; the only difference in the experimental procedure is that a pulse disturbance is used in place of a step disturbance. The pulse is introduced as a variation in valve top pressure as was done for step testing (see Fig. 19.3). In applying the pulse, the open-loop system is allowed to reach steady state, after which the valve top pressure is displaced from its steady-state value for a short time and then returned to its original value. The response is recorded at the output of the measuring element (B in Fig. 19.3). An arbitrary pulse and a typical response are shown in Fig. 19.14. Usually the pulse shape is rectangular in experimental work, but other well defined shapes are also used.

The input-output data obtained in a pulse test are converted to a frequency response diagram, which can be used to tune a controller. The transfer function of the valve, process, and measuring element (referred to as the process transfer function, for convenience) is given by:

$$G_p(s) = \frac{Y(s)}{X(s)} \tag{19.15}$$

where $Y(s)$ = Laplace transform of the function representing the recorded output response

$X(s)$ = Laplace transform of the function representing the pulse input

Applying the definition of the Laplace transform [Eq. (2.1)] to the numerator and denominator of Eq. (19.15) and replacing s by $j\omega$ gives

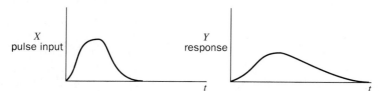

FIGURE 19-14
Typical process response to a pulse input.

$$G_p(j\omega) = \frac{\int_0^\infty Y(t)e^{-j\omega t}dt}{\int_0^\infty X(t)e^{-j\omega t}dt} \tag{19.16}$$

or

$$G_p(j\omega) = \frac{\int_0^\infty Y(t)\cos\omega t\ dt - j\int_0^\infty Y(t)\sin\omega t\ dt}{\int_0^\infty X(t)\cos\omega t\ dt - j\int_0^\infty X(t)\sin\omega t\ dt} \tag{19.17}$$

Reducing the input-output data of Fig. 19.14 to an expression for $G_p(j\omega)$ by means of Eq. (19.17) is a difficult, tedious task that has been described in the literature [Hougen (1964); Smith and Corripio (1985)]. The integration is done numerically: the time axis is divided into equal increments and the function $Y(t)$ is represented linearly over successive time increments. A computer is necessary to evaluate the integrals. Since the input-output variables (X and Y) are deviation variables that return to zero as t progresses, the integrals in Eq. (19.17) converge. If the input is a rectangular pulse, the integral in the denominator of Eq. (19.16) can be determined analytically. After the integrals in Eq. (19.17) are evaluated for several values of ω, $G_p(j\omega)$ for each value of ω can be expressed as

$$G_p(j\omega) = \frac{A + jB}{C + jD} = \alpha + j\beta$$

The magnitude and angle of $\alpha + j\beta$ can be found easily and used in plotting a frequency response diagram.

This brief outline describing pulse testing may appear deceptively simple. In practice, the data on the response must be very accurate and noise-free in order for the method to succeed. This means that the recorder used to measure the response must be very sensitive. The selection of the pulse height and width is also critical. If the pulse height and width are too small, the disturbance to the system will be too small to produce a transient that can be measured accurately by the recorder. If the pulse height is too large, the system may be operating too far from the linear range of interest. Obtaining the proper pulse height and width can be determined by some preliminary open-loop experiments. The pulse test is the least disruptive to plant operation among the process identification methods we have considered. The pulse disturbance does not cause the process output to depart far from its normal operating point. Also, the length of time that the process is tied up for an open-loop test is short compared to the frequency response method.

SUMMARY

In the practical application of process control, some methods for tuning and process identification are needed. The selection of controller modes depends on the process to be controlled. Proportional control is simple, but the response exhibits offset. The derivative action in PD control makes it possible to increase the controller gain with the result that the response has less offset and responds more quickly compared to proportional control. To eliminate offset, integral action must be present in the controller in the form of PI and PID control. PI control often causes the response to have large overshoot and a slow return to the set point especially for high-order processes. The presence of derivative action in a PID controller gives less overshoot and a faster return to the set point, compared to the response for PI control.

To compare the quality of control on a numerical basis, several criteria that integrate some function of the error with respect to time have been proposed. These include the integral of the square of the error (ISE), the integral of the absolute error (IAE), and the integral of the time-weighted absolute error (ITAE).

In the first part of this chapter two well known tuning methods are presented: the Ziegler-Nichols (Z-N) method (a closed-loop method) and the Cohen-Coon (C-C) method (an open-loop method). These two methods were applied to several examples and the transients for each compared. The lesson to be learned through these examples is that the controller parameters obtained from a tuning rule should be considered as first guesses; fine on-line tuning is usually needed to get a satisfactory transient.

The Z-N and C-C methods actually require information about the process model. The Z-N method is based on the ultimate gain at the crossover frequency, which is equivalent to knowing one point on the open-loop frequency response diagram. The C-C method requires the use of an open-loop step response (process reaction curve).

In the advanced control strategies discussed in Chap. 18, a process model is often needed to apply the strategy. When a process model cannot be found by application of theoretical principles, one must obtain a model experimentally. The experimental approach to obtaining a model is called process identification. The three methods of process identification discussed in this chapter are step testing, frequency testing, and pulse testing. The frequency method is seldom used because of the time it takes to test a system over a wide range of frequencies. Step testing is easy to apply and ties the process up for only enough time to obtain one transient. Pulse testing is also simple to apply, but the analysis of the input-output data require extensive calculations that must be done by a computer.

PROBLEMS

19.1. Use the semi-log graphical method to determine the process model for the following unit-step response data:

time. t	response. Y
0	0
0.25	0.07
0.50	0.20
0.75	0.34
1.00	0.47
1.25	0.57
1.50	0.66
2.00	0.79
2.50	0.87
3.00	0.92
3.50	0.94
4.00	0.96
∞	1.00

20

CONTROL
VALVES

One of the basic components of any control system is the final control element, which comes in a variety of forms depending on the specific control application. The most common type of final control element in chemical processing is the pneumatic control valve, which regulates the flow of fluids. Some other types include the variable speed pump and the power controller (used in electrical heating).

Since the pneumatic control valve is so widely used in chemical processing, this chapter will be devoted to the description, selection, and sizing of control valves.

CONTROL VALVE CONSTRUCTION

The control valve is essentially a variable resistance to the flow of a fluid, in which the resistance and therefore the flow, can be changed by a signal from a process controller.

As shown in Fig. 20.1, the control valve consists of an actuator and a valve. The valve itself is divided into the body and the trim. The body consists of a housing for mounting the actuator and connections for attachment of the valve to a supply line and a delivery line. The trim, which is enclosed within the body, consists of a plug, a valve seat, and a valve stem. The actuator moves the valve stem as the pressure on a spring-loaded diaphragm changes. The stem moves a plug in a valve seat in order to change the resistance to flow through the valve. When a valve is supplied by the manufacturer, the actuator and the valve are attached to each other to form one unit.

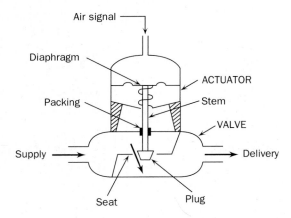

Air signal

Diaphragm

Packing

Supply

Seat

Plug

ACTUATOR

Stem

VALVE

Delivery

FIGURE 20-1
Pressure-to-close valve with single seating.

For most actuators, the motion of the stem is proportional to the pressure applied on the diaphragm. In general, this type of actuator can be used for functions other than moving a valve stem. For example, it can be used to adjust dampers, variable speed drives, rheostats, and other devices. As the pressure to the valve varies over its normal range of operation (3 to 15 psig) the range of motion of the stem varies from a fraction of an inch to several inches depending on the size of the actuator. Manufacturers provide a range of actuators for various valve sizes.

The valves available vary over a wide range of sizes. The size is usually referred to by the size of the end connectors. For example, a one-inch valve would have connectors (threaded or flanged) to fit into a one-inch pipe line. In general, the larger the valve size the larger the flow capacity of the valve.

For the control valve shown in Fig. 20.1, an increase in signal pressure above the diaphragm exerts a force on the diaphragm and back plate, which causes the stem to move down; this causes the cross-sectional area for flow between the plug and the seat to decrease, thereby reducing or throttling the flow. Such valve action as shown in Fig. 20.1 is called pressure-to-close action. The reverse action, pressure-to-open, can be accomplished by designing the actuator so that pressure is applied to the under side of the diaphragm, for which case an increase in pressure to the valve raises the stem. An alternate method to reverse the valve action is to leave the actuator as shown in Fig. 20.1 and to invert the plug on the stem and place it under the valve seat.

The valve shown in Fig. 20.1 is single-seated, meaning the valve contains one plug with one seating surface. For a single-seated valve, the plug must open against the full pressure drop across the valve. If the pressure drop is large, this means that a larger, more expensive actuator will be needed. To overcome this problem, valves are also constructed with double seating as shown in Fig. 20.2. In this type valve, two plugs are attached to the valve stem and each one has a seat. The flow pattern through the valve is designed so that the pressure drop across the seat at A tends to open the plug and the pressure drop across the seat at B tends to close the plug. This counterbalancing of forces on the plugs reduces

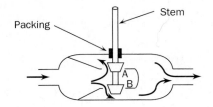

FIGURE 20-2
Double-seated valve.

the effort needed to open the valve with the result that a smaller, less expensive actuator is needed.

In a double-seated valve, it is difficult to have tight shut-off. If one plug has tight closure, there is usually a small gap between the other plug and its seat. For this reason, single-seated valves are recommended if the valve is required to be shut tight. In many processes, the valve is used for throttling flow and is never expected to operate near its shut-off position. For these conditions, the fact that the valve has a small leakage at shut-off position does not create a problem.

VALVE SIZING

In order to specify the size of a valve in terms of its capacity to provide flow when fully open, the following equation is used:

$$q = C_v \sqrt{\frac{\Delta p_v}{G}} \qquad \text{—incompressible fluid — fully open valve} \qquad (20.1)$$

where
q = flow rate, gpm
Δp_v = pressure drop across the wide-open valve, psi
G = specific gravity of fluid at stream temperature relative to water; for water $G = 1$.
C_v = factor associated with capacity of valve

Equation (20.1) applies to the flow of an incompressible fluid through a fully open valve. Manufacturers rate the size of a valve in terms of the factor C_v. Sometimes the C_v is defined as the flow (gpm) of a fluid of unit specific gravity through a fully open valve, across which a pressure drop of 1.0 lb_f/in^2 exists. This verbal definition is, of course, obtained directly from Eq. (20.1) by letting $q = 1$, $\Delta p_v = 1$, and $G = 1$. Equation (20.1) is based on the well-known Bernoulli equation for determining the pressure drop across valves and resistances. It is important to emphasize that C_v must be determined from Eq. (20.1) using the units listed. Since so many valves in use are rated in terms of C_v, Eq. (20.1) is of practical importance; however, some industries now are defining a valve coefficient K_v that is defined by the equation

$$q = K_v \sqrt{\frac{\Delta p_v}{G}}$$

where q = flow rate, m^3/hr
Δp_v = pressure drop across valve, Kg_f/cm^2
G = specific gravity relative to water

The relation between K_v and C_v is:

$$K_v = 0.856 C_v$$

For gases and steam, modified versions of Eq. (20.1) are used in which C_v is still used as a factor. Manufacturers of valves provide brochures, nomographs, and special slide rules for sizing valves for use with gases and steam.

In general, as the physical size of a valve body (i.e., size of pipe connectors) increases, the value of C_v increases. For a sliding stem and plug type of control valve, the value of C_v is roughly equal to the square of the pipe size multiplied by ten. Using this rule, a three-inch control valve should have a C_v of about 90, with units corresponding to those of Eq. (20.1).

Example 20.1. A valve with a C_v rating of 4.0 is used to throttle the flow of glycerine for which $G = 1.26$. Determine the maximum flow through the valve for a pressure drop of 100 psi.

$$q = 4.0 \sqrt{\frac{100}{1.26}} = 35.6 \text{ gpm}$$

The coefficient C_v varies with the design of the valve (shape, size, roughness) and the Reynolds number for the flow through the valve. This relationship is analogous to the relationship between friction factor and roughness and Reynolds number for flow through a pipe. For relatively nonviscous fluids, C_v in Eq. (20.1) can be taken as a constant for a valve of given size and type. The reason for this is that at high Reynolds numbers, the friction factor changes very little with flow rate. Except for very viscous fluids, the flow through a valve, which involves sudden contraction and expansion, is in the turbulent regime of fluid flow; turbulence in the valve exists even if the flow in the supply pipe is near the critical Reynolds number of 2100.

Consequently, for relatively nonviscous fluids, Eq. (20.1) is satisfactory for sizing a valve for any fluid. For the control of flow of very viscous fluids, such as tar or molasses, the value of C_v found from Eq. (20.1) must be multiplied by a correction factor that depends on viscosity, density, flow rate, and valve size (i.e., on the Reynolds number). Methods for determining the viscosity correction factor are provided by manufacturers for their valves. If one does not apply the correction factor for a very viscous fluid, the value of C_v will be too low and the valve will be undersized.

VALVE CHARACTERISTICS

The function of a control valve is to vary the flow of fluid through the valve by means of a change of pressure to the valve top. The relation between the flow through the valve and the valve stem position (or *lift*) is called the *valve characteristic*, which can be conveniently described by means of a graph as shown in Fig. 20.3 where three types of characteristics are illustrated.

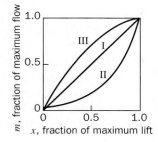

FIGURE 20-3
Inherent valve characteristics (pressure drop across valve is constant) I linear, II increasing sensitivity (e.g. equal percentage valve), III decreasing sensitivity.

In general, the flow through a control valve for a specific fluid at a given temperature can be expressed as:

$$q = f_1(L, p_0, p_1) \tag{20.2}$$

where q = volumetric flow rate

L = valve stem position (or lift)

p_0 = upstream pressure

p_1 = downstream pressure

The inherent valve characteristic is determined for fixed values of p_0 and p_1, for which case, Eq. (20.2) becomes

$$q = f_2(L) \tag{20.3}$$

For convenience let:

$$m = q/q_{max} \qquad \text{and} \qquad x = L/L_{max}$$

where q_{max} is the maximum flow when the valve stem is at its maximum lift
L_{max} (valve is full-open)

x is the fraction of maximum lift

m is the fraction of maximum flow.

Equation (20.3) may now be written

$$m = q/q_{max} = f(L/L_{max})$$

or

$$m = f(x) \tag{20.4}$$

The types of valve characteristics can be defined in terms of the sensitivity of the valve, which is simply the fractional change in flow to the fractional change in stem position for fixed upstream and downstream pressures; mathematically, sensitivity may be written

$$\text{sensitivity} = dm/dx$$

In terms of valve characteristics, valves can be divided into three types: decreasing sensitivity, linear, and increasing sensitivity. These types are shown in Fig. 20.3 where the fractional flow m is plotted against fractional lift x. For the decreasing sensitivity type, the sensitivity (or slope) decreases with m. For the linear type, the sensitivity is constant and the characteristic curve is a straight line. For the increasing sensitivity type, the sensitivity increases with flow.

Valve characteristic curves, such as the ones shown in Fig. 20.3, can be obtained experimentally for any valve by measuring the flow through the valve as a function of lift (or valve-top pressure) under conditions of constant upstream and downstream pressures. Two types of valves that are widely used are the linear valve and the logarithmic (or equal percentage) valve. The linear valve is one for which the sensitivity is constant and the relation between flow and lift is linear. The equal percentage valve is of the increasing sensitivity type.

It is useful to derive mathematical expressions for these types of valves. For the linear valve,

$$dm/dx = \alpha \tag{20.5}$$

where α is a constant.

Assuming that the valve is shut tight when the lift is at lowest position, we have that $m = 0$ at $x = 0$. For a single-seated valve that is not badly worn, the valve can be shut off for $x = 0$. Integrating Eq. (20.5) and introducing the limits $m = 0$ at $x = 0$ and $m = 1$ at $x = 1$ gives

$$\int_0^1 dm = \int_0^1 \alpha\,dx$$

Integrating this equation and inserting limits gives

$$\alpha = 1$$

Recall that the definitions of x and m require that $m = 1$ at $x = 1$. For $\alpha = 1$, Eq. (20.5) can now be integrated to give

$$m = x \qquad \text{(linear valve)} \tag{20.6}$$

For the equal percentage valve, the defining equation is

$$dm/dx = \beta m \tag{20.7}$$

where β is constant. Integration of this equation gives

$$\int_{m_0}^m \frac{dm}{m} = \int_0^x \beta\,dx \tag{20.8}$$

or

$$\ln \frac{m}{m_0} = \beta x \tag{20.9}$$

where m_0 is the flow at $x = 0$. Equation (20.9) shows that a plot of m versus x on semi-log paper gives a straight line. A convenient way to determine if a valve is of the equal percentage type is to plot the flow versus lift on semi-log paper. The relation expressed by Eq. (20.9) is the basis for calling the valve characteristic logarithmic. The basis for calling the valve characteristic equal percentage can be seen by rearranging Eq. (20.7) in the form

$$dm/m = \beta\,dx \qquad \text{or} \qquad \Delta m/m = \beta\,\Delta x$$

In this form it can be seen that an equal fractional (or percentage) change in flow ($\Delta m/m$) occurs for a specified increment of change in stem position (Δx), regardless of where the change in stem position occurs along the characteristic curve.

The term β can be expressed in terms of m_0 by inserting $m = 1$ at $x = 1$ into Eq. (20.9). The result is

$$\beta = \ln(1/m_0)$$

Solving Eq. (20.9) for m gives

$$m = m_0 e^{\beta x} \qquad \text{(equal percentage valve)} \qquad (20.10)$$

In integrating Eq. (20.7), the flow was assumed to be m_0 at $x = 0$. Mathematically this is necessary, because m_0 cannot be taken as zero at $x = 0$ because the term on the left side of Eq. (20.9) becomes infinite. In practice, there may be some leakage (hence $m_0 \neq 0$) when the stem is at its lowest position for a double-seated valve or for a valve in which the plug and seat have become worn.

For some valves, especially large ones, the valve manufacturer intentionally allows some leakage at minimum lift ($x = 0$) to prevent binding and wearing of the plug and seat surfaces. For a valve that does shut tight and is also classified as an equal percentage valve, the equal percentage characteristic will not be followed when the valve is nearly shut. In practice, the control valve serves as a throttling valve and is not intended to be wide-open or completely closed during normal operation.

In order to express the range over which an equal percentage valve will follow the equal percentage characteristic, the term *rangeability* is used. Rangeability is defined as the ratio of maximum flow to minimum controllable flow over which the valve characteristic is followed.

$$\text{Rangeability} = \frac{m_{\max}}{m_{\min, \text{ controllable}}}$$

For example, if m_0 is 0.02, the rangeability is 50. It is not uncommon for a control valve to have a rangeability as high as 50.

In practice, the ideal characteristics for linear and equal percentage valves are only approximated by commercially available valves. These discrepancies cause no difficulty because the inherent characteristics are changed considerably when the valve is installed in a line having resistance to flow, a situation that usually prevails in practice. In the next section, the effect of line loss on the effective valve characteristic will be discussed.

Effective Valve Characteristic

When a valve is placed in a line that offers resistance to flow, the inherent characteristic of the valve will be altered. The relation between flow and stem position (or valve-top pressure) for a valve installed in a process line will be called the *effective valve characteristic*.

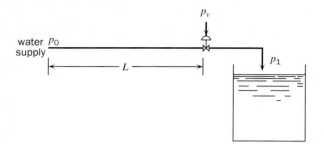

FIGURE 20-4
Control valve with supply line.

Consider a control valve having an inherent linear characteristic to be attached to the end of a pipeline that delivers water to an open tank. A diagram of the system is shown in Fig. 20.4. If the pipe is of large diameter relative to the size of the control valve, the pressure drop in the line will be negligible and the full pressure drop $p_0 - p_1$ will be across the valve as the lift varies between zero and one. In this case a plot of flow versus lift will give a linear relation as shown by Curve I of Fig. 20.5. This curve is for the flow of water at 5° C through a control valve for which $C_v = 4.0$ and the overall pressure drop, $p_0 - p_1$, is 100 psi. To show the effect of line loss, Curve II is constructed for the same conditions as Curve I, with the exception that 100 ft of 1.0 in. (inside diameter) pipe is used to supply the valve.

Example 20.2 will give the detailed calculations used to obtain the results in Fig. 20.5. For 100 ft of pipe, the plot of flow versus lift gives Curve II, shown in Fig. 20.5, in which the curve falls away or droops from the linear relation that holds for no line loss. Since line loss is proportional to the square of the velocity, the line loss is very small when the valve is nearly closed, for which case the total pressure drop is across the valve. For this reason, Curves I and II in Fig. 20.5 are close together at low rates. A rule often followed in industrial application of control valves is that the pressure drop across the wide-open valve should be greater that 25 percent of the pressure drop across the closed valve. A valve not selected according to this rule will lose its effectiveness to control at high flow rates.

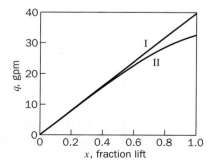

FIGURE 20-5
Effect of line loss on effective control valve characteristics from Example 20.2. I no pressure drop in supply line to valve, II pressure drop present in supply line to valve.

Example 20.2. Determine the flow versus lift relation for the linear control valve installed in the flow system of Fig. 20.4. The fluid is water at 5° C. The following data apply:

pipe length	100 ft
inside pipe diameter	1.0 in
density of water	62.4 lb/ft^3
viscosity of water	1.5 cp
C_v of control valve	4.0
total pressure drop, $p_0 - p_1$	100 psi

If there is no line loss as is the case for a large diameter line, the maximum flow can be calculated from Eq. (20.1):

$$q = C_v \sqrt{\frac{\Delta p_v}{G}} = 4.0 \sqrt{\frac{100}{1}} = 40.0 \text{ gpm}$$

To determine the flow/lift relation for the case of line loss, we arbitrarily start the calculation with a flow of 30 gpm. The pressure drop in the 100 ft pipe can be calculated from the well known expression from fluid mechanics:

$$\Delta p = \frac{32 f L \rho q^2}{144 \pi^2 g_c D^5} \tag{20.11}$$

where Δp = pressure loss in line, psi
   q = flow through pipe, ft^3/sec
   g_c = 32.174 (lb$_m$/lb$_f$)(ft/sec^2)
   L = pipe length, ft
   ρ = density of fluid, lb$_m$/ft^3
   D = inside pipe diameter, ft
   f = fanning friction factor, dimensionless

The fanning friction factor is a function of the Reynolds number and the pipe roughness. Equation (20.11) and a correlation for the fanning friction factor can be found in the literature (Perry and Chilton, 1973). We now calculate the Reynolds number (Re):

$$\text{Re} = D u \rho / \mu$$

Replacing the velocity u with $q/[(\pi/4)D^2]$ gives

$$\text{Re} = \frac{4q\rho}{\pi \mu D} \tag{20.12}$$

$$q = 30/(60)(7.48) = 0.0668 \text{ ft}^3/\text{sec or } 240.6 \text{ ft}^3/\text{hr}$$

$$\text{Re} = \frac{(4)(240.6)(62.4)}{(\pi)(1.50)(2.42)(1/12)} = 63,224$$

For this value of Reynolds number and for smooth pipe, the fanning friction factor f is 0.005. Equation (20.11) may now be used to calculate line loss:

$$\Delta p = \frac{(32)(0.005)(100)(62.4)(0.0668)^2}{(144)(\pi^2)(32.2)(1/12)^5} = 24.2 \text{ psi}$$

therefore Δp across valve $= 100 - 24.2 = 75.8$ psi

We next calculate the flow through the wide-open valve for a pressure drop of 75.8 psi:

$$q_{max} = C_v \sqrt{\frac{\Delta p_v}{G}} = 4.0 \sqrt{\frac{75.8}{1}} = 34.8 \text{ gpm}$$

Since the flow through the wide-open valve of 34.8 gpm at a pressure drop across the valve of 75.8 psi is greater than the selected value of 30 gpm, which was used to begin the calculation, we know the valve must be partially closed. Since the valve is linear, we calculate the lift x as follows:

$$x = 30/34.8 = 0.86$$

By means of similar calculations, several points on the effective characteristic curve of Fig. 20.5 can be found; the results are summarized in Table 20.1. The results shown in this table were used to obtain Curve II in Fig. 20.5.

Example 20.3. A control valve is to be installed in the flow system of Fig. 20.4. The valve is supplied by water at 5°C through 200 ft of pipe having an inside diameter of 1.0 in. The total pressure drop, $p_0 - p_1$, is 100 psi. When the valve is wide-open, the flow is to be 30 gpm. Determine C_v for the valve. Plot the effective characteristic curve for the valve as flow versus lift. Do this problem for a linear valve and for an equal percentage valve. The equal percentage valve has an m_0 of 0.03.

Linear Valve. To obtain the pressure drop in the line, use is made of Eqs. (20.11) and (20.12) as was done in Example 20.2. From Eq. (20.12), we obtain the Reynolds number as follows:

$$q = 30/(60)(7.48) = 0.0668 \text{ ft}^3/\text{sec or } 240.6 \text{ ft}^3/\text{hr}$$

$$Re = \frac{4q\rho}{\pi\mu D} = \frac{(4)(240.6)(62.4)}{\pi(1.5)(2.42)(1/12)} = 63,200$$

TABLE 20.1
Effective characteristic for a linear valve with supply line loss (Example 20.2).

q, gpm	x, fraction lift	Δp in line, psi
0	0	0
20	0.53	10.8
30	0.86	24.2
33	1.0	30.0

From a correlation for the fanning friction factor, we obtain $f = 0.005$. From Eq. (20.11), the line loss is calculated to be:

$$\Delta p = \frac{(32)(0.005)(200)(62.4)(0.0668)^2}{(144)(\pi^2)(32.2)(1/12)^5} = 48.5 \, \text{psi}$$

$$\Delta p_v = 100 - 48.5 = 51.5 \, \text{psi}$$

From knowledge of the maximum flow through the wide-open valve (30 gpm) and Δp_v, we calculate C_v from Eq. (20.1) as follows:

$$C_v = \frac{q}{\sqrt{\Delta p_v / G}} = \frac{30}{\sqrt{51.5}} = 4.18$$

From C_v, one can now calculate the stem position x needed for various flow rates m.

For $q = 20$ gpm, one obtains from Eq. (20.11)

$$\Delta p = 21.6 \text{ psi}$$

and

$$\Delta p_v = 100 - 21.6 = 78.4 \text{ psi}$$

For a wide-open valve ($x = 1$), across which the pressure drop is 78.4 psi, we obtain

$$q_{max} = C_v \sqrt{\frac{\Delta p_v}{G}} = 4.18 \sqrt{\frac{78.4}{1}} = 37.0 \text{ gpm}$$

The fraction of lift needed to reduce the flow to 20 gpm is

$$x = 20/37 = 0.54$$

For other flow rates, one can repeat this calculation to obtain values of x. The results are shown in Table 20.2 and in Fig. 20.6. The latter also shows the inherent characteristic of the linear valve for comparison with the effective characteristic of the valve when line loss is present.

TABLE 20.2
Effective characteristics for a linear valve and an equal percentage valve (Example 20.3).

Maximum flow of water at 5° C: 30 gpm
Pressure drop across flow system: 100 psi
Pipe length: 200 ft, Inside pipe diameter: 1.0 in.
$C_v = 4.18$, $m_0 = 0.03$

q gpm	Δp_v psi	x linear	x equal percentage
0	100	0	*
10	94.6	0.25	0.60
20	78.5	0.54	0.82
30	51.4	1.00	1.00

* For the equal percentage valve, $m_0 = 0.03$ when $x = 0$.

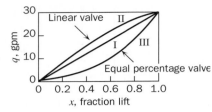

FIGURE 20-6
Comparison of effective valve characteristics of a linear valve and an equal percentage valve from Example 20.3. I ideal linear characteristic, II linear valve, III equal percentage valve.

Equal percentage valve. Calculation of the effective characteristic will now be made for an equal percentage valve having the same C_v of 4.18 as calculated for the linear valve in the first part of this example.

For $m_0 = 0.03$, the value of β is calculated to be

$$\beta = \ln(1/m_0) = \ln(1/0.03) = 3.51$$

For a flow rate of 20 gpm,

$$m = q/q_{max} = 20/37 = 0.54$$

Solving Eq. (20.9) for x and inserting the values for β, m_0, and m give

$$x = (1/3.51)\ln(0.54/0.03) = 0.82$$

For other values of flow, corresponding values of x are calculated and the results are shown in Table 20.2 and Fig. 20.6.

BENEFIT OF AN EQUAL PERCENTAGE VALVE. It is often stated in the control literature that the benefit derived from an equal percentage valve arises from its inherent nonlinear characteristic that compensates for the line loss to give an effective valve characteristic that is nearly linear. A study of Fig. 20.6 shows that in this example an equal percentage valve overcompensates for line loss and produces an effective characteristic that is not linear, but is bowed in the opposite direction to that of the effective characteristic of the linear valve. In summary, neither valve in this example produces an effective characteristic that is linear. One can show that as the line loss increases, the linear valve will depart more from the ideal linear relation and the equal percentage valve will move more closely toward the linear relation.

In practice, a valve designated as linear will not give a linear characterisitic exactly as defined in this chapter. To achieve a truly linear characteristic would require very careful design and precision machining of the valve plug and seat. The same comment can be made for an equal percentage valve, as defined by Eq. (20.10). In order to know the effective characteristic of a valve, one must test it experimentally.

VALVE POSITIONER

The friction in the packing and guiding surfaces of a control valve causes a control valve to exhibit hysteresis as shown in Fig. 20.7, in which stem position is

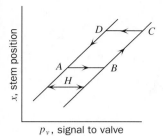

FIGURE 20-7
Control valve hysteresis.

plotted against valve-top pressure. When the pressure increases, the stem position increases along the lower curve. When the pressure decreases, the stem position decreases along the upper curve. At the moment the air pressure signal reverses, the stem position stays in the last position until the dead band H is exceeded, after which the pressure begins to decrease or increase along the paths shown by the arrows. If the valve is subjected to a slow periodic variation in pressure, a typical path taken by the stem position is shown by the closed curve ABCDA in Fig. 20.7.

The hysteresis described in the previous paragraph should be distinguished from the dynamic lag of a valve discussed in Chap. 10. The dynamic lag discussed in Chap. 10 is caused by the volume of space above the valve diaphragm, the resistance to flow of air to the valve top, and the inertia of the valve stem and plug; such a lag is expressed by a first-order or second-order transfer function. On the other hand, hysteresis, which is caused by the friction between the stem and the packing, is a nonlinear phenomenon and cannot be expressed by a transfer function. A valve can exhibit both dynamic lag and hysteresis.

The presence of hysteresis in the valve can cause the controlled signal to exhibit an oscillation or ripple called a *limit cycle*. Since this limit cycle is usually considered objectionable and contributes to wear of the valve, a method is needed to eliminate it. Since the limit cycle is a nonlinear phenomenon related to the hysteresis, controller tuning is not a solution to the problem.

To reduce the deleterious effect of hysteresis and to also speed up the response of the valve, one can attach to the control valve a *positioner* which acts as a high-gain proportional controller that receives a set-point signal from the primary controller and a measurement from the valve stem position. In this sense, the addition of a valve positioner introduces a form of cascade control, which was discussed in a previous chapter. A sketch of a control valve with a positioner attached is shown in Fig. 20.8. The positioner, bolted to the valve actuator, has an arm that is clamped to the valve stem to detect the stem position.

Notice that the valve positioner shown in Fig. 20.8, has the usual connections for a controller: a set point that calls for a desired stem position in the form of a signal from the primary controller p_c, a measurement in the form of stem position x, and a pneumatic output in the form of a pressure to the valve top p_v. The mechanical details of an actual valve positioner involve a pneumatic mechanism functioning as a high-gain proportional controller. The gain is built

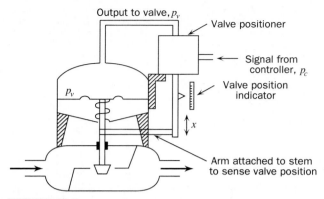

FIGURE 20-8
Control valve with positioner (Compare with Fig. 20.1).

into the design of the positioner and cannot be adjusted. The valve positioner is especially important for speeding up the valve motion, and eliminating hysteresis and valve stem friction.

SUMMARY

The control valve is a component of a control system often overlooked in a course on process control. In this chapter, the description, selection, and sizing of pneumatic control valves were presented. Valves may be of the pressure-to-close or the pressure-to-open type; the selection of the type is often related to safety considerations. If the air pressure fails, the valve should return to a position which ensures safe operating conditions for a process.

The flow capacity of a valve is based on an equation relating flow to the square root of the pressure drop across the valve; the proportionality constant C_v in this equation is a measure of the valve's capacity for flow. The larger C_v, then the larger the flow.

Valves are classified according to their inherent flow characteristics such as linear or equal percentage. A linear valve produces a flow (for constant pressure drop across the valve) that is proportional to the valve stem position, which in turn is proportional to the valve-top pressure.

The presence of a long, small-diameter line supplying a valve causes the pressure drop across the valve to decrease with the increase of flow, for a fixed, overall pressure drop across the system. If the pressure drop in the line is excessive, the characteristic of the linear valve will become nonlinear and in terms of control theory, the steady-state gain K_v of the valve decreases with flow. As a result of the change in valve gain, the controller in the loop must be readjusted for different flow rates in order to maintain the same degree of stability. To overcome this limitation of the linear valve, an equal percentage (or logarithmic) valve is available for which the gain of the valve increases with flow rate. Such a valve compensates for the line loss and produces an effective charateristic that approaches a linear relation. The basis for the name equal percentage (or logarithmic) is related to one form of the mathematical expression that describes the valve. In this form, an equal percentage change in flow occurs for a specified change in stem position, regardless of the stem position.

In order to eliminate hysteresis, which can produce cycling and cause wear of the valve plug and seat, a valve positioner may be attached to a control valve. The positioner also speeds up the motion of the valve in response to a signal from the controller.

PROBLEMS

20.1. A linear control valve having a C_v of 0.1 is connected to a source of water. If the pressure drop across the valve is 400 psi and if the pneumatic pressure to the valve top is 12 psig, what is the flow rate through the valve? The valve goes from completely shut to completely open as the valve-top pressure varies from 3 to 15 psig.

20.2. (*a*) Under what conditions would an equal percentage valve be used instead of a linear valve?

(*b*) What are some reasons to use a valve positioner?

CHAPTER
21

THEORETICAL ANALYSIS OF COMPLEX PROCESSES

In order to investigate theoretically the control of a process, it is necessary first to know the dynamic character of the process that is being controlled. In the previous chapters, the processes have been very simple for the purpose of illustrating control theory. Many physical processes are extremely complicated, and it requires considerable effort to construct a mathematical model that will adequately simulate the dynamics of the actual system. In this chapter, we shall analyze several complex systems to indicate some of the types of problems that can be encountered. In these examples, the technique of linearization, first presented in Chap. 6, will be applied to a function of several variables. One example will lead to a multiloop control system. In the last section, distributed-parameter systems will be discussed.

CONTROL OF A STEAM-JACKETED KETTLE

The dynamic response and control of the steam-jacketed kettle shown in Fig. 21.1 are to be considered. The system consists of a kettle through which water flows at a variable rate w lb/time. The entering water is at temperature T_i, which may vary with time. The kettle water, which is well agitated, is heated by steam condensing in the jacket at temperature T_v and pressure p_v. The temperature of the water in the kettle is measured and transmitted to the controller. The output signal from the controller is used to change the stem position of the valve, which adjusts the flow of steam to the jacket. The major problem in this example is to determine

318

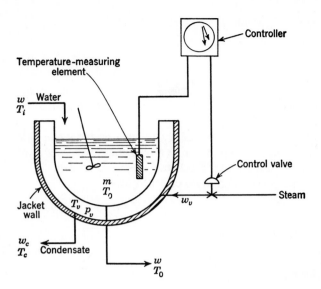

Temperature-measuring element

w T_i Water

Controller

Control valve

Steam

m T_0

Jacket wall

T_v P_v

w_v

w_c T_c Condensate

w T_0

FIGURE 21-1
Control of a steam-jacketed kettle.

the dynamic characteristics of the kettle. The kettle is actually a nonlinear system, and in order to obtain a linear model a number of simplifying assumptions are needed.

Analysis of Kettle

The following assumptions are made for the kettle:

1. The heat loss to the atmosphere is negligible.
2. The holdup volume of water in the kettle is constant.
3. The thermal capacity of the kettle wall, which separates steam from water, is negligible compared with that of the water in the kettle.
4. The thermal capacity of the outer jacket wall, adjacent to the surroundings, is finite, and the temperature of this jacket wall is uniform and equal to the steam temperature at any instant.
5. The kettle water is sufficiently agitated to result in a uniform temperature.
6. The flow of heat from the steam to the water in the kettle is described by the expression

$$q = U(T_v - T_o)$$

where q = flow rate of heat, Btu/(hr)(ft^2)
   U = overall heat-transfer coefficient, Btu/(hr)(ft^2)(°F)
   T_v = steam temperature, °F
   T_o = water temperature, °F

The overall heat-transfer coefficient U is constant.

7. The heat capacities of water and the metal wall are constant.

8. The density of water is constant.

9. The steam in the jacket is saturated.

The assumptions listed here are more or less arbitrary. For a specific kettle operating under a particular set of conditions, some of these assumptions may require modification.

     The approach to this problem is to make an energy balance on the water side and another energy balance on the steam side. In order to aid the development of the transfer functions, a schematic diagram of the kettle is shown in Fig. 21.2. The symbols used throughout this analysis are defined as follows:

T_i = temperature of inlet water, °F

T_o = temperature of outlet water, °F

T_v = temperature of jacket steam, °F

T_c = temperature of condensate, °F

w = flow rate of inlet water, lb/time

w_v = flow rate of steam, lb/time

w_c = flow rate of condensate from kettle, lb/time

m = mass of water in kettle, lb

m_1 = mass of jacket wall, lb

V = volume of jacket steam space, ft^3

C = heat capacity of water, Btu/(lb)(°F)

C_1 = heat capacity of metal in jacket wall, Btu/(lb)(°F)

A = cross-sectional area for heat exchange, ft^2

t = time

H_v = specific enthalpy of steam entering, Btu/lb

H_c = specific enthalpy of condensate leaving, Btu/lb

U_v = specific internal energy of steam in jacket, Btu/lb

ρ_v = density of steam in jacket, lb/ft^3

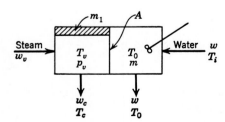

FIGURE 21-2
Schematic diagram of kettle.

An energy balance on the water side gives

$$wC(T_i - T_o) + UA(T_v - T_o) = mC\frac{dT_o}{dt} \tag{21.1}$$

In Eq. (21.1), the terms C, U, A, and m are constants. The first term in Eq. (21.1) is nonlinear, since it contains the product of flow rate and temperature, that is, wT_i and wT_o. In order to obtain a transfer function from Eq. (21.1), these nonlinear terms must be linearized. Before continuing the analysis, we shall digress briefly to discuss the general problem of linearization of a function of several variables.

Consider a function of two variables, $z(x,y)$. By means of a Taylor series expansion, the function can be expanded* around an operating point x_s,y_s as follows:

$$z = z(x_s,y_s) + \frac{\partial z}{\partial x}\Big|_{x_s,y_s}(x - x_s) + \frac{\partial z}{\partial y}\Big|_{x_s,y_s}(y - y_s)$$
$$+ \text{higher-order terms in } (x - x_s) \text{ and } (y - y_s) \tag{21.2}$$

The subscript s stands for steady state.

In control problems, the operating point (x_s,y_s), around which the expansion is to be made, is selected at steady-state values of the variables before any disturbance occurs. Linearization of the function z consists of retaining only the linear terms, on the basis that the deviations $(x - x_s)$, etc., will be small. Thus,

$$z \cong z_s + z_{x_s}(x - x_s) + z_{y_s}(y - y_s) \tag{21.3}$$

where z_{x_s} and z_{y_s} are the partial derivatives in Eq. (21.2). If z is a function of three or more variables, the linearized form is the same as that of Eq. (21.3) with an additional term for each variable.

The linearization expressed by Eq. (21.3) may be applied to the terms wT_i and wT_o in Eq. (21.1) to obtain

$$wT_i = w_sT_{i_s} + w_s(T_i - T_{i_s}) + T_{i_s}(w - w_s) \tag{21.4}$$

and

$$wT_o = w_sT_{o_s} + w_s(T_o - T_{o_s}) + T_{o_s}(w - w_s) \tag{21.5}$$

Notice that for these cases the nonlinear terms are wT_i and wT_o. The first partial derivatives, evaluated at the operating point, are

$$\frac{\partial(wT_i)}{\partial w}\Big|_{w_s,T_{i_s}} = T_{i_s}$$

$$\frac{\partial(wT_i)}{\partial T_i}\Big|_{w_s,T_{i_s}} = w_s$$

and so on.

---

*The reader may refer to I. S. Sokolnikoff and R. M. Redheffer (1966) for further discussion of this expansion.

Introducing Eq. (21.4) and (21.5) into (21.1) gives the following linearized equation:

$$[(T_{i_s} - T_{o_s})(w - w_s) + w_s(T_i - T_o)]C + UA(T_v - T_o) = mC\frac{dT_o}{dt} \quad (21.6)$$

At steady state, $dT_o/dt = 0$, and Eq. (21.1) can be written

$$w_sC(T_{i_s} - T_{o_s}) + UA(T_{v_s} - T_{o_s}) = 0 \quad (21.7)$$

Subtracting Eq. (21.7) from (21.6) and introducing the deviation variables

$$T_i' = T_i - T_{i_s}$$
$$T_o' = T_o - T_{o_s}$$
$$T_v' = T_v - T_{v_s}$$
$$W = w - w_s$$

and rearranging give the result

$$C[(T_{i_s} - T_{o_s})W + w_s(T_i' - T_o')] + UA(T_v' - T_o') = mC\frac{dT_o'}{dt} \quad (21.8)$$

Taking the transform of Eq. (21.8) and solving for $T_o'(s)$ give

$$T_o'(s) = \frac{K_1}{\tau_w s + 1}T_i'(s) + \frac{K_2}{\tau_w s + 1}T_v'(s) - \frac{K_3}{\tau_w s + 1}W(s) \quad (21.9)$$

where $K_1 = \dfrac{w_s C}{UA + w_s C}$

$$K_2 = \frac{UA}{UA + w_s C}$$

$$K_3 = \frac{C(T_{o_s} - T_{i_s})}{UA + w_s C}$$

$$\tau_w = \frac{mC}{UA + w_s C}$$

From Eq. (21.9), we see that the response of T_o' to T_i', T_v', or W is first-order with a time constant τ_w. The steady-state gains (Ks) in Eq. (21.9) are all positive.

The following energy balance can be written for the steam side of the kettle:

$$w_v H_v - w_c H_c = UA(T_v - T_o) + \frac{Vd(\rho_v U_v)}{dt} + m_1 C_1 \frac{dT_v}{dt} \quad (21.10)$$

Notice that we have made use of assumption 4 in writing the last term of Eq. (21.10), which implies that the metal in the outer jacket wall is always at the steam temperature.

A mass balance on the steam side of the kettle yields

$$w_v - w_c = V\frac{d\rho_v}{dt} \quad (21.11)$$

Combining Eqs. (21.10) and (21.11) to eliminate w_c gives

$$w_v(H_v - H_c) = (U_v - H_c)V\frac{d\rho_v}{dt} + m_1C_1\frac{dT_v}{dt} + UA(T_v - T_o)$$
$$+ V\rho_v\frac{dU_v}{dt} \tag{21.12}$$

The variables ρ_v, U_v, H_v, and H_c are functions of the steam and condensate temperatures and can be approximated by expansion in Taylor series and linearization as follows:

$$\begin{aligned}
\rho_v &= \rho_{v_s} + \alpha(T_v - T_{v_s})\\
U_v &= U_{v_s} + \phi(T_v - T_{v_s})\\
H_v &= H_{v_s} + \gamma(T_v - T_{v_s})\\
H_c &= H_{c_s} + \sigma(T_c - T_{c_s})
\end{aligned} \tag{21.13}$$

where $\alpha = \dfrac{d\rho_v}{dT_v}\Big|_s$

$\phi = \dfrac{dU_v}{dT_v}\Big|_s$

$\gamma = \dfrac{dH_v}{dT_v}\Big|_s$

$\sigma = \dfrac{dH_c}{dT_c}\Big|_s$

The parameters α, ϕ, γ, and σ in these relationships can be obtained from the steam tables once the operating point is selected.*

Introducing the relationships of Eq. (21.13) into Eq. (21.12) and assuming the condensate temperature T_c to be the same as the steam temperature T_v give the following result:

---

*For example, if the operating point is at $212°$ F and the deviation in steam temperature is $10°$ F, we obtain the following estimate of γ from the steam tables:

$T_{v_s} = 212°$ F

$H_{v_s} = 1150.4$ Btu/lb

At $T_v = 222°$ F,

$H_v = 1154.1$

At $T_v = 202°$ F,

$H_v = 1146.6$

$\gamma \approx \dfrac{1154.1 - 1146.6}{222 - 202} = 0.375$

and

$H_v = 1150.4 + 0.375(T_v - 212)$

In a similar manner, the properties of saturated steam can be used to evaluate α, ϕ, and σ.

$$[H_{v_s} - H_{c_s} + (\gamma - \sigma)(T_v - T_{v_s})]w_v = \left[(U_{v_s} - H_{c_s})\right.$$

$$\left. + (2\phi - \sigma)(T_v - T_{v_s}) + \frac{\phi}{\alpha}\rho_{v_s} + \frac{m_1 C_1}{\alpha V}\right]\alpha V \frac{dT_v}{dt} + UA(T_v - T_o) \quad (21.14)$$

Some of the terms in Eq. (21.14) can be neglected. The term

$$(\gamma - \sigma)(T_v - T_{v_s})$$

can be dropped because it is negligible compared with $(H_{v_s} - H_{c_s})$. For example, for steam at atmospheric pressure, a change of 10°F gives a value of $(\gamma - \sigma)(T_v - T_{v_s})$ of about 7 Btu/lb while $(H_{v_s} - H_{c_s})$ is 970 Btu/lb. Similarly, the term $(2\phi - \sigma)(T_v - T_{v_s})$ can be neglected. For example, this term is about -4 Btu/lb for a change in steam temperature of 10°F for steam at about 1 atm pressure; the term $(U_{v_s} - H_{c_s})$ is 897 Btu/lb under these conditions. Also, the term $\phi\rho_{v_s}/\alpha$ is about 15 Btu/lb and can be neglected. Discarding these terms, writing the remaining terms in deviation variables, and transforming yield

$$T_v'(s) = \frac{1}{\tau_v s + 1}T_o'(s) + \frac{K_5}{\tau_v s + 1}W_v(s) \quad (21.15)$$

where $T_v' = T_v - T_{v_s}$

$W_v = w_v - w_{v_s}$

$K_5 = \dfrac{H_{v_s} - H_{c_s}}{UA}$

$\tau_v = \dfrac{(U_{v_s} - H_{c_s})\alpha V + m_1 C_1}{UA}$

From Eq. (21.15), we see that the steam temperature T_v' depends on the steam flow rate W_v and the water temperature T_o'. The combination of Eqs. (21.9) and (21.15) give the dynamic response of the water temperature to changes in water flow rate, inlet water temperature, and steam flow rate. These equations are represented by a portion of the block diagram of Fig. 21.4. Before completing the analysis of the control system, we must consider the effect of valve-stem position on the steam flow rate.

Analysis of Valve

The flow of steam through the valve depends on three variables: steam supply pressure, steam pressure in the jacket, and the valve-stem position, which we shall assume to be proportional to the pneumatic valve-top pressure p. For simplicity, assume the steam supply pressure to be constant with the result that the steam flow rate is a function of only the two remaining variables; thus

$$w_v = f(p, p_v) \quad (21.16)$$

Because of the assumption that the steam in the jacket is always saturated, we know that p_v is a function of T_v; thus

$$p_v = g(T_v) \quad (21.17)$$

This functional relation can be obtained from the saturated steam tables. Equations (21.16) and (21.17) can be combined to give

$$w_v = f[p,g(T_v)] = f_1(p, T_v)$$

The function $f_1(p, T_v)$ is in general nonlinear, and if an analytic expression* is available, the function can be linearized as described previously. In this example, we shall assume that an analytic expression is not available. The linearized form of $f_1(p, T_v)$ can be obtained by making some experimental tests on the valve. If the valve-top pressure is fixed at its steady-state (or average) value and w_v is measured for several values of T_v (or p_v), a curve such as the one shown in Fig. 21.3a can be obtained. If the steam temperature T_v (or p_v) is held constant and the flow rate is measured at several values of valve-top pressure, a curve such as that shown in Fig. 21.3b can be obtained. These two curves can now be used to evaluate the partial derivatives in the linear expansion of $f_1(p,T_v)$ as we shall now demonstrate.

Expanding w_v about the operating point p_s,T_{v_s} and retaining only the linear terms give

$$w_v = w_{v_s} + \frac{\partial w_v}{\partial p}\bigg|_{p_s,T_{v_s}} (p - p_s) + \frac{\partial w_v}{\partial T_v}\bigg|_{p_s,T_{v_s}} (T_v - T_{v_s})$$

This equation can be written in the form

$$W_v = K_v P - \frac{1}{R_v}T_v' \tag{21.20}$$

where $W_v = w_v - w_{v_s}$

$P = p - p_s$

$T_v' = T_v - T_{v_s}$

$K_v = \dfrac{\partial w_v}{\partial p}\bigg|_{p_s,T_{v_s}}$

$\dfrac{1}{R_v} = -\dfrac{\partial w_v}{\partial T_v}\bigg|_{p_s,T_{v_s}}$

---

*The flow of steam through a control valve can often be represented by the relationship

$$w_v = A_0 C_v \sqrt{p_s - p_v} \tag{21.18}$$

where p_s = supply pressure of steam

p_v = pressure downstream of valve

A_0 = cross-sectional area for flow of steam through valve

C_v = constant of the valve

For a linear valve, A_0 is proportional to stem position and the stem position is proportional to valve-top pressure p; under these conditions, Eq. (21.18) takes the form

$$w_v = C_v' p \sqrt{p_s - p_v} \tag{21.19}$$

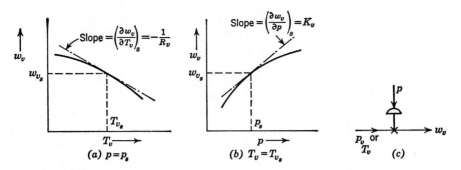

FIGURE 21-3
Linearization of valve characteristics from experimental tests.

The coefficients K_v and $-1/R_v$ in Eq. (21.20) are the slopes of the curves of Fig. 21.3 at the operating point p_s, T_{v_s}. This follows from the definition of a partial derivative. Notice that $1/R_v$ has been defined as the negative of the slope so that R_v is a positive quantity. The experimental approach described here for obtaining a linear form for the flow characteristics of a valve is always possible in principle. However, it must be emphasized that the linear form is useful only for small deviations from the operating point. If the operating point is changed considerably, the coefficients K_v and $1/R_v$ must be reevaluated. Notice that, in writing Eq. (21.20), we have assumed the valve to have no dynamic lag between p and stem position. This assumption is valid for a system having large time constants, such as a steam-jacketed kettle, as was demonstrated in Chap. 10.

Block Diagram of Control System

We have now completed the analysis of the kettle and valve. A block diagram of the control system, based on Eqs. (21.9), (21.15), and (21.20) is shown in Fig. 21.4.

The controller action is not specified but merely denoted by G_c in the block diagram. Also, the feedback element is denoted as H. From Fig. 21.4, we see that the steam-jacketed kettle is a multiloop control system. Furthermore, the loops overlap. The block diagram can be used to obtain the overall transfer function between any two variables by applying the methods of Chap. 12. After considerable algebraic manipulation, the following result is obtained:

$$T_o' = \frac{G_c G_2 G_5 K_v}{D(s)} R + \frac{G_1(1 + G_5/R_v)}{D(s)} T_i' - \frac{G_3(1 + G_5/R_v)}{D(s)} W \qquad (21.21)$$

where $D(s) = 1 + G_5/R_v + G_c G_2 G_5 K_v H - G_2 G_4$. The terms G_1, G_2, G_3, G_4, G_5, G_c, and H are defined in Fig. 21.4. For example, if $G_c = K_c$ and $H = 1$, one obtains from Eq. (21.21) the transfer function

$$\frac{T_o'}{R} = \frac{K}{\tau^2 s^2 + 2\zeta \tau s + 1} \qquad (21.22)$$

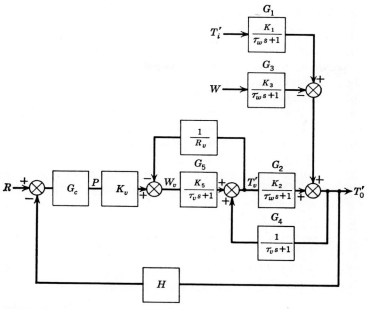

FIGURE 21-4
Block diagram for control of steam-jacketed kettle.

where $K = \dfrac{K_c K_v K_2 K_5}{D_1}$

$\tau^2 = \dfrac{\tau_v \tau_w}{D_1}$

$2\zeta\tau = \dfrac{\tau_v + \tau_w + K_5\tau_w/R_v}{D_1}$

$D_1 = 1 + \dfrac{K_5}{R_v} + K_c K_v K_2 K_5 - K_2$

It is seen that the response of the control system is second-order when proportional control is used and the measuring element does not have dynamic lag. Notice that the parameters K, τ^2, and $2\zeta\tau$ in Eq. (21.22) are positive. This follows from the fact that the parameters K_c, K_v, K_2, K_5, R_v, τ_v, and τ_w are all positive and that $K_2 < 1$. When a block diagram of a control system becomes very complicated, such as the one in this example, it is convenient to simulate the control system with a computer. When computer simulation is selected as the means of studying the transient response of the control system, the block diagram can be translated directly into a computer program. This computer-simulation technique will be covered in detail in Chap. 34.

DYNAMIC RESPONSE OF A GAS ABSORBER

Another example of a complex system is the plate absorber* shown in Fig. 21.5. In this process, air containing a soluble gas such as ammonia is contacted with fresh water in a two-plate column in order to remove part of the ammonia from the gas. The action of gas bubbling through the liquid causes thorough mixing of the two phases on each plate. During the mixing process, ammonia diffuses from the bubbles into the liquid. In an industrial operation, many plates may be used; however, for simplicity, we consider only two plates in this example, since the basic principles are unaffected by the number of plates.

Our problem is to analyze the system for its dynamic response. In other words, we want to know how the concentrations of liquid and gas change as a result of change in inlet composition or flow rate.

Throughout the analysis, the following symbols are used:

L_n = flow of liquid leaving nth plate, moles/min

V_n = flow of gas leaving nth plate, moles/min

x_n = concentration of liquid leaving nth plate, mole fraction NH_3

y_n = concentration of gas leaving nth plate, mole fraction NH_3

H_n = holdup (or storage) of liquid on nth plate, moles

---

*The reader who has not studied gas absorption may find this subject presented in any textbook on chemical engineering unit operations. For example, see Bennett and Myers (1982).

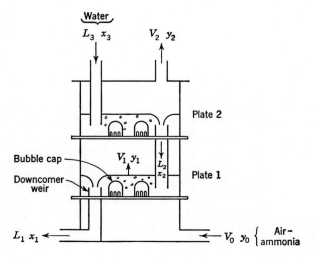

FIGURE 21-5
Bubble-cap gas absorber.

In order to avoid too many complicating details, the following assumptions will be used:

1. The temperature and total pressure throughout the column are uniform and do not vary with changes in flow rates.
2. The entering gas stream is dilute (say 5 mole percent NH_3) with the consequence that we can neglect the decrease in total molar flow rate of gas as ammonia is removed. Likewise, we can assume that the molar flow rate of liquid does not increase as ammonia is added.
3. The plate efficiency is 100 percent,[†] which means that the vapor and liquid streams leaving a plate are in equilibrium. Such a plate is called an *ideal* equilibrium stage.
4. The equilibrium relationship is linear and is given by the expression

$$y_n = mx_n^* + b \tag{21.23}$$

where m and b are constants that depend on the temperature and total pressure of the system, and x_n^* is the concentration of liquid in equilibrium with gas of concentration y_n. For an ideal plate

$$x_n = x_n^*$$

5. The holdup of liquid H_n on each plate is constant and independent of flow rate. Furthermore, the holdup is the same for each plate, that is, $H_1 = H_2 = H$.
6. The holdup of gas between plates is negligible. As a consequence of this assumption and assumption 2, the flow rate of gas from each plate is the same and equal to the entering gas flow rate; that is,

$$V_0 = V_1 = V_2 = V$$

In this list of assumptions, the one which is most likely to be invalid for a practical process is that the plate is an ideal equilibrium stage.

[†] If the efficiency of the plate is not 100 percent, we can introduce an individual tray efficiency of the Murphree type, defined as

$$E_n = \frac{x_n - x_{n+1}}{x_n^* - x_{n+1}}$$

where x_n^* is the concentration of the liquid in equilibrium with gas of composition y_n. Notice that for an ideal plate $E_n = 1$ and $x_n = x_n^*$. In general the efficiency of a plate depends on the design of the plate, the properties of the gas and liquid streams, and the flow rates. We could include efficiency in our mathematical model; however, to do so would greatly increase the complexity of the problem. To account properly for the variation in efficiency with flow rates would require empirical relationships for a specific plate design.

Analysis

We begin the analysis of this process by writing an ammonia balance around each plate. A mass balance on ammonia around plate 1 gives

$$H\frac{dx_1}{dt} = L_2x_2 + Vy_0 - L_1x_1 - Vy_1 \tag{21.24}$$

This last equation states that the accumulation of NH_3 on plate 1 is equal to the flow of NH_3 into the plate minus the flow of NH_3 out of the plate. Notice that V and H do not have subscripts because of assumptions 5 and 6.

A mass balance on ammonia around plate 2 gives

$$H\frac{dx_2}{dt} = Vy_1 - L_2x_2 - Vy_2 \tag{21.25}$$

The last equation does not contain a term L_3x_3, since we have assumed that $x_3 = 0$.

For an ideal plate $x_n = x_n^*$, and the equilibrium relation of Eq. (21.23) becomes

$$y_n = mx_n + b$$

Substituting the equilibrium relationship into (21.24) and (21.25) gives

$$H\frac{dx_1}{dt} = L_2x_2 - L_1x_1 + Vm(x_0 - x_1)$$

and

$$H\frac{dx_2}{dt} = Vm(x_1 - x_2) - L_2x_2$$

where $x_0 = (y_0 - b)/m$ is the composition of liquid that would be in equilibrium with the entering gas of composition y_0. Solving these last two equations for the derivatives gives

$$\frac{dx_1}{dt} = \frac{1}{H}(L_2x_2 - L_1x_1) + \frac{Vm}{H}(x_0 - x_1) \tag{21.26}$$

$$\frac{dx_2}{dt} = \frac{Vm}{H}(x_1 - x_2) - \frac{1}{H}L_2x_2 \tag{21.27}$$

Thus far the analysis has resulted in two nonlinear first-order differential equations. The nonlinear terms in Eqs. (21.26) and (21.27) are L_2x_2 and L_1x_1. The forcing functions in this process, which must be specified as functions of t, are the inlet gas concentration $[x_0 = (y_0 - b)/m]$ and the inlet liquid flow rate L_3. In order to solve for $x_1(t)$ and $x_2(t)$, we must have two more equations, obtained by considering the liquid-flow dynamics on each plate. Assume that each plate can be considered as a first-order system for which the following equations hold:*

$$\tau_2 \frac{dL_2}{dt} = L_3 - L_2 \tag{21.28}$$

and

$$\tau_1 \frac{dL_1}{dt} = L_2 - L_1 \tag{21.29}$$

The time constants in these equations (τ_1 and τ_2) can be determined experimentally by the methods of Chap. 19. The first-order representation for liquid dynamics was found to be adequate by Nobbe (1961). We now have four differential equations [Eqs. (21.26) to (21.29)], and six variables (x_1, x_2, x_0, L_1, L_2, L_3). Since x_0 and L_3 are the forcing functions, which are specified functions of time, these four equations can be solved for $x_1(t)$, $x_2(t)$, $L_1(t)$, and $L_2(t)$ in terms of x_0 and L_3.

We shall now divide the problem into two cases. The first case requires that we find the response of y_2 to a change in the inlet gas concentration only, the liquid flow rate remaining constant. In this case, the problem is linear and only Eqs. (21.26) and (21.27) are needed.

In the second case, it is assumed that we want to know the change in outlet concentration y_2 for a change in both inlet flow and inlet gas concentration. For this case, four simultaneous differential equations must be solved, two of which contain nonlinear terms. One approach to this problem is to linearize the nonlinear terms as was done in the case of the steam-jacketed kettle of the previous example; however, since this technique has already been illustrated, we shall not repeat it here.

---

*The assumption that the plate behaves as a first-order system with respect to liquid-flow dynamics would have to be justified experimentally. For the common bubble-cap plate, liquid builds up on the plate and flows over a weir, which may consist of a circular pipe or a vertical plate. The resistance to flow from the plate is therefore a weir, for which flow-head relationships are known (see footnote in Chap. 6). However, these flow-head relationships for weirs have been developed for the flow of liquids that are not aerated. In the case of flow of liquid over a bubble-cap plate, the liquid is very turbulent as a result of the agitation of the bubbles rising through the liquid. For this reason, one cannot expect the flow-head relations developed for quiescent flow to apply to the turbulent conditions present in the liquid on a plate. The true flow-head relation should be determined experimentally.

The fact that the flow rate is assumed to vary without change in holdup on the plate (assumption 5) appears to be contradictory. Actually, to increase the flow rate, a slight increase in level (and therefore holdup volume) above the crest of the weir is required. However, for the example under consideration, it will be assumed that the change in level needed to produce a substantial increase in flow is so small that the change in the amount of liquid on the plate is a small fraction of the total liquid holdup.

For the first case where the inlet liquid flow rate remains constant ($L_1 = L_2 = L$), Eqs. (21.26) and (21.27) can be written

$$\frac{dx_1}{dt} = -ax_1 + bx_2 + cx_0 \tag{21.30}$$

$$\frac{dx_2}{dt} = cx_1 - ax_2 \tag{21.31}$$

where $a = \dfrac{L}{H} + \dfrac{Vm}{H}$

$\quad\quad b = \dfrac{L}{H}$

$\quad\quad c = \dfrac{Vm}{H}$

At steady state, $dx_1/dt = dx_2/dt = 0$, and Eqs. (21.30) and (21.31) can be written

$$0 = -ax_{1_s} + bx_{2_s} + cx_{0_s}$$

$$0 = cx_{1_s} - ax_{2_s}$$

Subtracting these steady-state equations from Eqs. (21.30) and (21.31) and introducing the deviation variables $X_1 = x_1 - x_{1_s}$, $X_2 = x_2 - x_{2_s}$, and $X_0 = x_0 - x_{0_s}$ give

$$\frac{dX_1}{dt} = -aX_1 + bX_2 + cX_0 \tag{21.32}$$

$$\frac{dX_2}{dt} = cX_1 - aX_2 \tag{21.33}$$

Notice that $X_0 = Y_0/m$ because

$$X_0 = x_0 - x_{0_s}$$

$$X_0 = \frac{y_0 - b}{m} - \frac{y_{0_s} - b}{m} = \frac{y_0 - y_{0_s}}{m} = \frac{Y_0}{m}$$

Equations (21.32) and (21.33) can be transformed to give

$$sX_1 = -aX_1 + bX_2 + cX_0$$

$$sX_2 = cX_1 - aX_2$$

We now have two algebraic equations and three unknowns (X_1, X_2, and X_0). Solving this pair of equations to eliminate X_1 and replacing X_2 by Y_2/m and X_0 by Y_0/m give the transfer function

$$\frac{Y_2(s)}{Y_0(s)} = \frac{c^2/(a^2 - bc)}{[1/(a^2 - bc)]s^2 + [2a/(a^2 - bc)]s + 1} \tag{21.34}$$

This result shows that the response of outlet gas concentration to a change in inlet gas concentration is second-order. One can show* that ζ for this system is greater than 1, meaning that the response is overdamped. If the analysis is repeated for a gas absorber containing n plates, it will be found that the response between inlet gas concentration and outlet gas concentration is nth-order.

DISTRIBUTED-PARAMETER SYSTEMS

Heat Conduction into a Solid

In Chap. 5, the analysis of the mercury thermometer was based on a "lumped-parameter" model. At that time, reference was made to a distributed-parameter model of the thermometer. To illustrate the difference between a lumped-parameter system and a distributed-parameter system, consider a slab of solid conducting material of infinite thickness, as shown in Fig. 21.6. Let the input to this system be the temperature at the left face ($x = 0$), which is some arbitrary function of time. The output will be the temperature at the position $x = L$. For convenience, we may consider this system to represent the response of a bare thermocouple embedded in a thick wall, as the surface of the wall experiences a variation in temperature. The conductivity k, heat capacity C, and density ρ of the conducting material are constant, independent of temperature. Initially ($t < 0$), the slab is at a uniform steady-state temperature. Therefore in deviation variables, which will be used henceforth, the initial temperature is zero. The cross-sectional area of the slab is A.

ANALYSIS. In this problem the temperature in the slab is a function of position and time and is indicated by $T(x,t)$. The temperature at the surface is indicated by $T(0,t)$, and that at $x = L$ by $T(L,t)$. To derive a differential equation that

---

*Equation (21.34) is of the standard second-order form, $K/(\tau^2 s^2 + 2\zeta\tau s + 1)$, with the parameters

$$\tau^2 = \frac{1}{a^2 - bc} \qquad \text{and} \qquad 2\zeta\tau = \frac{2a}{a^2 - bc}$$

Solving these two equations to eliminate τ gives

$$\zeta = \frac{1}{\sqrt{1 - bc/a^2}}$$

Writing a and b in terms of the original system parameters (L,H,V,m) gives

$$\zeta = \left[1 - \frac{(L/H)(Vm/H)}{(L/H + Vm/H)^2} \right]^{-1/2}$$

Simplifying this expression gives

$$\zeta = \left[1 - \frac{Vm/L}{(1 + Vm/L)^2} \right]^{-1/2}$$

Since $Vm/L > 0$, we see that $\zeta > 1$.

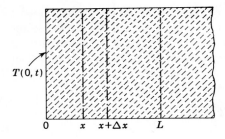

FIGURE 21-6
Heat conduction in a solid.

describes the heat conduction in the slab, we first write an energy balance over a differential length Δx of the slab. This energy balance can be written

$$\left\{ \begin{array}{c} \text{Flow of heat} \\ \text{into left face} \\ \text{by conduction} \end{array} \right\} - \left\{ \begin{array}{c} \text{Flow of heat out} \\ \text{of right face} \\ \text{by conduction} \end{array} \right\} = \left\{ \begin{array}{c} \text{Rate of accumulation} \\ \text{of internal energy in} \\ \text{the volume element} \end{array} \right\} \quad (21.35)$$

The flow of heat by conduction follows Fourier's law:

$$q = -k\frac{\partial T}{\partial x} \tag{21.36}$$

where q = heat flux by conduction
$\partial T/\partial x$ = temperature gradient
k = thermal conductivity

Applying Eq. (21.36) to Eq. (21.35) gives

$$-Ak\frac{\partial T}{\partial x}\bigg|_x - \left(-Ak\frac{\partial T}{\partial x}\bigg|_{x+\Delta x}\right) = \frac{\partial}{\partial t}[C\rho A\Delta x(T - T_r)] \tag{21.37}$$

where T_r is the reference temperature used to evaluate internal energy. The term $\partial T/\partial x\big|_{x+\Delta x}$ can be written

$$\frac{\partial T}{\partial x}\bigg|_{x+\Delta x} = \frac{\partial T}{\partial x}\bigg|_x + \frac{\partial}{\partial x}\frac{\partial T}{\partial x}\Delta x \tag{21.38}$$

Substituting Eq. (21.38) into (21.37) and simplifying give the fundamental equation describing conduction in a solid

$$k\frac{\partial^2 T}{\partial x^2} = \rho C\frac{\partial T}{\partial t}$$

This is often written as

$$K\frac{\partial^2 T}{\partial x^2} = \frac{\partial T}{\partial t} \tag{21.39}$$

where $K = k/\rho C$ = thermal diffusivity.

Several points are worth noticing at this time. In this analysis, we have allowed the capacity for storing heat ($\rho C A$ per unit length of x) and the resistance to heat conduction ($1/kA$ per unit length of x) to be "spread out" or distributed uniformly throughout the medium. This distribution of capacitance and resistance is the basis for the term *distributed parameter*. The analysis has also led to a partial differential equation, which in general is more difficult to solve than the ordinary differential equation that results from a lumped-parameter model.

TRANSFER FUNCTION. We are now in a position to derive a transfer function from Eq. (21.39). First notice that, since T is a function of both time t and position x, a transfer function may be written for an arbitrary value of x. In this problem, the temperature is to be observed at $x = L$; hence the transfer function will relate $T(L,t)$ to the temperature at the left surface $T(0,t)$, which is taken as the forcing function.

Equation (21.39) will be solved by the method of Laplace transforms. Taking the Laplace transform of both sides of Eq. (21.39) with respect to t gives

$$K \int_0^\infty \frac{\partial^2 T}{dx^2}(x, t)e^{-st}\,dt = \int_0^\infty \frac{\partial T}{\partial t}(x, t)e^{-st}\,dt \tag{21.40}$$

Consider first the integral on the left side of Eq. (21.40). Interchanging the order of integration and differentiation* results in

$$\int_0^\infty \frac{\partial^2 T}{\partial x^2}(x, t)e^{-st}\,dt = \frac{\partial^2}{\partial x^2}\int_0^\infty T(x, t)e^{-st}\,dt = \frac{d^2\overline{T}(x, s)}{dx^2} \tag{21.41}$$

where $\overline{T}(x,s)$ is the Laplace transform of $T(x,t)$.[†] It should be noted that the presence of x has no effect on the second integral of Eq. (21.41) because the integration is with respect to t. Also note that the derivative on the right side of Eq. (21.41) is taken as an ordinary derivative because $T(x,s)$ will later be seen to be a function of only one independent variable x and a parameter s. Next consider the integral on the right side of Eq. (21.40). Again, the presence of x has no effect on the integration with respect to t, and the rule for the transform of a derivative may be applied directly to yield

$$\int_0^\infty \frac{\partial T}{\partial t}(x,t)e^{-st}\,dt = s\overline{T}(x,s) - T(x,0) \tag{21.42}$$

where $T(x,0)$ is the initial temperature distribution in the solid. Introducing the results of the transformation into Eq. (21.40) gives

---

*This interchange is allowed for most functions of engineering interest. See R. V. Churchill (1972).

[†] In this chapter the overbar will often be used to indicate the Laplace transform of a function of two variables.

$$K\frac{d^2\overline{T}(x,s)}{dx^2} = s\overline{T}(x,s) - T(x,0) \tag{21.43}$$

The partial differential equation has now been reduced to an ordinary differential equation, which can usually be solved without difficulty. It should be clear that s in Eq. (21.43) is merely a parameter, with the result that this equation is an *ordinary* second-order differential equation in the independent variable x. This follows because there are no derivatives with respect to s in Eq. (21.43). Since we have taken $T(x,0) = 0$ for the example under consideration, Eq. (21.43) becomes

$$\frac{d^2\overline{T}}{dx^2} - \frac{s}{K}\overline{T} = 0 \tag{21.44}$$

Equation (21.44) is a linear differential equation and can be solved to give

$$\overline{T} = A_1 e^{-\sqrt{s/K}\,x} + A_2 e^{\sqrt{s/K}\,x} \tag{21.45}$$

The arbitrary coefficients A_1 and A_2 may be evaluated as follows: In order that $\overline{T}$ may be finite as $x \to \infty$, it is necessary that $A_2 = 0$. Equation (21.45) then becomes

$$\overline{T} = A_1 e^{-\sqrt{s/K}\,x} \tag{21.45a}$$

The transformed forcing function at $x = 0$ is $\overline{T}(0,s)$, which can be substituted into Eq. (21.45a) to determine A_1; then

$$\overline{T}(0, s) = A_1 e^0$$

or

$$A_1 = \overline{T}(0, s)$$

Substituting A_1 into Eq. (21.45a) gives

$$\frac{\overline{T}(x,s)}{\overline{T}(0,s)} = e^{-\sqrt{s/K}\,x} \tag{21.46}$$

By specifying a particular value of x, say $x = L$, the transfer function is

$$\frac{\overline{T}(L,s)}{\overline{T}(0,s)} = e^{-\sqrt{s/K}\,L} \tag{21.47}$$

STEP RESPONSE. To illustrate the use of this transfer function, consider a forcing function that is the unit-step function; thus

$$T(0, t) = u(t)$$

for which case $\overline{T}(0,s) = 1/s$. Substituting this into Eq. (21.47) gives

$$\overline{T}(L,s) = \frac{1}{s}e^{-\sqrt{s/K}\,L} \tag{21.48}$$

To obtain the response in the time domain, we must invert Eq. (21.48). A table of transforms* gives the following transform pair:

$$L\left\{\frac{1}{s}e^{-\sqrt{s/K}\,x}\right\} = \text{erfc}\frac{x}{\sqrt{4Kt}} \qquad (21.49)$$

where erfc x is the error-function complement of x defined as

$$\text{erfc } x = 1 - \frac{2}{\sqrt{\pi}}\int_{0}^{x} e^{-u^2}\,du$$

This function is tabulated in many textbooks[†] and mathematical tables.
Using this transform pair, Eq. (21.48) becomes

$$T(L, t) = \text{erfc}\frac{L}{\sqrt{4Kt}} = \text{erfc}\left[\frac{1}{2}\left(\frac{Kt}{L^2}\right)^{-1/2}\right] \qquad (21.50)$$

A plot of T versus the dimensionless group Kt/L^2 is shown in Fig. 21.7.

SINUSOIDAL RESPONSE. It is instructive to consider the response in temperature at $x = L$ for the case where the forcing function is a sinusoidal variation; thus

$$T(0, t) = A\sin\omega t$$

Using the substitution rule of Chap. 16, in which s is replaced by $j\omega$, Eq. (21.47) becomes

$$\frac{\overline{T}(L, j\omega)}{\overline{T}(0, j\omega)} = e^{-\sqrt{j\omega/K}\,L} \qquad (21.51)$$

---

*Tables of transforms that include transform pairs frequently encountered in the solution of partial differential equations may be found in many textbooks on heat conduction and applied mathematics. For example, see Mickley, Sherwood, and Reed (1957).
    Inversion of complicated transforms such as that of Eq. (21.48) can be achieved systematically by the method of complex residues, which is also discussed in the above reference.
[†] See Carslaw and Jaeger (1959), p. 485.

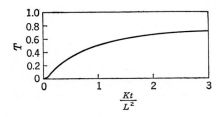

FIGURE 21-7
Response of temperature in the interior of a solid to a unit-step change in temperature at the surface.

To obtain the AR and phase angle requires that the magnitude and argument of the right side of Eq. (21.51) be evaluated. This can be done as follows: First write j in polar form; thus

$$j = e^{j\pi/2}$$

from which we get

$$\sqrt{j} = [e^{j(\pi/2)}]^{1/2} = \pm e^{j\pi/4} = \pm\frac{1}{\sqrt{2}}(1 + j)$$

Substituting the positive form* of $\sqrt{j}$ into Eq. (21.51) gives

$$\frac{\overline{T}(L, j\omega)}{\overline{T}(0, j\omega)} = e^{-\sqrt{\omega/2K}\,L}\,e^{-j\,\sqrt{\omega/2K}\,L}$$

From this form, we can write by inspection

$$\text{AR} = \left|\frac{\overline{T}(L, j\omega)}{\overline{T}(0, j\omega)}\right| = e^{-\sqrt{\omega/2K}\,L} \tag{21.52}$$

$$\text{Phase angle} = \sphericalangle\frac{\overline{T}(L, j\omega)}{\overline{T}(0, j\omega)} = -\sqrt{\frac{\omega}{2K}}\,L \qquad \text{rad} \tag{21.53}$$

From these results, it is seen that the AR approaches zero as $\omega \to \infty$ and the phase angle decreases without limit as $\omega \to \infty$. Such a system is said to have *nonminimum* phase lag characteristics. With the exception of the distance-velocity lag, all the systems that have been considered up to now have given a limited value of phase angle as $\omega \to \infty$. These are called minimum phase systems and always occur for lumped-parameter systems. The nonminimum phase behavior is typical of distributed-parameter systems.

Transport Lag as a Distributed-parameter System

We can demonstrate that the transport lag (distance-velocity lag) is, in fact, a distributed-parameter system as follows: Consider the flow of an incompressible fluid through an insulated pipe of uniform cross-sectional area A and length L, as shown in Fig. 21.8a. The fluid flows at velocity v, and the velocity profile is flat. We know from Chap. 8 that the transfer function relating outlet temperature T_o to the inlet temperature T_i is

$$\frac{T_o(s)}{T_i(s)} = e^{-(L/v)s}$$

---

*Notice that the substitution of $-(1 + j)/\sqrt{2}$ into Eq. (21.51) leads to a result in which the AR is greater than 1 and the phase angle leads. This is contrary to the response of the physical system and is not admitted as a useful solution.

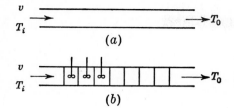

FIGURE 21-8
Obtaining the transfer function of a transport lag from a lumped-parameter model.

Let the pipe be divided into n zones as shown in Fig. 21.8b. If each zone of length L/n is considered to be a well-stirred tank, then the pipe is equivalent to n noninteracting first-order systems in series, each having a time constant*

$$\tau = \frac{L}{n}\frac{1}{v}$$

(Note that taking each zone to be a well-stirred tank is called lumping of parameters.) The overall transfer function for this lumped-parameter model is therefore

$$\frac{T_o(s)}{T_i(s)} = \left(\frac{1}{\tau s + 1}\right)^n = \left[\frac{1}{(L/v)s/n + 1}\right]^n$$

To "distribute" the parameters, we let the size of the individual lumps go to zero by letting $n \to \infty$.

$$\frac{T_o(s)}{T_i(s)} = \lim_{n \to \infty}\left[\frac{1}{(L/v)s/n + 1}\right]^n$$

The thermal capacitance is now distributed over the tube length. It can be shown by use of the calculus that the limit is

$$e^{-(L/v)s}$$

which is the transfer function derived previously. This demonstration should provide some initial insight into the relationship between a distributed-parameter system and a lumped-parameter system and indicates that a transport lag is a distributed system.

Heat Exchanger

As our last example[†] of a distributed-parameter system, we consider the double-pipe heat exchanger shown in Fig. 21.9. The fluid that flows through the inner

---

*This expression for τ is equivalent to that appearing in Eq. (9.10). Since the transfer function for flow through a tank was developed in Chap. 9, the analysis will not be repeated here.

[†]The analysis presented here essentially follows that of W. C. Cohen and E. F. Johnson (1956). These authors also present the experimental results of frequency response tests on a double-pipe, steam-to-water heat exchanger.

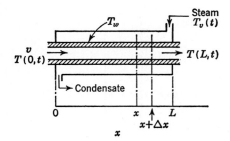

FIGURE 21-9
Double-pipe heat exchanger.

pipe at constant velocity v is heated by steam condensing outside the pipe. The temperature of the fluid entering the pipe and the steam temperature vary according to some arbitrary functions of time. The steam temperature varies with time, but not with position in the exchanger. The metal wall separating steam from fluid is assumed to have significant thermal capacity that must be accounted for in the analysis. The heat transfer from the steam to the fluid depends on the heat-transfer coefficient on the steam side (h_o) and the convective transfer coefficient on the water side (h_i). The resistance of the metal wall is neglected. The goal of the analysis will be to find transfer functions relating the exiting fluid temperature $T(L,t)$ to the entering fluid temperature $T(0,t)$ and the steam temperature $T_v(t)$.

The following symbols will be used in this analysis:

$T(x,t) =$ fluid temperature

$T_w(x,t) =$ wall temperature

$T_v(t) =$ steam temperature

$T_r =$ reference temperature for evaluating enthalpy

$\rho =$ density of fluid

$C =$ heat capacity of fluid

$\rho_w =$ density of metal in wall

$C_w =$ heat capacity of metal in wall

$A_i =$ cross-sectional area for flow inside pipe

$A_w =$ cross-sectional area of metal wall

$D_i =$ inside diameter of inner pipe

$D_o =$ outside diameter of inner pipe

$h_i =$ convective heat-transfer coefficient inside pipe

$h_o =$ heat-transfer coefficient for condensing steam

$v =$ fluid velocity

ANALYSIS. We begin the analysis by writing a differential energy balance for the fluid inside the pipe over the volume element of length Δx (see Fig. 21.9). This balance can be stated

$$
\left\{ \begin{array}{c} \text{Flow of} \\ \text{enthalpy in} \end{array} \right\} - \left\{ \begin{array}{c} \text{Flow of} \\ \text{enthalpy out} \end{array} \right\} + \left\{ \begin{array}{c} \text{Heat transferred} \\ \text{through film on} \\ \text{inside wall} \end{array} \right\}
$$

$$
= \left\{ \begin{array}{c} \text{Rate of accum-} \\ \text{ulation of} \\ \text{internal energy} \end{array} \right\} \tag{21.54}
$$

The terms in this balance can be evaluated as follows:

$$
\text{Flow of enthalpy in at } x = vA_i\rho C(T - T_r)
$$

$$
\text{Flow of enthalpy out at } x + \Delta x = vA_i\rho C\left[\left(T + \frac{\partial T}{\partial x}\Delta x\right) - T_r\right]
$$

$$
\text{Heat transfer through film} = \pi D_i h_i \Delta x(T_w - T)
$$

$$
\text{Accumulation of internal energy} = \frac{\partial}{\partial t}[A_i\rho\Delta x C(T - T_r)]
$$

Introducing these terms into Eq. (21.54) gives, after simplification,

$$
\frac{\partial T}{\partial t} = -v\frac{\partial T}{\partial x} + \frac{1}{\tau_1}(T_w - T) \tag{21.55}
$$

where $\dfrac{1}{\tau_1} = \dfrac{\pi D_i h_i}{A_i\rho C}$

An energy balance is next written for the metal in the wall, over the volume element of length Δx. This can be stated as follows:

$$
\left\{ \begin{array}{c} \text{Heat transfer in} \\ \text{through steam} \\ \text{condensate film} \end{array} \right\} - \left\{ \begin{array}{c} \text{Heat transfer out} \\ \text{through fluid film} \end{array} \right\} = \left\{ \begin{array}{c} \text{Accumulation of} \\ \text{energy in wall} \end{array} \right\}
$$

Expressing each term in this balance by symbols gives

$$
\pi D_o h_o \Delta x(T_v - T_w) - \pi D_i h_i \Delta x(T_w - T) = A_w \Delta x \rho_w C_w \frac{\partial T_w}{\partial t} \tag{21.56}
$$

Simplifying this expression gives

$$
\frac{\partial T_w}{\partial t} = \frac{1}{\tau_{22}}(T_v - T_w) - \frac{1}{\tau_{12}}(T_w - T) \tag{21.57}
$$

where $\dfrac{1}{\tau_{12}} = \dfrac{\pi D_i h_i}{A_w \rho_w C_w} \qquad \dfrac{1}{\tau_{22}} = \dfrac{\pi D_o h_o}{A_w \rho_w C_w}$

We now have obtained the differential equations that describe the dynamics of the system. As in previous problems, the dependent variables will be transformed to deviation variables. At steady state, the time derivatives in Eqs. (21.55) and (21.57) are zero, and it follows that

$$0 = -v\frac{dT_s}{dx} + \frac{1}{\tau_1}(T_{w_s} - T_s) \tag{21.58}$$

and

$$0 = \frac{1}{\tau_{22}}(T_{v_s} - T_{w_s}) - \frac{1}{\tau_{12}}(T_{w_s} - T_s) \tag{21.59}$$

where the subscript s is used to denote the steady-state value. Note that to determine the steady-state values of the temperature requires the solution of two simultaneous equations, the first of which is an ordinary differential equation. Thus, the steady-state temperature T_s is a function of x and may be obtained by solution of Eqs. (21.58) and (21.59) as

$$T_s = T_{v_s} + (T_{s_0} - T_{v_s})\exp\left[-\frac{x}{v\tau_1}\bigg/\left(1 + \frac{\tau_{22}}{\tau_{12}}\right)\right]$$

where T_{s_0} is the normal entrance temperature. All equations for T' to be derived below should be recognized as deviations from this expression.

Subtracting Eq. (21.58) from (21.55) and Eq. (21.59) from (21.57) and introducing deviation variables give

$$\frac{\partial T'}{\partial t} = -v\frac{\partial T'}{\partial x} + \frac{1}{\tau_1}(T_w' - T') \tag{21.60}$$

and

$$\frac{\partial T_w'}{\partial t} = \frac{1}{\tau_{22}}(T_v' - T_w') - \frac{1}{\tau_{12}}(T_w' - T') \tag{21.61}$$

where $T' = T - T_s$
$\quad\quad T_w' = T_w - T_{w_s}$
$\quad\quad T_v' = T_v - T_{v_s}$

Equations (21.60) and (21.61) may be transformed with respect to t to yield

$$s\overline{T}' = -v\frac{d\overline{T}'}{dx} + \frac{1}{\tau_1}(\overline{T}_w' - \overline{T}') \tag{21.62}$$

and

$$s\overline{T}_w' = \frac{1}{\tau_{22}}(\overline{T}_v' - \overline{T}_w') - \frac{1}{\tau_{12}}(\overline{T}_w' - \overline{T}') \tag{21.63}$$

where $\overline{T}' = \overline{T}'(x,s)$
$\quad\quad \overline{T}_w' = \overline{T}_w'(x,s)$
$\quad\quad \overline{T}_v' = \overline{T}_v'(s)$

In Eqs. (21.62) and (21.63) it has been assumed that the exchanger is initially at steady state, so that $T(x,0) = T_s$, $T_w(x,0) = T_{w_s}$, and $T_v(0) = T_{v_s}$.

Eliminating $\overline{T}'_w$ from Eqs. (21.62) and (21.63) gives, after considerable simplification

$$\frac{d\overline{T}'}{dx} + \frac{a}{v}\overline{T}' = \frac{b}{v}\overline{T}'_v \tag{21.64}$$

where $a(s) = s + \dfrac{1}{\tau_1} - \dfrac{\tau_{22}}{\tau_1(\tau_{12}\tau_{22}s + \tau_{12} + \tau_{22})}$

$$b(s) = \frac{\tau_{12}}{\tau_1(\tau_{12}\tau_{22}s + \tau_{12} + \tau_{22})}$$

Equation (21.64) is an ordinary first-order differential equation with boundary condition $\overline{T}'(x,s) = \overline{T}'(0,s)$ at $x = 0$.

It can be readily solved to yield

$$\overline{T}'(x, s) = \overline{T}'(0, s) + [1 - e^{-(a/v)x}]\left[\frac{b}{a}\overline{T}'_v(s) - \overline{T}'(0, s)\right] \tag{21.65}$$

where $\overline{T}'(0,s)$ is the transform of the fluid temperature at the entrance to the pipe and $\overline{T}'_v(s)$ is the transform of the steam temperature. From Eq. (21.65), the transfer functions can be obtained as follows:

If the steam temperature does not vary, $T'_v(s) = 0$; the transfer function relating temperature at the end of the pipe $(x = L)$ to temperature at the entrance is

$$\frac{\overline{T}'(L, s)}{\overline{T}'(0, s)} = e^{-(a/v)L} \tag{21.66}$$

Setting $1/\tau_1$ to zero in the expression for $a(s)$ [Eq. (21.64)] shows that $a(s) = s$ and hence the response is simply that of a transport lag. This is in agreement with the physical situation where h_i approaches zero [Eq. (21.55)], for which case the wall separating cold fluid from hot fluid acts as a perfect insulator. We saw in Chap. 8 that this situation is represented by a transport lag.

If the temperature of the fluid entering the pipe does not vary, the transfer function relating the exit fluid temperature to the steam temperature is

$$\frac{\overline{T}'(L, s)}{\overline{T}'_v(s)} = \frac{b}{a}[1 - e^{-(a/v)L}] \tag{21.67}$$

In principle, the response in the temperature of the fluid leaving the exchanger can be found for any forcing function, $T(0,t)$ or $T_v(t)$, by introducing the corresponding transforms into Eq. (21.66) or (21.67). However, the resulting expression is very complex and cannot be easily inverted. For the case of sinusoidal inputs, the substitution rule discussed in Chap. 16 can be used to determine the AR and phase angle of the frequency response. Cohen and Johnson give a Bode diagram corresponding to Eq. (21.67) for a specific set of heat-exchanger parameters. This diagram is shown in Fig. 21.10.

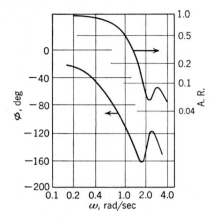

FIGURE 21-10
Bode diagram of heat exchanger for variation in steam temperature (Cohen and Johnson).

Notice that the theory predicts an interesting resonance effect at higher frequencies. The resonance effect has been observed experimentally in a steam-to-water exchanger. See Lees and Hougen (1956). Unfortunately, the experimental data of Cohen and Johnson do not extend to sufficiently high frequencies to exhibit resonance. The reader is referred to the original article for further details.

SUMMARY

In this chapter, several complex systems have been analyzed mathematically. The result of each analysis was a set of equations (algebraic and/or differential) that presumably describe the dynamic response of the system to one or more disturbances. The process of obtaining the set of equations is often called *modeling,* and the set of equations is referred to as the *mathematical model* of the system. In general, the model is based on the physics and chemistry of the system. For example, in the analysis of a heat exchanger, one may write that the heat flux through a wall is equal to a convective transfer coefficient times a temperature driving force.

For a process not well understood, there is little chance that an accurate model can be obtained from the theoretical approach used here. For such systems, a direct dynamic test can be made. To do this, a known disturbance such as a pulse, step, or sinusoidal input is applied and the response recorded. This approach was discussed in Chap. 19. On the other hand, a model based on a theoretical analysis is extremely valuable, for it means that the system is well understood and that the effect of changes in system design and operation can be predicted.

The analysis of a steam-jacketed kettle provided an example of a nonlinear system containing nonlinear functions of several variables. The problem was handled by linearizing these functions about an operating point and ultimately obtaining a block diagram of the system from which the transfer function of the control system could be obtained. Although this approach is relatively straightforward, the resulting linear model can only be used over a narrow range of variables.

The analysis of the gas absorber gave some insight into the dynamic character of a typical multistage process that is widely used in the chemical process industries. A linear analysis of an n-plate column leads to n ordinary differential equations, which combine to give an overdamped nth-order response. Nonlinearities may be present in this system in such forms as a product of flow and concentration or a nonlinear equilibrium relationship. When changes in inlet flow occur, a set of differential equations describing the dynamics of the liquid flow must be added to those describing mass transfer. When the change of plate efficiency with flow is considered, the model of a gas absorber becomes even more complex. Most of the design techniques developed in the past for multistage operations (gas absorption, distillation, etc.) have applied to steady-state operation. The dynamic analysis of such processes calls for dynamic parameters that are usually unavailable. For example, the liquid-flow dynamics of trays used in distillation towers are relatively unknown.

The discussion of distributed-parameter systems further illustrated the complexities that can arise in physical systems. The distributed-parameter systems lead to partial differential equations, which may be very difficult to solve for most of the forcing functions of practical interest. However, we saw that the response of distributed-parameter systems to sinusoidal forcing functions can be obtained directly by application of the substitution rule, in which s is replaced by $j\omega$. A distributed-parameter system features nonminimum phase lag characteristics. This is in sharp contrast to the lumped-parameter systems for which the phase angle approaches a limit at infinite frequency.

As systems are analyzed in more detail and with fewer assumptions, the models that describe them become more complex, although more accurate. To predict the response of the system from the model requires that equations of the model be solved for some specific input disturbance. The only practical way to solve a complex model is to use a computer. This method of solving the mathematical model is often called computer simulation. The computer response will resemble that of the physical system if the model is accurate. In the last section of this text, the computer and its use to simulate control systems will be discussed in considerable detail.

PART
VI

SAMPLED-DATA CONTROL SYSTEMS

CHAPTER
22

SAMPLING
AND
Z-TRANSFORMS

The wide use of digital computers in control makes the inclusion of a section on sampled-data control imperative. Sampled-data control was actually established over 25 years ago by electrical engineers before the digital computer was widely used to control chemical processes. A sampled-data system is one in which the flow of signals in the control system is interrupted at one or more points. In this book, the interruption or sampling will occur every T units of time. Such sampling is called uniform sampling and is the usual type in practical applications.

To understand the nature of a sampled-data system, consider a typical, single-loop continuous control system, shown in Fig. 22.1a. The system is referred to as continuous because the signal flow between blocks is continuous or without interruption; i.e., at any instant of time and at any location in the loop, one can observe a changing value of the signal during a transient. For example, the response from the measuring element varies in a continuous manner from moment to moment. A typical temperature transmitter would provide such a continuous signal. In chemical processes, some measurements cannot be made continuously. Chemical composition is a measurement that may not be continuous. For example, a sample of a process stream may have to be subjected to a chemical analysis that takes some finite period of time. An example would be an automatic chromatograph that must process a sample of fluid in a packed column for a fixed time T. For this example the measured value of composition is known only at the end of the processing time T. If a new sample of fluid is taken successively every T units of time and the result of the chemical analysis is held constant between sampling instants, one can represent the control system as shown in Fig 22.1*b*. In this case, the measured signal *B* is held constant between sampling instants and delayed by an amount *T*. The delay occurs because of the time needed to process the sample of fluid in the chromatograph. Note that for simplicity, the chromatograph processing time and the sampling time have been taken to be the same value *T*. In Fig. 22.1*b* this disruption in the flow of signal *B* is represented by a sampling switch and the holding of the value of the signal is obtained by the block labeled "hold." The nature of the signals from the measuring element *B* and the sampled signal B_c are shown in Fig. 22.1*b*. The output from the hold is a stair-step response, which approximates the continuous signal *B* more accurately as the sampling period *T* decreases.

Another reason for studying sampled-data control is to be able to describe the operation of a digital computer as a controller. The output of a continuous electronic or pneumatic controller changes in a continuous manner. When a digital computer is used to implement a control law, a calculation must be performed to calculate the new value of the controller output every *T* units of time. The calculation, which is based on a numerical expression of the control law, will be developed in detail later. At each sampling instant, the computation of the controller output is made and then held at a constant value until the next sampling

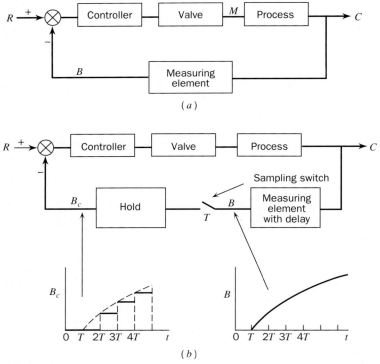

FIGURE 22-1
Comparison of (*a*) continuous control and (*b*) sampled-data control.

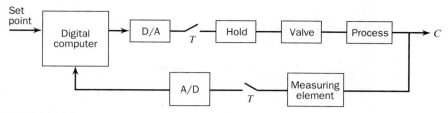

FIGURE 22-2
Computer control system.

instant occurs, at which time a new controller output is computed and held. This computation and holding is repeated every T units of time. Figure 22.2 shows a diagram that represents a digital control system. In this figure, the typical elements of the system (valve, process, and measuring element) are continuous and behave the same as those in a continuous control system. The usual continuous controller is replaced by a digital computer, which is programmed to implement a control law, such as PI control. Since the digital computer works with digital information, an analog-to-digital converter (A/D) is needed to convert the continuous (analog) signal to a digital signal that can be used by the computer. Since the output from the calculation is a digital signal, a digital-to-analog converter (D/A) is needed on the output of the computer so that a continuous (analog) signal is available to operate the valve. Typical analog signals associated with the A/D and D/A converters range from 4–20 ma or 1–5 V. The sampling switches are shown to indicate the sampled nature of the signals. These switches are purely symbolic; there are no mechanical switches in the hardware used to implement a controller. The hold block, which is shown in the figure, holds the value of the controller output constant between sampling instants. The output of the hold is a stair-step function.

CLAMPING

The continuous function $f(t)$ is said to be clamped to produce the function $f_c(t)$ as shown in Fig. 22.3. The period of sampling is T and the frequency is $\omega_s = 2\pi/T$ radians/time. The clamping can be described mathematically by the combination of impulse modulation and the application of a zero-order hold as will soon be shown.

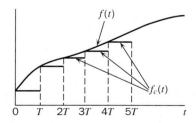

FIGURE 22-3
Clamping.

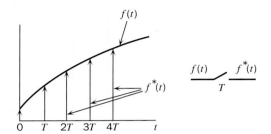

FIGURE 22-4
Impulse modulation.

Impulse Modulation

As shown in Fig. 22.4, an impulse-modulated function consists of a sequence of impulse functions, the magnitudes of which equal the values of the continuous function at sampling instants. The impulse-modulated function is given the symbol $f^*(t)$. A convenient symbol for the impulse modulation sampler is shown as a switch ($\diagup$), which closes momentarily every T sec. It should be noted that the impulse modulation switch is purely symbolic, for there is no switch of this type present in the hardware used in sampling signals.

Zero-order Hold

The zero-order hold is defined by the transfer function

$$G_{h0}(s) = \frac{1 - e^{-Ts}}{s} \tag{22.1}$$

Combining impulse modulation with the zero-order hold, as shown in Fig. 22.5, provides the clamping. To illustrate how the zero-order hold shapes the sequence of impulses into a clamped signal, a block diagram of the hold is shown in Fig. 22.6 in which an integrator and a transport lag are connected to implement the zero-order hold. The operation of a zero-order hold can be understood by expressing $F_c(s)$ as follows (see Fig. 22.5):

$$F_c(s) = F^*(s)\frac{1 - e^{-Ts}}{s} = \frac{F^*(s)}{s} - \frac{F^*(s)}{s}e^{-Ts} \tag{22.2}$$

where $F^*(s)$ is the Laplace transform of the impulse-modulated function, $f^*(t)$. This expression shows that the clamped function $f_c(t)$ is obtained by a combination of integration, transport lag, and subtraction. Recall that integration in Eq. (22.2) is represented by $1/s$ and transport lag by e^{-Ts}. To understand the signals at the output of the integrator, one must recall that the integration of an impulse

$$\underset{F(s)}{\overset{f(t)}{}} \quad \diagup \quad \underset{\substack{F^*(s) \\ F(z)}}{\overset{f^*(t)}{}} \quad \boxed{\dfrac{1 - e^{-Ts}}{s}} \quad \underset{F_c(s)}{\overset{f_c(t)}{}}$$

FIGURE 22-5
Clamping.

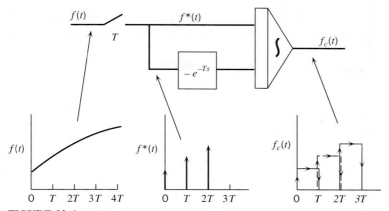

FIGURE 22-6
Construction of a clamped signal.

function is a step function. The impulse occurring at $t = 0$ results in a pulse*
of width T. (See Fig. 22.6.) The impulse occurring at $t = T$ results in a pulse of
width T starting at $t = T$. The remaining impulses each contribute a pulse of width
T starting at successive values of T. The height of each pulse equals the magnitude of the impulse that produced it. The combination of these pulse functions
provide the stair-step function associated with clamping. This rather "mechanical"
description of clamping through implementation of the zero-order hold transfer
function may give the reader an intuitive feel for the abstract mathematical
expression involved.

LAPLACE TRANSFORM OF THE
IMPULSE-MODULATED FUNCTION

Let $i(t)$ be a sequence, or train, of unit impulses that are separated by period T.
This may be expressed as follows:

$$i(t) = \delta(t) + \delta(t - T) + \delta(t - 2T) + \cdots = \sum_{n=0}^{\infty} \delta(t - nT) \qquad (22.3)$$

The "starred" function $f^*(t)$ may be written as the product of $f(t)$ and $i(t)$:

$$f^*(t) = f(t)i(t) \qquad (22.4)$$

Introducing the expression for $i(t)$ from Eq. (22.3) into this equation gives

$$f^*(t) = f(t) \sum_{n=0}^{\infty} \delta(t - nT)$$

---

*The pulse occurs because the integration of the impulse at $t = 0$ is combined with the integration
of a delayed impulse of the same magnitude, but opposite sign, at $t = T$.

Since $i(t)$ has a non-zero value only at sampling instants ($t = nT$), $f(t)$ may be replaced by $f(nT)$ and then placed within the summation to produce the result

$$f^*(t) = \sum_{n=0}^{\infty} f(nT)\delta(t - nT) \qquad (22.5)$$

Applying the Laplace transform of a unit impulse, which is unity, and applying the theorem on translation of a function (See Chap. 4) to this expression for $f^*(t)$ give the Laplace transform of the impulse-modulated function:

$$F^*(s) = L\{f^*(t)\} = \sum_{n=0}^{\infty} f(nT)e^{-nTs} \qquad (22.6)$$

The function $F^*(s)$ is referred to as the *starred transform* of $f(t)$.

An alternate form for $F^*(s)$ which is useful in proofs and derivations is given in the appendix of this chapter. This alternate form is based on the Fourier series representation of a periodic function.

THE Z-TRANSFORM

We have now reached the point where the Z-transform can be introduced. The Z-transform is simply the Laplace transform of the impulse-modulated function, $f^*(t)$, in which $z = e^{Ts}$. Keeping this point in mind, we may write Eq. (22.6) in the form

$$F(z) = Z\{f(t)\} = L\{f^*(t)\} = \sum_{n=0}^{\infty} f(nT)e^{-nTs}\Big|_{z=e^{Ts}} \qquad (22.7)$$

or

$$F(z) = Z\{f(t)\} = \sum_{n=0}^{\infty} f(nT)z^{-n} \qquad (22.8)$$

In the definition of the Z-transform given by either of these equations, we have expressed the Z-transform by $F(z)$ or $Z\{f(t)\}$. Two examples of the use of this definition of the Z-transform will be given.

Example 22.1.   The unit step.

$$f(t) = u(t) = 1$$
$$\text{therefore, } f(nT) = 1 \qquad \text{for } n \geq 0$$

From Eq. (22.8) we write

$$Z\{u(t)\} = \sum_{0}^{\infty} z^{-n}$$

This infinite series has the sum $z/(z - 1)$; therefore the result is

$$Z\{u(t)\} = \frac{z}{z - 1}$$

Example 22.2. The exponential function.

$$f(t) = e^{-t/\tau} u(t)$$

$$Z\{e^{-t/\tau} u(t)\} = \sum_0^\infty e^{-nT/\tau} z^{-n} = \sum_0^\infty (e^{T/\tau} z)^{-n}$$

This infinite series has the sum as shown below

$$Z\{e^{-t/\tau} u(t)\} = \frac{z}{z - e^{-T/\tau}}$$

Table of Transform Pairs

Tables have been prepared that give the Z-transforms of various functions. A short table of transform pairs is given in Table 22.1. This table was adapted from an extensive table in Tou(1959). Table 22.1 includes for each function of t listed, the Laplace transform, $F(s)$, the Z-transform, $F(z)$, and the modified Z-transform, $F(z,m)$. The modified Z-transform will be discussed later. An example of a Z-transform pair from this table is

$$e^{-at} : \frac{z}{z - e^{-aT}}$$

Note that this is the same as Example 22.2 with $a = 1/\tau$. Tables of Z-transforms are very useful in obtaining the transients for sampled-data systems and they are used in much the same way as tables of ordinary Laplace transforms are used for continuous systems.

SUMMARY

One reason for studying sampled-data control is to be able to describe mathematically a process in which the flow of signals is interrupted periodically. An example of such a system is one that contains a chemical analyzer (e.g., a chromatograph) that produces a measured value of composition after a fixed processing time. Another reason for studying sampled-data control is to be able to describe the operation of a microprocessor-based controller.

The form of sampling used in practical applications is clamping, a process of sampling that holds a signal constant between sampling instants. It was shown that clamping is produced by sending an impulse modulated signal through a zero-order hold.

Two forms of the Laplace transform of the impulse-modulated function $f^*(t)$ were presented. One of these forms was used to define the Z-transform in which the Laplace variable s is replaced by z through use of the transformation $z = e^{Ts}$. A short table of Z-transforms was provided. The Z-transform will be used in the next chapter to compute the response of sampled-data systems at sampling instants.

TABLE 22.1
Z-transforms and modified Z-transforms (adapted from Tou, 1959)

No.	$F(s)$	$f(t)$	$F(z)$	$F(z,m)$
1	ϵ^{-kTs}	$\delta(t - kT)$	z^{-k}	z^{m-1-k}
2	1	$\delta(t)$	1 or z^{-0}	0
3	$\dfrac{1}{s}$	$u(t)$	$\dfrac{z}{z-1}$	$\dfrac{1}{z-1}$
4	$\dfrac{1}{s^2}$	t	$\dfrac{Tz}{(z-1)^2}$	$\dfrac{mT}{z-1} + \dfrac{T}{(z-1)^2}$
5	$\dfrac{1}{s^3}$	$\dfrac{1}{2!}t^2$	$\dfrac{T^2 z(z+1)}{2(z-1)^3}$	$\dfrac{T^2}{2}\left[\dfrac{m^2}{z-1} + \dfrac{2m+1}{(z-1)^2} + \dfrac{2}{(z-1)^3}\right]$
6	$\dfrac{1}{s^{k+1}}$	$\dfrac{1}{k!}t^k$	$\displaystyle\lim_{a\to 0}\dfrac{(-1)^k}{k!}\dfrac{\partial^k}{\partial a^k}\left(\dfrac{z}{z - \epsilon^{-aT}}\right)$	$\displaystyle\lim_{a\to 0}\dfrac{(-1)^k}{k!}\dfrac{\partial^k}{\partial a^k}\left(\dfrac{\epsilon^{-amT}}{z - \epsilon^{-aT}}\right)$
7	$\dfrac{1}{s - (1/T)\ln a}$	$a^{1/T}$	$\dfrac{z}{z-a}$	$\dfrac{a^m}{z-a}$
8	$\dfrac{1}{s+a}$	ϵ^{-at}	$\dfrac{z}{z - \epsilon^{-aT}}$	$\dfrac{\epsilon^{-amT}}{z - \epsilon^{-aT}}$
9	$\dfrac{1}{(s+a)^2}$	$t\epsilon^{-at}$	$\dfrac{Tz\epsilon^{-aT}}{(z - \epsilon^{-aT})^2}$	$\dfrac{T\epsilon^{-amT}[\epsilon^{-aT} + m(z - \epsilon^{-aT})]}{(z - \epsilon^{-aT})^2}$

	$F(s)$	$f(t)$	$F(z)$	$F(z, m)$
10	$\dfrac{1}{(s+a)^3}$	$\dfrac{t^2}{2}\epsilon^{-at}$	$\dfrac{T^2\epsilon^{-aT}z}{2(z-\epsilon^{-aT})^2} + \dfrac{T^2\epsilon^{-2aT}z}{(z-\epsilon^{-aT})^3}$	$\dfrac{T^2\epsilon^{-amT}}{2}\left[\dfrac{m^2}{z-\epsilon^{-aT}} + \dfrac{(2m+1)\epsilon^{-aT}}{(z-\epsilon^{-aT})^2} + \dfrac{2\epsilon^{-2aT}}{(z-\epsilon^{-aT})^3}\right]$
11	$\dfrac{1}{(s+a)^{k+1}}$	$\dfrac{t^k}{k!}\epsilon^{-at}$	$\dfrac{(-1)^k}{k!}\dfrac{\partial^k}{\partial a^k}\left(\dfrac{z}{z-\epsilon^{-aT}}\right)$	$\dfrac{(-1)^k}{k!}\dfrac{\partial^k}{\partial a^k}\left(\dfrac{\epsilon^{-amT}}{z-\epsilon^{-aT}}\right)$
12	$\dfrac{a}{s(s+a)}$	$1-\epsilon^{-at}$	$\dfrac{(1-\epsilon^{-aT})z}{(z-1)(z-\epsilon^{-aT})}$	$\dfrac{1}{z-1} - \dfrac{\epsilon^{-amT}}{z-\epsilon^{-aT}}$
13	$\dfrac{a}{s^2(s+a)}$	$t - \dfrac{1-\epsilon^{-at}}{a}$	$\dfrac{Tz}{(z-1)^2} - \dfrac{(1-\epsilon^{-aT})z}{a(z-1)(z-\epsilon^{-aT})}$	$\dfrac{T}{(z-1)^2} + \dfrac{mT-1/a}{z-1} + \dfrac{\epsilon^{-amT}}{a(z-\epsilon^{-aT})}$
14	$\dfrac{\omega_0}{s^2+\omega_0^2}$	$\sin\omega_0 t$	$\dfrac{z\sin\omega_0 T}{z^2-2z\cos\omega_0 T+1}$	$\dfrac{z\sin m\omega_0 T + \sin(1-m)\omega_0 T}{z^2-2z\cos\omega_0 T+1}$
15	$\dfrac{s}{s^2+\omega_0^2}$	$\cos\omega_0 t$	$\dfrac{z(z-\cos\omega_0 T)}{z^2-2z\cos\omega_0 T+1}$	$\dfrac{z\cos m\omega_0 T - \cos(1-m)\omega_0 T}{z^2-2z\cos\omega_0 T+1}$
16	$\dfrac{\omega_0^2}{s(s^2+\omega_0^2)}$	$1-\cos\omega_0 t$	$\dfrac{z}{z-1} - \dfrac{z(z-\cos\omega_0 T)}{z^2-2z\cos\omega_0 T+1}$	$\dfrac{1}{z-1} - \dfrac{z\cos m\omega_0 T - \cos(1-m)\omega_0 T}{z^2-2z\cos\omega_0 T+1}$
17	$\dfrac{b-a}{(s+a)(s+b)}$	$\epsilon^{-at} - \epsilon^{-bt}$	$\dfrac{z}{z-\epsilon^{-aT}} - \dfrac{z}{z-\epsilon^{-bT}}$	$\dfrac{\epsilon^{-amT}}{z-\epsilon^{-aT}} - \dfrac{\epsilon^{-bmT}}{z-\epsilon^{-bT}}$
18	$\dfrac{(b-a)(s+c)}{(s+a)(s+b)}$	$(c-a)\epsilon^{-at} + (b-c)\epsilon^{-bt}$	$\dfrac{(c-a)z}{z-\epsilon^{-aT}} + \dfrac{(b-c)z}{z-\epsilon^{-bT}}$	$\dfrac{(c-a)\epsilon^{-amT}}{z-\epsilon^{-aT}} + \dfrac{(b-c)\epsilon^{-bmT}}{z-\epsilon^{-bT}}$
19	$\dfrac{ab}{s(s+a)(s+b)}$	$1 + \dfrac{b}{a-b}\epsilon^{-at} - \dfrac{a}{a-b}\epsilon^{-bt}$	$\dfrac{z}{z-1} + \dfrac{bz}{(a-b)(z-\epsilon^{-aT})} - \dfrac{az}{(a-b)(z-\epsilon^{-bT})}$	$\dfrac{1}{z-1} + \dfrac{b\epsilon^{-amT}}{(a-b)(z-\epsilon^{-aT})} - \dfrac{a\epsilon^{-bmT}}{(a-b)(z-\epsilon^{-bT})}$

An Alternate Form of $F^*(s)$

Another form for $F^*(s)$ which is useful in proofs and derivations can be obtained by the application of a Fourier series expansion. (See Tou, 1959 for more detail). An outline of the derivation is given below.

A Fourier series representation of a periodic function $g(t)$ may be written

$$g(t) = \sum_{k=-\infty}^{\infty} C_k e^{jk\omega_s t}$$

where $j = \sqrt{-1}$ and the coefficients C_k are obtained from the following equation in which the integration is done over one period T. For this application, it is convenient to choose the period from $-T/2$ to $T/2$.

$$C_k = \frac{1}{T} \int_{-T/2}^{T/2} g(t) e^{-jk\omega_s t} dt$$

Applying this to $i(t)$, the sequence of unit impulses given by Eq. (22.3), one obtains

$$i(t) = \sum_{k=-\infty}^{\infty} C_k e^{jk\omega_s t}$$

where

$$C_k = \frac{1}{T} \int_{-T/2}^{T/2} \sum_{n=0}^{\infty} \delta(t - nT) e^{-jk\omega_s t} dt$$

In the range of integration from $-T/2$ to $T/2$, the only term in the summation of delayed unit-impulse functions that contributes to the integrand is $\delta(t)$; therefore, we may write the equation for C_k as

$$C_k = \frac{1}{T} \int_{-T/2}^{T/2} \delta(t) e^{-jk\omega_s t} dt$$

One can show that this integral becomes 1; therefore

$$C_k = \frac{1}{T}$$

Equation (22.4) can now be written

$$f^*(t) = f(t) \frac{1}{T} \sum_{k=-\infty}^{\infty} e^{jk\omega_s t}$$

After placing $f(t)$ inside the summation, we take the Laplace transform of each side of the above equation and apply the theorem on the translation of a transform from Chap. 4 to each term on the right side; the result is

$$F^*(s) = L\{f^*(t)\} = \frac{1}{T} \sum_{k=-\infty}^{\infty} F(s + jk\omega_s) \qquad (22.9)$$

Use will be made of this expression in the next chapter.

CHAPTER
23

OPEN-LOOP AND CLOSED-LOOP RESPONSE

To calculate the open-loop response of a sampled-data system, one can develop a pulse transfer function that is the counterpart of the transfer function for continuous systems.

OPEN-LOOP RESPONSE

Pulse Transfer Function

Consider the block diagram in Fig. 23.1 in which an impulse-modulated signal enters a block having the transfer function $G(s)$. We may write

$$C(s) = G(s)F^*(s) \tag{23.1}$$

Let there be a fictitious sampler attached to the output of $G(s)$. Using the alternate definition for a starred transform from the previous chapter [Eq. (22.9)], the sampled function $C^*(s)$ in Fig. 23.1 may be written:

$$C^*(s) = \frac{1}{T} \sum_{n=-\infty}^{\infty} C(s + jn\omega_s) = \frac{1}{T} \sum_{n=-\infty}^{\infty} G(s + jn\omega_s)F^*(s + jn\omega_s) \tag{23.2}$$

As shown in the appendix of this chapter, $F^*(s)$ is periodic in s with frequency ω_s, which means that

$$F^*(s) = F^*(s + jn\omega_s) \tag{23.3}$$

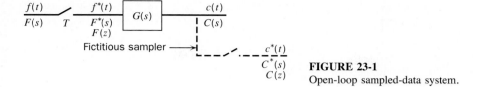

FIGURE 23-1
Open-loop sampled-data system.

Equation (23.2) may be written:

$$C^*(s) = F^*(s) \left\{ \frac{1}{T} \sum_{n=-\infty}^{\infty} G(s + jn\omega_s) \right\} \tag{23.4}$$

Recognizing the term within braces on the right side to be simply $G^*(s)$ according to the alternate definition of the starred transform given by Eq. (22.9), we may write

$$C^*(s) = F^*(s)G^*(s) \tag{23.5}$$

Recalling that the Z-transform is simply the Laplace transform of the starred function in which $z = e^{Ts}$, Eq. (23.5) may be written

$$C(z) = F(z)G(z) \tag{23.6}$$

The term $G(z)$ is called the *pulse transfer function*. Equation (23.6) states that the sampled output is equal to the product of the sampled input and the pulse transfer function. This is analogous to the continuous case where we write $C(s) = F(s)G(s)$. Note that the inverse of $C(z)$ gives information about $c(t)$ only at sampling instants, $0, T, 2T, 3T, \dots, nT$.

Example 23.1. Use of the pulse transfer function. To see how Eq. (23.6) may be used, consider the example shown in Fig. 23.2 in which a triangular wave signal enters the sampler. For this example

$$G(s) = \frac{1}{\tau s + 1} = \frac{1/\tau}{s + \frac{1}{\tau}} \tag{23.7}$$

From a table of Z-transforms (Table 22.1) we obtain for this $G(s)$

$$G(z) = \frac{1}{\tau} \frac{z}{z - e^{-T/\tau}} \tag{23.8}$$

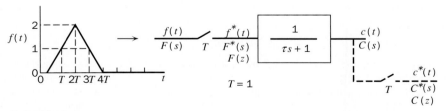

FIGURE 23-2
Example of an open-loop system.

Using the basic definition of a Z-transform in Eq. (22.8), we may express the output of the sampler as:

$$F(z) = \sum_{0}^{\infty} f(nT)z^{-n} = 0 + z^{-1} + 2z^{-2} + z^{-3} + 0 \qquad (23.9)$$

Applying Eq. (23.6) gives

$$C(z) = F(z)G(z) = \left(z^{-1} + 2z^{-2} + z^{-3}\right)\left\{\frac{1}{\tau}\frac{z}{z - e^{-T/\tau}}\right\} \qquad (23.10)$$

or

$$C(z) = \frac{1}{\tau}\frac{1 + 2z^{-1} + z^{-2}}{z - e^{-T/\tau}} \qquad (23.11)$$

For the purpose of having a numerical result, let $T = 1$ and $\tau = 1$. Then $T/\tau = 1$ and $e^{-T/\tau} = 0.368$.

The problem now facing us is the inversion of Eq. (23.11). Two methods will be discussed: (1) long division and (2) use of a table of Z-transforms. To apply the method of long division, we simply divide the denominator of Eq. (23.11) into its numerator as shown here.

$$
\begin{array}{r}
z^{-1} + 2.368z^{-2} + 1.871z^{-3} + \cdots \\
z - 0.368 \enclose{longdiv}{1 \quad + 2z^{-1} \quad + \quad z^{-2}} \\
\underline{1 \quad - 0.368z^{-1}} \\
2.368z^{-1} + \quad z^{-2} \\
\underline{2.368z^{-1} - 0.871z^{-2}} \\
1.871z^{-2}
\end{array}
$$

From this result, we may write

$$C(z) = z^{-1} + 2.368z^{-2} + 1.871z^{-3} + \cdots \qquad (23.12)$$

Interpretation of Eq. (23.12) in the time domain may be done with the aid of the basic definition of the Z-transform of Eq. (22.8) by recognizing the coefficients of the terms on the right side of Eq. (23.12) to be the values of $c(t)$ at sampling instants; thus

$$c(0) = 0$$
$$c(T) = 1.0$$
$$c(2T) = 2.368$$
$$c(3T) = 1.871$$

and so on

Recall that the inversion of the Z-transform gives information about the continuous function $c(t)$ only at sampling instants. The values of $c(t)$ at times other than sampling instants must be obtained by some other means. Later, we shall show that the modified Z-transform can be used to obtain intersample information.

One may also apply basic knowledge of the response of the system to determine $c(t)$ at times between sampling instants. For a first-order response, this

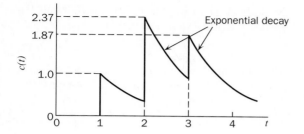

FIGURE 23-3
Response of open-loop system in Fig. 23.2 ($T = 1$).

approach is quite easy. In Figure 23.3 is shown the continuous response of $c(t)$. Between sampling instants, the response decays exponentially.

Inversion of Z-transform

Two methods often used to invert Z-transforms are:

1. Method of long division. (This method was just covered in the previous example.)
2. Method of partial fraction expansion.

The method of partial fraction expansion follows the same procedure as that for inversion of ordinary Laplace transforms. To illustrate this method, consider the following example.

Example 23.2. Invert the following $C(z)$:

$$C(z) = \frac{z}{(z - a)(z - b)}$$

This may be written:

$$\frac{C(z)}{z} = \frac{1}{(z - a)(z - b)} = \frac{A}{z - a} + \frac{B}{z - b} \tag{23.13}$$

The reason for placing z in the denominator on the left side is for mathematical convenience, as will be shown later. Evaluating the constants A and B gives:

$$A = 1/(a - b) \qquad \text{and} \qquad B = -1/(a - b)$$

We may now write

$$C(z) = \frac{1}{a - b} \left[\frac{z}{z - a} - \frac{z}{z - b} \right] \tag{23.14}$$

Each term within the brackets can be inverted by referring to a table of transform pairs. From Table 22.1, we have the transform pair

$$z/(z - a) : a^{t/T}$$

At sampling instants, $a^{t/T}$ becomes $a^{nT/T}$ or a^n. In a similar manner, the inverse of $z/(z - b)$ is b^n. Using these results gives

$$c(nT) = \frac{1}{a - b} [a^n - b^n] \tag{23.15}$$

For the problem solved by long division in Example 23.1, we obtain the following result if the method of partial fraction expansion is used:

$$c(nT) = e^{-(n-1)} + 2e^{-(n-2)} + e^{-(n-3)} \qquad (23.16)$$

The three terms in brackets do not apply until $n = 1$, $n = 2$, and $n = 3$, respectively. To obtain this result, we need to use a theorem on the Z-transform of a translated function, which will be discussed later.

Comparison of Methods of Inversion

1. *Long division.* This method is good when one is interested in the solution for only the first few sampling instants. One must be careful not to make errors in performing the long division, for an error at one sampling instant will propagate errors at later sampling instants. The method of long division can be programmed for a computer to obtain an error-free solution for as many sampling instants as desired. A computer program written in BASIC for long division is given in the appendix of this chapter for use by the interested reader.
2. *Partial fraction expansion.* This method requires the usual algebra of partial fraction expansion. However, once the solution is obtained, the response at any sampling instant can be found without relying on values at previous sampling instants.

CLOSED-LOOP RESPONSE

The closed-loop response for a sampled-data system can be obtained in a manner similar to that for continuous closed-loop systems. However, there are some differences that will be explained in this section.

Consider the sampled-data negative feedback control system shown in Fig. 23.4. In this process, clamping is provided by the combination of an impulse modulator and a zero-order hold. To obtain expressions that relate C to R or C to U, we proceed as follows. From the diagram, we can write:

$$C(s) = G_c(s)G_p(s)R(s) - G_c(s)G_p(s)G_h(s)C^*(s) + G_p(s)U(s) \qquad (23.17)$$

This expression is obtained by combining the signals resulting from $R(s)$, $C^*(s)$, and $U(s)$ after they move through their respective paths to the output C. This expression may also be written as

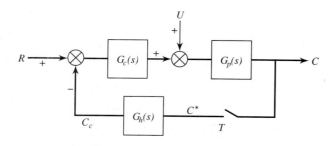

FIGURE 23-4
Closed-loop sampled-data system.

$$C(s) = \overline{G_cG_pR}(s) - \overline{G_cG_pG_h}(s)C^*(s) + \overline{G_pU}(s) \qquad (23.18)$$

where the overbar above several terms means that the functions of s corresponding to each term are multiplied together to form one function of s. Further discussion of the usefulness of the overbar will be given later.

Taking the starred transform of both sides of Eq. (23.18) gives

$$C^*(s) = \overline{G_cG_pR}^*(s) - \overline{G_cG_pG_h}^*(s)C^*(s) + \overline{G_pU}^*(s) \qquad (23.19)$$

Again, it must be pointed out that the overbar above several terms means that these terms are multiplied together before taking the starred transform. The middle term on the right side of Eq. (23.18) requires special attention; taking the starred transform of $\overline{G_cG_pG}_h(s)C^*(s)$ gives $\overline{G_cG_pG}_h^*(s)C^*(s)$. The explanation of this result is given in the appendix.

Solving for $C^*(s)$ gives

$$C^*(s) = \frac{\overline{G_cG_pR}^*(s)}{1 + \overline{G_cG_pG_h}^*(s)} + \frac{\overline{G_pU}^*(s)}{1 + \overline{G_cG_pG_h}^*(s)} \qquad (23.20)$$

The starred functions of s appearing in Eq. (23.20) can be converted to functions of z formally by letting $z = e^{Ts}$ as was done in the previous chapter to obtain the definition of a Z-transform. The result is

$$C(z) = \frac{\overline{G_cG_pR}(z)}{1 + G(z)} + \frac{\overline{G_pU}(z)}{1 + G(z)} \qquad (23.21)$$

where $G(z) = \overline{G_cG_pG_h}(z)$

The overbar above a group of terms in Eq. (23.21), such as G_pU, is useful in reminding one that the functions of s corresponding to each term in the group are multiplied together to form one function of s before the starred transform or the equivalent Z-transform of the group of terms is taken. For example, one cannot obtain $G_pU(z)$ by looking up the Z-transform of $G_p(s)$ and then looking up the Z-transform of $U(s)$ in tables and multiplying these two functions of z together, i.e., it is wrong to write

$$\overline{UG}_p(z) = U(z)G_p(z) \qquad \textit{Wrong}$$

The correct procedure is to obtain one function of s, written as $\overline{G_pU}(s)$, and to use this combined function of s to look up its Z-transform. Although the overbar is useful to remind one that the terms must be multiplied together before taking the Z-transform, this convention is not always used in the literature or in this book, for it is often difficult and inconvenient to place a bar above a group of terms. For this reason, the overbar will not be used in the equations following this section. If a group of terms are multiplied together followed by the argument z, the overbar above the terms will be understood. The examples to follow should clear up this rather subtle, mysterious taking of a Z-transform of a group of terms.

The two expressions on the right side of Eq. (23.21) may be said to contain the transfer functions relating C to R and C to U. However, this is not strictly true, for R cannot be separated from G_cG_p as in the case of a continuous system.

To clarify this important point, the expression for $C(z)$ for a change in set point only (i.e., $U = 0$) is

$$C(z) = \frac{\overline{G_c G_p R(z)}}{1 + G(z)} \tag{23.22}$$

It is wrong to write

$$\frac{C(z)}{R(z)} = \frac{\overline{G_c G_p(z)}}{1 + G(z)} \qquad Wrong$$

In other words, the term $G_c G_p R(z)$ must be worked out for each $R(s)$ to be studied. A similar comment applies to the term $G_p U(z)$. These subtle points in the correct use of Eq. (23.21) can be made much clearer by the example shown later.

Table Relating Z-Transform Outputs to Sampled-Data Systems

Obtaining the expression $C(z)$ in Eq. (23.22) for the sampled-data block diagram of Fig. 23.4 requires considerable effort. As we shall see in later chapters, other sampled-data block diagrams occur for which an expression for $C(z)$ is needed. In the literature (Tou, 1959), one can find tables of various types of sampled-data block diagrams with the corresponding expressions for $C(z)$. A short table, which will be useful later, is shown in Table 23.1. This table also lists the modified Z-transform $C(z,m)$, which will be discussed in Chap. 25. (Notice that the overbar is not used in this table.) It is important to know how to interpret the entries in the table. For the diagram in Fig. 23.4, we see that item 2 in Table 23.1 applies. For this case

$$C(z) = \frac{GR(z)}{1 + GH(z)} \tag{23.23}$$

The expression $GR(z)$ in Eq. (23.23) is equal to $G_c G_p R(z)$ in Fig. 23.4 and $GH(z)$ is equal to $G_c G_p G_h(z)$. Using these equivalent expressions for $GR(z)$ and $GH(z)$, we write directly from Table 23.1

$$C(z) = \frac{G_c G_p R(z)}{1 + G_c G_p G_h(z)}$$

This agrees with Eq. (23.22).

Example 23.3.   Closed-loop response.   For the diagram shown in Fig. 23.4, let

$$G_p = 1/(\tau s + 1) \qquad G_c = K$$

The process is now equivalent to the sampled-data control of a first-order process with proportional control. For this specific example, we obtain for $G(s) = G_c G_p G_h(s)$

$$G(s) = \frac{K\left(1 - e^{-Ts}\right)}{s\left(\tau s + 1\right)} = K\left(1 - e^{-Ts}\right)\frac{1/\tau}{s\left(s + \frac{1}{\tau}\right)} \tag{23.24}$$

TABLE 23.1
Output Z-transforms for sampled-data systems

Sampled-data system	$C(z)$ and $C(z, m)$
1.	$C(z) = G(z)R(z)$ $C(z, m) = G(z, m)R(z)$
2.	$C(z) = \dfrac{GR(z)}{1 + GH(z)}$ $C(z, m) = RG(z, m)$ $\qquad - \dfrac{GH(z, m)RG(z)}{1 + GH(z)}$
3.	$C(z) = \dfrac{G(z)R(z)}{1 + GH(z)}$ $C(z, m) = \dfrac{G(z, m)R(z)}{1 + GH(z)}$
4.	$C(z) = \dfrac{G_2 U(z)}{1 + G_1 G_2 H(z)}$ $C(z, m) = U G_2(z, m)$ $\qquad - \dfrac{U G_2(z) G_1 G_2 H(z, m)}{1 + G_1 G_2 H(z)}$
5.	$C(z) = \dfrac{G_1(z)G_2(z)R(z)}{1 + G_1(z)G_2 H(z)}$ $C(z, m) = \dfrac{G_1(z)G_2(z, m)R(z)}{1 + G_1(z)G_2 H(z)}$
6.	$C(z) = \dfrac{G_2 U(z)}{1 + G_1(z)G_2 H(z)}$ $C(z, m) = U G_2(z, m)$ $\qquad - \dfrac{G_1(z)U G_2 H(z)G_2(z, m)}{1 + G_2 H(z)G_1(z)}$

$G(s)$ may now be written as two terms:

$$G(s) = K\left[\frac{1/\tau}{s[s + (1/\tau)]} - \frac{(1/\tau)e^{-Ts}}{s[s + (1/\tau)]}\right] \tag{23.25}$$

To obtain $G(z)$ for Eq. (23.25), the Z-transform of each term on the right must be found. The first term can be transformed easily by using the following transform pair from Table 22.1:

$$\frac{a}{s(s + a)} : \frac{z\left(1 - e^{-aT}\right)}{(z - 1)\left(z - e^{-aT}\right)}$$

The second term can be transformed with the use of the following important theorem.

Theorem. If $g(t)$ is Z-transformable and has the Z-transform $G(z)$, then the Z-transform of the delayed function $g(t - nT)$ is given by

$$Z\{g(t - nT)\} = z^{-n}G(z) \tag{23.26}$$

This theorem applies when the delay time nT is an integral number of sampling periods. This theorem applies only when $g(t) = 0$ for $t < 0$, a condition that will always apply in this book. The proof of this theorem can be found in other references (Tou, 1959). Note that this theorem is similar to the one for the transform of a delayed continuous function (i.e., $L\{g(t - \tau)\} = e^{-\tau s}G(s)$).

Applying the transform pair and the theorem just given to the terms on the right side of Eq. (23.25) gives

$$G(z) = K\frac{z\left(1 - e^{-T/\tau}\right)}{(z - 1)\left(z - e^{-T/\tau}\right)} - K\frac{z^{-1}z\left(1 - e^{-T/\tau}\right)}{(z - 1)\left(z - e^{-T/\tau}\right)} \tag{23.27}$$

or

$$G(z) = \frac{Kz\left(1 - z^{-1}\right)\left(1 - e^{-T/\tau}\right)}{(z - 1)\left(z - e^{-T/\tau}\right)} \tag{23.28}$$

Simplifying gives

$$G(z) = K\frac{1 - b}{z - b} \tag{23.29}$$

where $b = e^{-T/\tau}$.

To obtain $G_cG_pR(z)$ for a unit-step change in R, we proceed as follows:

$$G_cG_pR(s) = \frac{K}{s(\tau s + 1)} = K\frac{1/\tau}{s[s + (1/\tau)]} \tag{23.30}$$

Taking the Z-transform

$$G_cG_pR(z) = \frac{Kz(1 - b)}{(z - 1)(z - b)} \tag{23.31}$$

For a unit-step change in R, Eq. (23.21) becomes

$$C(z) = \frac{G_cG_pR(z)}{1 + G(z)} = \frac{Kz(1 - b)/(z - 1)(z - b)}{1 + [K(1 - b)/(z - b)]} \tag{23.32}$$

or

$$C(z) = \frac{K(1 - b)z}{(z - 1)[z - b + K(1 - b)]} \tag{23.33}$$

INVERSION. The inversion of $C(z)$ may be obtained by long division or by partial fraction expansion. Using the latter method, we proceed as follows

$$\frac{C(z)}{z} = \frac{K(1 - b)}{(z - 1)\{z - [b - K(1 - b)]\}} \tag{23.34}$$

or

$$\frac{C(z)}{z} = \frac{K(1 - b)}{(z - 1)(z - \alpha)} = \frac{A}{z - 1} + \frac{B}{z - \alpha} \tag{23.35}$$

where $\alpha = b - K(1 - b)$. Solving for the constants A and B gives $A = K/(1 + K)$, $B = -K/(1 + K)$. Inverting by means of the table of transforms gives

$$c(nT) = \frac{K}{1 + K}(1^n - \alpha^n) = \frac{K}{1 + K}\{1 - [b - K(1 - b)]^n\} \tag{23.36}$$

Stability. From the result given by Eq. (23.36), the stability of this system can be studied as follows.

For $c(nT)$ to converge, $|b - K(1 - b)| < 1$

Note that $b < 1$ since $b = e^{-T/\tau}$ and T/τ is positive.
The inequality may be written in two ways:

I. $b - K(1 - b) < 1$
II. $b - K(1 - b) > -1$

For **I.**: $\quad -b + K(1 - b) > -1$
$$K(1 - b) > -1 + b$$
$$K > -(1 - b)/(1 - b)$$

or $\qquad\qquad K > -1$

Since K is always positive, this result is of no practical value.

For **II.**: $\quad b - K(1 - b) > -1$
$$-b + K(1 - b) < 1$$
$$K(1 - b) < (1 + b)$$

$$K < \frac{1 + b}{1 - b} \qquad \text{(requirement for stability)} \tag{23.37}$$

This is a useful result and shows how a sampled-data system differs from a continuous system. For a continuous system, proportional control of a first-order system is always stable. For the sampled-data case, there is a value of controller gain K, above which the system goes unstable.

Transient response. For this example, the transient response consists of connected arcs of exponentials. A typical response is shown in Fig. 23.5. The sampled-data

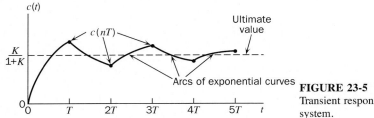

FIGURE 23-5
Transient response of a sampled-data system.

response, $c(nT)$, gives only values of $c(t)$ at sampling instants. The continuous response $c(t)$ is obtained from basic knowledge of the first-order system.

Offset. From Eq. (23.36), one can see that for a stable system, $c(\infty) = K/(1 + K)$. This is the same result that would be obtained for a continuous system under proportional control. In terms of offset, we have

$$\text{Offset} = r(\infty) - c(\infty) = 1 - K/(1 + K) = 1/(1 + K)$$

This result should not be surprising, for once the transient terms have disappeared, which is always the condition under which offset is determined, the sampled-data system is equivalent to the continuous system.

SUMMARY

The methods used to obtain the response of open-loop and closed-loop sampled-data systems are similar to those used for continuous systems. The block diagram for a sampled-data system contains one or more sampling switches. For an open-loop system, the response at sampling instants is obtained by expressing the Z-transform of the output $C(z)$ as the product of the pulse transfer function $G(z)$ and the Z-transform of the forcing function $F(z) : C(z) = G(z)F(z)$. This expression is analogous to the one used for continuous systems: $C(s) = G(s)F(s)$. The inversion of $C(z)$ can be obtained by (1) partial fraction expansion and use of a table of Z-transforms or (2) by long division.

   The method using partial fraction expansion gives an analytical result that can be used to find the response at any sampling instant. The process of long division must be continued until the particular output term of interest is reached; for this reason long division is better suited for obtaining the response during the first few sampling periods. Because of the sampling switches present in a sampled-data system, obtaining the expression for the closed-loop response $C(z)$ requires considerable effort. To assist in this effort, a table relating Z-transform outputs to a variety of closed-loop sampled-data configurations was presented. The response of a sampled-data system containing a transport lag can be obtained easily as an analytical expression; this is in contrast to the difficulty one has for continuous systems that contain transport lag.

AI. Derivation of Eq. (23.3)

The basic definition of $F^*(s)$ given by Eq. (22.9) is

$$F^*(s) = \frac{1}{T} \sum_{k=-\infty}^{\infty} F(s + jk\omega_s)$$

Replacing s by $s + jn\omega_s$, where n is an integer, gives

$$F^*(s + jn\omega_s) = \frac{1}{T} \sum_{k=-\infty}^{\infty} F(s + j(n + k)\omega_s)$$

Let $\mu = n + k$, then the above equation becomes

$$F^*(s + jn\omega_s) = \frac{1}{T} \sum_{\mu=-\infty}^{\infty} F(s + j\mu\omega_s)$$

Since the limits on k are ∞ and $-\infty$, the limits on μ are the same. By the definition of Eq. (22.9), the term on the right is simply $F^*(s)$ and we may write

$$F^*(s + jn\omega_s) = F^*(s)$$

which is Eq. (23.3). We describe this relation by stating that $F^*(s)$ is periodic in s with frequency ω_s.

AII. Taking the Starred Transform of $G_cG_pG_h(s)C^*(s)$ in Eq. (23.18)

In obtaining the closed-loop transfer function for the system in Fig. 23.4, we took the starred transform of $G_cG_pG_h(s)C^*(s)$ in Eq. (23.18) to get $\overline{G_cG_pG_h}^*(s)C^*(s)$. An explanation of this step is as follows.

For convenience, let $G_cG_pG_h(s) = G_1(s)$. Consider the block diagram in Fig. 23.6 in which $G_1(s)$ operates on the sampled value of C, which is $C^*(s)$. From this diagram, we write

$$Y(s) = G_1(s)C^*(s)$$

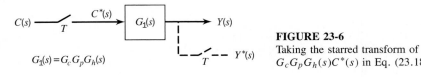

FIGURE 23-6
Taking the starred transform of $G_cG_pG_h(s)C^*(s)$ in Eq. (23.18).

Taking the starred transform of both sides of this equation and using the definition of a starred function given by Eq. (22.9) give

$$Y^*(s) = \frac{1}{T} \sum_{n=-\infty}^{\infty} Y(s + jn\omega_s) = \frac{1}{T} \sum_{n=-\infty}^{\infty} G_1(s + jn\omega_s)C^*(s + jn\omega_s)$$

From Eq. (23.3), we can write

$$C^*(s + jn\omega_s) = C^*(s)$$

therefore

$$Y^*(s) = C^*(s)\left\{ \frac{1}{T} \sum_{n=-\infty}^{\infty} G_1(s + jn\omega_s) \right\}$$

The term in brackets, according to the definition of Eq. (22.9), is $G_1^*(s)$. We can now write

$$Y^*(s) = C^*(s)G_1^*(s)$$

Converting G_1 to the original variables gives

$$Y^*(s) = C^*(s)\overline{G_cG_pG_h}^*(s)$$

AIII. BASIC Program for Long Division

The BASIC program given in Fig. 23.7 is useful to invert a Z-transform by the method of long division. To use the program, one must arrange the Z-transform $C(z)$ in the form

$$C(z) = \frac{a_0 + a_1z^{-1} + a_2z^{-2} + \cdots a_mz^{-m}}{1 + b_1z^{-1} + b_2z^{-2} + \cdots b_nz^{-n}} \tag{23.38}$$

When the computer program is run, one enters the values of $a_0, a_1, \ldots, b_1, b_2, \ldots, m$, and n when requested by the program. One also enters the number of terms desired in the long division. To illustrate the use of the program, we shall do the long division that was done in Example 23.1. After introducing the parameters of Example 23.1 into Eq. (23.11), $C(z)$ becomes

$$C(z) = \frac{1 + 2z^{-1} + z^{-2}}{z - 0.368}$$

To put this in the form of Eq. (23.38), multiply numerator and denominator by z^{-1} and the result is

$$C(z) = \frac{z^{-1} + 2z^{-2} + z^{-3}}{1 - 0.368z^{-1}}$$

From the numerator of this expression, we see that $a_0 = 0$, $a_1 = 1$, $a_2 = 2$, $a_3 = 1$, and $m = 3$. From the denominator, we see that $b_1 = -0.368$ and $n = 1$. Introducing these values into the sample run shown in Table 23.2 gives a result that agrees with that of Example 23.1.

```
5   DIM A(20), B(20), X(20), D(20)
10  PRINT "Z-TRANSFORM INVERSION BY LONG DIVISION"
15  PRINT
20  PRINT "ORDER OF NUMERATIOR, M": INPUT M
25  PRINT
30  PRINT "INPUT NUMERATOR COEFFICIENTS A OF THE FORM:"
40  PRINT "A0 + A1/Z**1 + A2/Z**2 + ... + AM/Z**M"
50  FOR I = 0 TO M
60  PRINT "COEFFICIENT", I: INPUT A(I)
70  NEXT I
75  PRINT
80  PRINT "ORDER OF DENOMINATOR,N": INPUT N
85  PRINT
90  PRINT "INPUT OF DENOMINATOR COEFFICIENTS,B, OF THE FORM:"
100 PRINT "1 + B1/Z**1 + B2/Z**2 + ... + BN/Z**N"
110 PRINT "NOTE THAT B0 = 1"
120 FOR I = 1 TO N
130 PRINT "COEFFICIENT", I: INPUT B(I)
140 NEXT I
142 PRINT
143 PRINT "HOW MANY TERMS DO YOU WANT IN THE INVERSE FORM?"
145 PRINT "C0 + C1/Z**1 + C2/Z**2 + ... + CJ/Z**J"
147 INPUT N3
150 N1 = N: IF M>N THEN N1=M
160 FOR I = 1 TO N1
170 D(I) = A(I)
180 IF I>N THEN B(I) = 0
190 IF I>M THEN D(I) =0
200 NEXT I
210 D(N1+1) = 0
220 IF A(0) = 0 GOTO 270
230 X(0) = A(0)
240 FOR I = 1 TO N: D(I) = D(I) - X(0)*B(I): NEXT I
250 X(1) = D(1)
260 GOTO 280
270 X(0) = 0: X(1) = A(1)
280 PRINT X(0),"/Z** 0 +"
282 PRINT X(1),"/Z** 1 +"
288 FOR J = 2 TO N3
290 FOR K = 1 TO N1: D(K) = D(K+1) - X(J-1)*B(K): NEXT K
300 X(J) = D(1)
310 PRINT X(J),"Z/**";J;"+"
320 NEXT J
330 END
```

FIGURE 23-7
BASIC program for long division.

TABLE 23.2
Output from a BASIC program for long division

```
RUN
Z-TRANSFORM INVERSION BY LONG DIVISION
ORDER OF NUMERATOR, M
? 3

INPUT NUMERATOR COEFFICIENTS A OF THE FORM:
A0 + A1/Z**1 + A2/Z**2 + ... +AM/Z**M
COEFFICIENT    0
? 0
COEFFICIENT    1
? 1
COEFFICIENT    2
? 2
COEFFICIENT    3
? 1

ORDER OF DENOMINATOR,N
? 1

INPUT OF DENOMINATOR COEFFICIENTS, B, OF THE FORM:
1 + B1/Z**1 + B2/Z**2 + ... + BN/Z**N
NOTE THAT B0 = 1
COEFFICIENT    1
? -.368

HOW MANY TERMS DO YOU WANT IN THE INVERSE FORM?
C0 + C1/Z**1 + C2/Z**2 + ... + CJ/Z**J
? 5
0                /Z** 0 +
1                /Z** 1 +
2.368            /Z** 2 +
1.871424         /Z** 3 +
.688684          /Z** 4 +
.2534357         /Z** 5 +
Ok
```

PROBLEMS

23.1. (*a*) For the open-loop system shown in Fig. P23.1, determine $c(nT)$ for $R = \delta(t)$, $u(t)$, $tu(t)$ when $T = 1$ and $T = 0.5$. Sketch the continuous response $c(t)$ for each disturbance.

(*b*) Repeat part *a* for the case in which the zero-order hold is removed.
Note: The complete solution to this problem requires the solution of 12 open-loop problems.

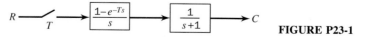

$$R \xrightarrow{\quad\;\; T \quad} \boxed{\frac{1-e^{-Ts}}{s}} \longrightarrow \boxed{\frac{1}{s+1}} \longrightarrow C$$

FIGURE P23-1

23.2. For the sampled-data process in Fig. P23.2, determine
(*a*) $C(z)$.
(*b*) $c(nT)$ for several values of n.
(*c*) Plot the continuous response, $c(t)$.

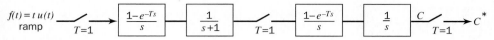

$f(t) = t\,u(t)$ ramp $\quad T=1 \quad \dfrac{1-e^{-Ts}}{s} \quad \dfrac{1}{s+1} \quad T=1 \quad \dfrac{1-e^{-Ts}}{s} \quad \dfrac{1}{s} \quad C \quad T=1 \quad C^*$

FIGURE P23-2

23.3. Consider the transfer function

$$G_{h_1} = \frac{(1 + Ts)}{T}\left(\frac{1 - e^{-Ts}}{s}\right)^2$$

If the function $X(t)$ shown in Fig. P23.3 is fed to an impulse sampler, which is followed by G_{h_1}, determine the output $Y(t)$. Present your results graphically. The term G_{h_1} is called a first-order hold.

$X \quad T=1 \quad G_{h_1} \quad Y$

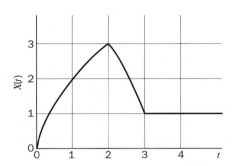

FIGURE P23-3

23.4. For the sampled-data control system shown in Fig. P23.4, determine $c(nT)$ for $K = 1$ and $K = 2$. Sketch the continuous response $c(t)$. Determine the ultimate controller gain.

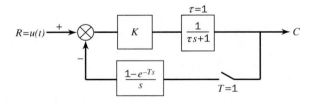

$R=u(t) \quad + \quad \otimes \quad - \quad K \quad \begin{matrix}\tau=1\\ \dfrac{1}{\tau s+1}\end{matrix} \quad C \qquad \dfrac{1-e^{-Ts}}{s} \quad T=1$

FIGURE P23-4

23.5. For the sampled-data control system shown in Fig. P23.5, determine $c(nT)$ for $K = 0.2$. Sketch the continuous response $c(t)$.

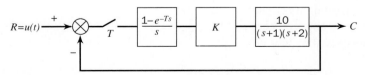

$R=u(t) \quad + \quad \otimes \quad - \quad T \quad \dfrac{1-e^{-Ts}}{s} \quad K \quad \dfrac{10}{(s+1)(s+2)} \quad C$

FIGURE P23-5

CHAPTER
24

STABILITY

We have seen in Example 23.3 of the previous chapter on the proportional, sampled-data control of a first-order system that the question of stability arises. By solving the response $c(nT)$ for this relatively easy problem [Eq. (23.36)] we were able to derive the conditions for stability [Eq. (23.37)]. Using this same approach for finding conditions for stability for higher-order processes can be quite complicated. Fortunately, one may develop general rules of stability that resemble those for continuous systems.

Consider the response of a sampled-data system to be of the form:

$$C(z) = \frac{F_1(z)}{1 + G(z)} = \frac{F_2(z)}{(z - z_1)(z - z_2) \cdots (z - z_n)} \tag{24.1}$$

To obtain the response $c(nT)$, we may expand the right side by the method of partial fractions to obtain

$$C(z) = \frac{F_1(z)}{1 + G(z)} = \left(A_1 \frac{z}{z - z_1} + A_2 \frac{z}{z - z_2} + \cdots \right) z^{-1} \tag{24.2}$$

In anticipation of an entry in the table of transform pairs, each term within the parentheses of Eq. (24.2) is written as $z/(z - z_i)$. The term z^{-1}, which has been placed outside the parentheses to balance the z placed inside, will simply shift the time response by T units and in no way affect the conclusion on stability to be given in the following discussion.

For the moment, consider the roots $z_1, z_2, \ldots$ to be real. We have seen from previous examples that the inverse of a typical term $z/(z - z_i)$ is:

$$Z^{-1}\left\{ \frac{z}{z - z_i} \right\} = z_i^n \tag{24.3}$$

For this term to contribute a bounded response to $c(nT)$ requires that $|z_i| < 1$. We shall now extend this special case of real roots, which has been presented to introduce the subject, to the general case of roots being complex.

GENERAL CONDITIONS FOR STABILITY

The general conditions for stability of a continuous system are that the roots of the characteristic equation fall in the left half of the complex plane. Before the sampled signal $C^*(s)$ is changed to the form $C(z)$ by introducing the transformation $z = e^{Ts}$, the characteristic equation of the sampled system is

$$1 + G^*(s) = 0$$

We may apply the general stability criterion and require that all roots of the characteristic equation be in the left half of the s-plane. When the characteristic equation expressed in the s-domain is transformed to the z-domain through the transformation $z = e^{Ts}$, we get

$$1 + G(z) = 0$$

Consider a typical stable root of the characteristic equation to have the value

$$s = -\sigma + j\omega \qquad \text{where } \sigma > 0$$

This root is shown in the complex s-plane in Fig. 24.1. By applying the transformation $z = e^{Ts}$, we may write

$$z = e^{Ts} = e^{-T\sigma}e^{j\omega T}$$

This expression for z, a complex number, is of the form

$$z = Me^{j\theta} \qquad \text{or} \qquad z = |z| \angle z$$

where M or $|z|$ is the magnitude of the complex number and θ or $\angle z$ is the angle associated with the complex number.

$$\text{Since } \sigma > 0, \ e^{-T\sigma} < 1$$

$$\text{therefore } |z| < 1$$

In terms of the complex z-plane, this result states that stability for a sampled-data control system requires that the roots of the characteristic equation $1 + G(z) = 0$ fall within the unit circle as shown in Fig. 24.2.

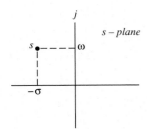

FIGURE 24-1
Root location in complex plane.

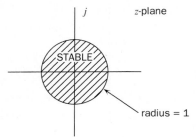

FIGURE 24-2
Region of stability in the z-plane.

Routh Test

The Routh test, which is often used to examine the roots of the characteristic equation of a continuous system (see Chap. 14), may also be used to examine the roots of the characteristic equation of a sampled-data system. Recall that the Routh test detects the presence of roots in the right half plane. Since the criterion of stability of a sampled-data system requires that all roots fall within the unit circle of the z-plane, one must first apply a transformation that will map the inside of the unit circle of the z-plane into the left half of the w-plane. One can then apply the Routh test to discover roots in the right half of the w-plane, and if none are found, we know that the roots of the characteristic equation $1 + G(z) = 0$ fall within the unit circle and that the sampled-data control system is stable.

A transformation that will map the inside of the unit circle of the z-plane into the left half of the w-plane is

$$z = \frac{w + 1}{w - 1} \tag{24.4}$$

This transformation is called the *bilinear* transformation. The regions involving the transformation are shown in Fig. 24.3. An alternate transformation is

$$z = \frac{1 + w}{1 - w} \tag{24.5}$$

The reader should check to see that the transformations given by Eqs. (24.4) and (24.5) do what is claimed. For example, if $w = -1 + j$, a point in the left half of the w-plane, then Eq. (24.4) becomes

$$z = \frac{-1 + j + 1}{-1 + j - 1} = \frac{j}{-2 + j}$$

Multiplying numerator and denominator by $-2 - j$, the complex conjugate of $-2 + j$, gives

$$z = \frac{j}{-2 + j} \frac{-2 - j}{-2 - j} = \frac{1 - 2j}{5} = \frac{1}{5} - j\frac{2}{5} \qquad \text{a point inside the unit circle}$$

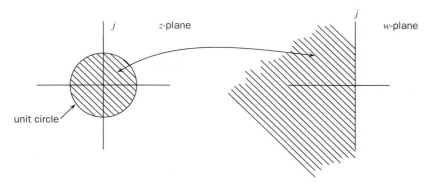

FIGURE 24-3
Transformation from the z-plane to the w-plane.

A general proof that the transformation maps the inside of the unit circle into the left half plane is given here. Solving Eq. (24.4) for w gives

$$w = \frac{z + 1}{z - 1} \tag{24.6}$$

Let $z = x + jy$. Equation (24.6) becomes

$$w = \frac{x + jy + 1}{x + jy - 1} = \frac{(x + 1) + jy}{(x - 1) + jy} \tag{24.7}$$

or

$$w = \frac{(x + 1) + jy \, (x - 1) - jy}{(x - 1) + jy \, (x - 1) - jy} \tag{24.8}$$

Multiplying out the factors in the numerator and the denominator gives, after algebraic rearrangement

$$w = \frac{x^2 + y^2 - 1}{(x - 1)^2 + y^2} - j\frac{2y}{(x - 1)^2 + y^2} \tag{24.9}$$

We may now use the analytical expression for a unit circle, $x^2 + y^2 = 1$, to complete the proof. If a point is inside the unit circle of the z-plane,

$$|z| < 1 \quad \text{and} \quad x^2 + y^2 < 1$$

Introducing this inequality into Eq. (24.9) leads to the result that the real part of w is negative; thus:

$$\text{Re}\{w\} < 0 \tag{24.10}$$

Since this is equivalent to stating that the values of w fall in the left half plane, we have completed the general proof.

Root Locus

One may determine the stability of a sampled-data system by plotting the root locus diagram for the characteristic equation. In this case, there is no need to use a transformation, as is needed in applying the Routh test. In general, the open-loop transfer function for the sampled-data system can be placed in the form

$$G(z) = K\frac{(z - v_1)(z - v_2)\cdots}{(z - p_1)(z - p_2)\cdots} \tag{24.11}$$

where $v_1, v_2, \ldots$ are the zeros of the open-loop transfer function and $p_1, p_2, \ldots$ are the poles of the open-loop transfer function.

To obtain the root locus plot for $1 + G(z) = 0$, one places the open-loop zeros and poles on the complex plane and applies the angle criterion used in root locus construction. The stability boundary occurs when one of the branches of the root locus diagram crosses the unit circle. To find the gain K at the stability boundary, one applies the magnitude criterion of root locus construction. (See Chap. 15.)

Example 24.1. The stability of proportional control of a first-order process will be examined. This same problem was presented as Example 23.3. Both the Routh test and the root locus method will be used. The system is shown in Fig. 24.4. For this system, we have shown in Eq. (23.29) that

$$G(z) = K\frac{1 - b}{z - b}$$

where $b = e^{-T/\tau}$.

Using the transformation given by Eq. (24.4), we obtain for $1 + G(z) = 0$

$$1 + \frac{K(1 - b)}{\dfrac{w + 1}{w - 1} - b} = 0 \tag{24.12}$$

or

$$1 + \frac{K(1 - b)(w - 1)}{w + 1 - b(w - 1)} = 0 \tag{24.13}$$

Rearranging this result in polynomial form for applying the Routh test gives

$$[(K + 1)(1 - b)]w + [(1 + b) - K(1 - b)] = 0 \tag{24.14}$$

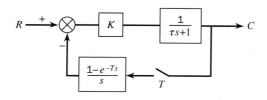

FIGURE 24-4
Block diagram for Example 24.1.

The Routh array for this expression becomes

Row	term
1	$(K + 1)(1 - b)$
2	$(1 + b) - K(1 - b)$

Since $b = e^{-T/\tau}$ is always positive and less than one and K is positive, the first element in the array is positive. For stability, the Routh test requires that all elements of the first column be positive. Therefore,

$$(1 + b) - K(1 - b) > 0$$

or

$$1 + b > K(1 - b)$$

or

$$K < \frac{1 + b}{1 - b} \qquad \text{for stability}$$

This is the same result given by Eq. (23.37), which was obtained from an expression for $c(nT)$.

We shall now use the root locus method on the same example. For this simple problem, there is only one pole of the open-loop transfer function, $G(z)$, which is located at b as shown in Fig. 24.5. The root locus consists of one branch that moves from the pole at b along the real axis to the left. The intersection of this branch with the unit circle at $z = -1$ gives us the stability boundary. Using the magnitude criterion of root locus construction gives

$$\frac{K(1 - b)}{|z_i - p_1|} = 1$$

We can obtain the value of K at the stability boundary by solving for K; thus

$$K = \frac{|z_i - p_1|}{1 - b} = \frac{1 + b}{1 - b}$$

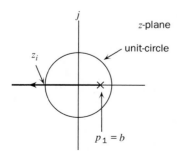

FIGURE 24-5
Root locus plot for Example 24.1.

Since the root locus branch moves to the left with increase of K, we see that

$$K < \frac{1+b}{1-b} \quad \text{for stability}$$

For this simple example, the root locus method is easier, for one does not need to use a transformation and the root locus diagram is very simple. However, for higher-order systems, the apparent advantage of the root locus method over the Routh test is lost. To appreciate the details of applying the stability criterion to sampled-data systems, the reader is encouraged to work a few of the more complex problems at the end of this chapter.

Other methods for determining stability of sampled-data systems include the Schur-Cohn criterion and the Jury criterion (see Jury, 1964 and Tou, 1959). The Jury criterion is a simplification of the Schur-Cohn criterion. These methods, which can be applied directly to the characteristic equation written in the z-domain, can detect roots outside the unit circle of the z-plane. Since these methods require the evaluation of high-order determinants, they are limited to simple systems.

SUMMARY

The presence of sampling in a control system contributes to instability. The criterion for stability of a sampled-data system requires that the roots of the characteristic equation, $1 + G(z) = 0$, fall within the unit circle of the complex z-plane. Based on this criterion, two methods were developed to determine stability: (a) the modified Routh test and (b) the root locus method. To use the Routh test, one must first apply the bilinear transformation, which maps the inside of the unit circle into the left half of the w-plane. The usual rules of the Routh test are then applied to the transformed characteristic equation. Using the root locus method is simpler, for one simply constructs the root locus diagram from the poles and zeroes of the open-loop transfer function $G(z)$. When a branch of the root locus diagram crosses the unit circle, the system becomes unstable. It is of interest to note that systems having transport lag can be analyzed easily for stability in sampled-data systems by either the modified Routh test or the root locus method; this was not the case for continuous systems having transport lag.

PROBLEMS

24.1. For the system shown in Fig. P23.5, determine the ultimate gain by use of the Routh test and by use of the root locus method.

24.2. For the control system shown in Fig. P24.2, determine
   (a) an expression for $C(z)$ when a unit-step change occurs in U, R remaining 0.
   (b) the stability criteria for the control system.
   (c) plot the continuous response $c(t)$ for at least a period of time equal to $2T$. Obtain this information from basic knowledge of first-order systems.
   Note: Clamping is not present in this system, for there is no zero-order hold in the block diagram.

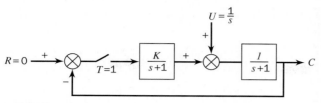

FIGURE P24-2

24.3. The sampled-data control system shown in Fig. P24.3 contains a first-order hold, for which the transfer function is

$$G_{h_1} = \frac{(1 + Ts)}{T} \left(\frac{1 - e^{-Ts}}{s} \right)^2$$

(a) Determine $G(z)$ for the closed-loop response.
(b) Determine the value of K for which the closed-loop response is on the verge of instability by means of the root locus method. Sketch the root locus diagram.
(c) If a zero-order hold were used in place of the first-order hold, what would be the value of K for the system to be on the verge of instability?

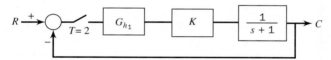

FIGURE P24-3

24.4. One can show that for the sampled-data system in Fig. P24.4

$$G(z) = \frac{K_1(z + \alpha)}{z(z - b)}$$

where $\alpha = 0.517$, $b = 0.607$, K_1 is proportional to K.

Draw accurately the root-locus diagram and from it determine the ultimate value of K, above which the system is unstable.

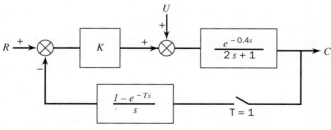

FIGURE P24-4

CHAPTER
25

MODIFIED
Z-TRANSFORMS

The inversion of the Z-transform $C(z)$ gives information about $c(t)$ only at sampling instants. This, of course, is a result of introducing the sampling switch. The "simplicity" of the mathematics of Z-transforms must be paid for by the limited information about $c(t)$. For some processes, knowing the response at sampling instants is quite sufficient. However, if one wants information between sampling instants (intersample information, as it is called), a procedure other than the use of Z-transforms is required.

One method, which can be very tedious except for first-order processes, is to compute the continuous response $c(t)$ by solving the differential equations describing the process. If one were to go through this much effort, there would be little reason to use Z-transforms in the first place.

Another method for finding intersample information is to use the modified Z-transform. This method is nearly as easy to use as the ordinary Z-transform and gives the intersample information about the response at any desired time between sampling instants.

Another reason for introducing the modified Z-transform is to have a method for obtaining the pulse transfer function of a system that includes a transport lag $(e^{-\tau s})$ for which τ is not equal to an integral number of sampling periods $(\tau \neq nT)$. The development of such a pulse transfer function will be described in the next chapter.

DEVELOPMENT OF MODIFIED Z-TRANSFORM

Consider the process shown in Fig. 25.1 in which a fictitious delay $e^{-\lambda Ts}$ has been placed after the block $G(s)$. The value of λ is between 0 and 1. The use of ordinary Z-transforms will give $c(nT)$, which is obtained by inverting $C(z)$. From Eq. (23.6), $C(z)$ can be found quite simply from the expression:

$$C(z) = G(z)F(z) \tag{25.1}$$

To obtain $c(t)$ at times other than sampling instants, $c(t)$ is delayed (or translated) by an amount λT before sampling. The choice of λ gives the desired intersample value of $c(t)$.

Before developing the definition of the modified Z-transform, Fig. 25.2 provides a simple example that will clarify the timing of signals in Fig. 25.1. In Fig. 25.2 $f(t) = u(t)$, a unit-step function, $G(s) = 1/(s + 1)$, and $\lambda = 0.7$. By studying Fig. 25.2, one can see the nature of the signal at each position in the diagram. Notice that the continuous signal $c_\lambda(t)$ from the delay block is the response $c(t)$ shifted by $0.7T$ to the right. The sampled response $c_\lambda^*(t)$ consists of a train of impulses, the magnitudes of which equal the values of $c(t)$ at $0.3T$ into each sampling period. As will be shown later $c_\lambda(nT)$ gives the intersample information that is provided by the modified Z-transform.

We may now turn to the general development of the modified Z-transform. From Fig. 25.1, we may write

$$C_\lambda(s) = G_\lambda(s)F^*(s) \tag{25.2}$$

where $G_\lambda(s) = G(s)e^{-\lambda Ts}$

Taking the starred transform of this expression:

$$C_\lambda^*(s) = G_\lambda^*(s)F^*(s) \tag{25.3}$$

The basis for performing this last step has been discussed in detail in connection with Eq. (23.5).

To develop the modified Z-transform, consider separately the processing of $c(t)$ as shown in Fig. 25.3. To find the Z-transform $C_\lambda(z)$ corresponding to $C_\lambda^*(s)$,

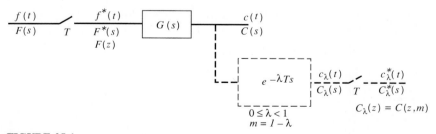

FIGURE 25-1
Development of modified Z-transform.

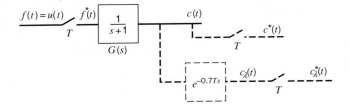

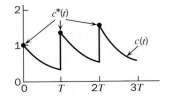

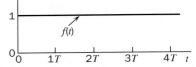

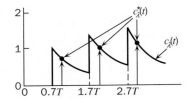

FIGURE 25-2
Example to illustrate the development of the modified Z-transform ($\lambda = 0.7$ or $m = 0.3$).

we may use the definition of Eq. (22.8) to write

$$C_\lambda(z) = \sum_{n=0}^{\infty} c(nT - \lambda T)z^{-n} \tag{25.4}$$

Since we work only with functions of t for which the function is zero for $t < 0$, we have for $n = 0$

$$c(0 - \lambda T) = c(-\lambda T) = 0$$

FIGURE 25-3
Delay and sampling of $c(t)$.

Equation (25.4) may now be written

$$C_\lambda(z) = \sum_{n=1}^{\infty} c(nT - \lambda T)z^{-n}$$
(25.5)

If we let $m = 1 - \lambda$, we may write for the argument of c in Eq. (25.5)

$$nT - \lambda T = nT - (1 - m)T = (n - 1)T + mT$$
(25.6)

Equation (25.5) may now be written

$$C_\lambda(z) = \sum_{n=1}^{\infty} c[(n - 1)T + mT]z^{-n}$$
(25.7)

If we let $n' = n - 1$, Eq. (25.7) may be written

$$C_\lambda(z) = \sum_{n'=0}^{\infty} c[(n' + m)T]z^{-1}z^{-n'}$$
(25.8)

or

$$C_\lambda(z) = z^{-1} \sum_{n'=0}^{\infty} c[(n' + m)T]z^{-n'}$$
(25.9)

This last expression is the definition of the modified Z-transform. Replacing the index n' with n, to avoid an awkward symbol in the definition, we have the expression for the modified Z-transform:

$$C_\lambda(z) = C(z,m) = z^{-1} \sum_{n=0}^{\infty} c[(n + m)T]z^{-n}$$
(25.10)

The symbol $C(z,m)$ has replaced $C_\lambda(z)$ and $m = 1 - \lambda$.

Tables of transform pairs have been developed that relate a function $f(t)$ to its modified Z-transform. Table 22.1 provides the modified Z-transforms for the functions of t listed in the table.

PULSE TRANSFER FUNCTION FOR MODIFIED Z-TRANSFORM

Returning to Eq. (25.3), we may write

$$C_\lambda^*(s) = G_\lambda^*(s)F^*(s)$$
(25.11)

Writing this in terms of Z-transforms, we have

$$C_\lambda(z) = G_\lambda(z)F(z)$$
(25.12)

The convention has been established to replace the Z-transform of the delayed function, such as $C_\lambda(z)$ in Eq. (25.12), with the symbol $C(z,m)$ where m is related to λ by the relation $m = 1 - \lambda$. Changing the subscripted symbols in Eq. (25.12) according to this convention gives the equivalent expression

$$C(z,m) = G(z,m)F(z) \tag{25.13}$$

where $G(z,m)$ is the Z-transform corresponding to $G_\lambda(s)$. Remember that $G_\lambda(s)$ is simply the transfer function for the process $G(s)$ multiplied by the transfer function for the fictitious delay; thus:

$$G_\lambda(s) = G(s)e^{-\lambda Ts} \tag{25.14}$$

To find $G(z,m)$ one may refer to a table of transforms and find the entry $G(z,m)$ corresponding to the desired $G(s)$.

SUMMARY OF USE OF THE MODIFIED Z-TRANSFORM.   To find the output of the block diagram in Fig. 25.4 at times other than sampling instants, one uses the modified Z-transform and writes

$$C(z,m) = G(z,m)F(z) \tag{25.15}$$

It should be realized that $C(z,m)$ is simply a Z-transform and that it can be inverted by the same procedures used for inverting ordinary Z-transforms. Furthermore, the inversion gives information about the response only at sampling instants. The result from the inversion of the modified Z-transform gives the values of $c(t)$ between sampling instants. By choosing m between 0 and 1, one can obtain the values of $c(t)$ at any desired time within the sampling intervals.

For convenience, one may apply the following rule to determine the effect of the size of m on the time into the sampling intervals.

> **Rule.**   Inversion of $C(z,m)$ gives a response that is equivalent to stepping back one sampling period in $c(t)$ and advancing mT units of time. Thus, for $m = 1$, there is no delay and the result is the same as that obtained from the ordinary Z-transform. This can be seen from the fact that $\lambda = 1 - m = 1 - 1 = 0$. On the other extreme, for $m = 0$, one has a delay that approaches a full sampling period.

The system presented earlier in Fig. 25.2 for the purpose of explaining the timing of the various signals will now be worked as an example using the method just discussed.

> **Example 25.1.**   Obtain by means of ordinary and modified Z-transforms the response for the system shown in Fig. 25.5. Determine the response $c(t)$ at sampling instants and at times positioned $0.3T$ from the beginning of sampling instants.

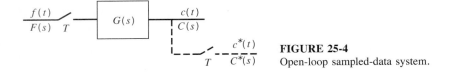

FIGURE 25-4
Open-loop sampled-data system.

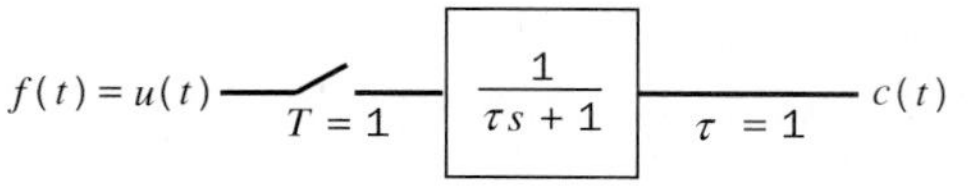

$$f(t) = u(t) \xrightarrow{\quad T = 1 \quad} \boxed{\dfrac{1}{\tau s + 1}} \xrightarrow{\quad \tau = 1 \quad} c(t)$$

FIGURE 25-5
Open-loop system of Example 25.1.

Using ordinary Z-transforms, one obtains:

$$C(z) = G(z)F(z) \tag{25.16}$$

From the table of transforms,

$$G(z) = \frac{1}{\tau} \frac{z}{z - e^{-T/\tau}} \qquad F(z) = \frac{z}{z - 1} \tag{25.17}$$

Introducing these expressions into Eq. (25.16) gives

$$C(z) = \frac{1}{\tau} \frac{z}{z - e^{-T/\tau}} \frac{z}{z - 1} = \frac{1}{\tau} \frac{z^2}{(z - e^{-T/\tau})(z - 1)} \tag{25.18}$$

We next determine the intersample response by using the modified Z-transform as expressed by Eq. (25.13). From the table of transforms, we obtain for $G(z,m)$

$$G(z, m) = \frac{1}{\tau} \frac{e^{-mT/\tau}}{z - e^{-T/\tau}} \tag{25.19}$$

$C(z,m)$ then becomes

$$C(z, m) = \frac{1}{\tau} \frac{e^{-mT/\tau}}{z - e^{-T/\tau}} \frac{z}{z - 1} \tag{25.20}$$

Inversion of $C(z)$. The inversion of $C(z)$ from Eq. (25.18) may be obtained by the method of partial fraction expansion, which has been discussed previously. The result is:

$$c(nT) = \frac{1/\tau}{(1 - b)}(1 - b^{n+1}) \tag{25.21}$$

where $b = e^{-T/\tau}$

Introducing the parameters $(T = \tau = 1)$ for this example gives

$$c(nT) = \frac{1}{(1 - 0.368)}(1 - 0.368^{n+1}) \tag{25.22}$$

Inversion of $C(z, m)$. The inversion of $C(z,m)$ is obtained in exactly the same manner as the inversion of $C(z)$. It should be remembered that $C(z,m)$ is simply a Z-transform of the output of a process in which some delay has been introduced. We may use either the method of long division or the method of partial fraction expansion to invert $C(z,m)$. Using the latter, we proceed as follows

$$\frac{C(z,m)}{\left\{ \dfrac{z e^{-mT/\tau}}{\tau} \right\}} = \frac{1}{(z - b)(z - 1)} = \frac{A}{z - b} + \frac{B}{z - 1} \tag{25.23}$$

where $b = e^{-T/\tau}$

Evaluating A and B gives $A = -1/(1 - b)$ and $B = 1/(1 - b)$, therefore

$$C(z,m) = \frac{e^{-mT/\tau}}{\tau(1 - b)}\left(\frac{z}{z - 1} - \frac{z}{z - b}\right) \tag{25.24}$$

From the table of transforms (Table 22.1), Eq. (25.24) becomes

$$c_\lambda(nT) = \frac{e^{-mT/\tau}}{\tau(1 - b)}(1 - b^n) \tag{25.25}$$

Using the parameters of this example, we get

$$c_\lambda(nT) = \frac{e^{-0.3}}{1 - 0.368}(1 - 0.368^n)$$

Note that

$$c_\lambda(nT) = e^{-0.3}c[(n - 1)T]$$

This result is to be expected for this process, i.e., the intersample value is a constant fraction, e^{-m}, of the peak value, which occurred at the previous sampling instant.

CLOSED-LOOP INTERSAMPLE RESPONSE

Consider the following closed-loop system in Fig. 25.6, which was discussed in Chap. 23. If $C(z)$ for this system is inverted, we obtain values of the response $c(t)$ at sampling instants, $c(nT)$. To obtain information about $c(t)$ at intersample positions, we may insert in the loop $e^{-\lambda Ts}$ and $e^{\lambda Ts}$ as shown in Fig. 25.7. This artificial insertion of lag and lead does not alter the system, but provides a means for calculating the desired information, $c_\lambda(t)$.

For simplicity, consider only the case of set-point disturbance. If load disturbance is also present, the same approach to be used here can be applied. From Fig. 25.7, we can write directly

$$C_\lambda(s) = RG_cG_pG_d(s) - C^*(s)G_hG_cG_pG_d(s) \tag{25.26}$$

where $G_d(s) = e^{-\lambda Ts}$.

Taking the starred transform of both sides of this equation gives

$$C_\lambda^*(s) = RG_cG_pG_d^*(s) - C^*(s)G_hG_cG_pG_d^*(s) \tag{25.27}$$

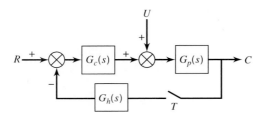

FIGURE 25-6
Closed-loop sampled-data system.

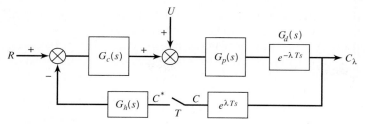

FIGURE 25-7
Modified closed-loop system for obtaining intersample information.

From Eq. (23.19), we may write for only a set-point disturbance:

$$C^*(s) = RG_cG_p^*(s) - G_cG_pG_h^*(s)C^*(s)$$

Solving for $C^*(s)$ gives

$$C^*(s) = \frac{RG_cG_P^*(s)}{1 + G_cG_pG_h^*(s)} \tag{25.28}$$

Introducing $C^*(s)$ from Eq. (25.28) into Eq. (25.27) gives

$$C_\lambda^*(s) = RG_cG_pG_d^*(s) - \frac{RG_cG_p^*(s)}{1 + G_cG_pG_h^*(s)}G_hG_cG_pG_d^*(s) \tag{25.29}$$

Expressing this as a modified Z-transform, we obtain

$$C(z, m) = RG_cG_p(z,m) - \frac{RG_cG_p(z)G_hG_cG_p(z,m)}{1 + G_cG_pG_h(z)} \tag{25.30}$$

In going from Eq. (25.29) to Eq. (25.30), any symbol with a subscript λ or any group of transfer functions containing G_d (the transport lag $e^{-\lambda Ts}$) is converted to the modified Z-transform symbol according to our previous discussion.

In using this last equation, one must calculate the terms on the right side carefully. For example, to find $G_hG_cG_p(z,m)$ one first obtains $G_hG_cG_p(s)$, which is obtained by multiplying together the individual transfer functions. One then obtains the desired result from a table of modified Z-transforms.

Table 23.1 gives the modified Z-transform outputs $C(z,m)$ for some sampled-data systems of interest in this book. The modified Z-transform output given by Eq. (25.30) for Fig. 25.6 corresponds to item 2 in Table 23.1.

SUMMARY

The modified Z-transform is needed to obtain the response of a sampled-data system at intersample positions. It is also needed to obtain the pulse transfer function of a system containing transport lag ($e^{-\tau s}$) for which τ is not an integral number of sampling periods. It should be remembered that a modified Z-transform is simply a Z-transform of a function in which a transport lag ($e^{-\lambda Ts}$) has been included. The inversion of a modified Z-transform is obtained in the same manner as the inversion of an ordinary Z-transform, by long division or by use of partial fraction expansion.

PROBLEMS

25.1. For the control system shown in Problem P23.4, determine the response between sampling for the case $m = 0.4$ by use of the modified Z-transform.

25.2. For the process shown in Fig. P25.2, determine $Y(z)$. By means of long division, determine $Y(t)$ for $t = 0$, 1, 2, 3, 4, and 5.

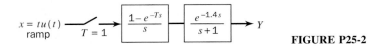

FIGURE P25-2

25.3. (*a*) For the control system shown in Fig. P25.3, obtain a general expression for $C(z)$ for the case where $R = 0$ (no set-point change) in terms of G_1, G_2, and U.

(*b*) One can show that for $R = 0$ and $U = 1/s$ that $C(z,m = 0.3) = 0.259z^{-1} + 0.587z^{-2} + 0.556z^{-3} + \ldots$.

From this result and any other information you wish to use, determine if possible the values of C at the following times: 0, 0.3, 0.5, 0.7, 1.0, 1.3, 1.5, 1.7, 2.0, 2.3, and 3.0. Present your results in a table.

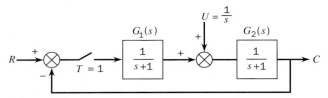

FIGURE P25-3

CHAPTER 26

SAMPLED-DATA CONTROL OF A FIRST-ORDER PROCESS WITH TRANSPORT LAG

The tools developed in the previous chapters for sampled-data systems will now be applied to a model found to fit a large class of systems in chemical processing. This model consists of a first-order process with transport lag (or delay). The transfer function may be written

$$G_p(s) = \frac{e^{-a\tau s}}{\tau s + 1} \tag{26.1}$$

where τ is the time constant and $a\tau$ is the transport lag parameter (a is a positive number). Consider the sampled-data control system shown in Fig. 26.1 in which this transfer function is used as a model of the process. In this discussion, the hold will be a zero-order hold for which the transfer function is

$$G_h(s) = \frac{1}{s}(1 - e^{-Ts}) \tag{26.2}$$

Recall that the combination of the sampling switch and the zero-order hold provides clamping. We shall take the control action to be proportional, for which

$$G_c = K \tag{26.3}$$

393

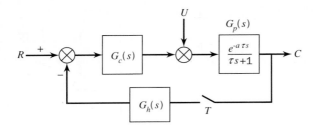

FIGURE 26-1
Sampled-data control of a first-order system with transport lag.

For the case of no transport lag ($a = 0$), the proportional control of this process was discussed in Chap. 23. As will be shown, the higher the value of a, the higher the order of the characteristic equation for the closed-loop system.

We shall consider a number of cases that are based on the size of a. For this purpose, let

$$nT < a\tau \le (n + 1)T \tag{26.4}$$

where $n = 0, 1, 2, 3, \ldots$

For convenience in obtaining $G(z)$, $a\tau$ may be written as follows:

$$a\tau = nT + (a\tau - nT) \tag{26.5}$$

In this form $a\tau$ is equal to an integral number of sampling periods (nT) plus a fraction of a sampling period [i.e., $(a\tau - nT) \le T$]. Using this expression for $a\tau$ in Eq. (26.1), the open-loop transfer function becomes

$$G(s) = K(1 - e^{-Ts})e^{-nTs}\frac{e^{-(a\tau-nT)s}}{s(\tau s + 1)} \tag{26.6}$$

Using the theorem on translation in Chap. 23 [Eq. (23.26)], we may write $G(z)$ as

$$G(z) = \frac{K(1 - z^{-1})}{z^n}Z\left\{\frac{e^{-(a\tau-nT)s}}{s(\tau s + 1)}\right\} \tag{26.7}$$

Note that the expression within braces is equivalent to delaying the response from $1/[s(\tau s + 1)]$ by $(a\tau - nT)$. We may apply the concepts used in developing the modified Z-transform to find the Z-transform of the expression within braces in Eq. (26.7) by equating $(a\tau - nT)$ to λT and recalling that $m = 1 - \lambda$. This leads to

$$\lambda T = a\tau - nT$$
$$m = 1 - \lambda = 1 - a\frac{\tau}{T} + n \tag{26.8}$$

Equation (26.7) may now be written

$$G(z) = \frac{K(1 - z^{-1})}{z^n}Z_m\left\{\frac{1}{s(\tau s + 1)}\right\} \tag{26.9}$$

From the table of modified Z-transforms (Table 22.1), we obtain

$$G(z) = \frac{K(1 - z^{-1})}{z^n} \left\{ \frac{1}{z - 1} - \frac{\exp[-\frac{1}{\tau}(1 - \frac{a\tau}{T} + n)T]}{z - e^{-T/\tau}} \right\} \tag{26.10}$$

Rearranging this expression gives

$$G(z) = \frac{K(z - 1)}{z^{n+1}} \left[\frac{1}{z - 1} - \frac{d}{z - b} \right] \tag{26.11}$$

where $d = \exp[a - (n + 1)T/\tau]$ and $b = e^{-T/\tau}$

The characteristic equation for the closed-loop response may be written

$$1 + G(z) = 0$$

or

$$1 + \frac{K(z - 1)}{z^{n+1}} \left[\frac{1}{z - 1} - \frac{d}{z - b} \right] = 0 \tag{26.12}$$

or

$$z^{n+2} - bz^{n+1} + K(1 - d)z + K(d - b) = 0 \tag{26.13}$$

Note that the order of the characteristic equation increases if (1) the delay time $a\tau$ is increased for a fixed sampling period T; (2) the sampling rate increases (lower T) for a fixed delay time.

STABILITY

The stability of the sampled-data system represented by Eq. (26.13) can be examined by applying the Routh test as discussed in Chap. 24. As the order of the characteristic equation increases, the effort involved in determining stability criteria greatly increases. A few cases corresponding to various values of n in Eq. (26.13) are presented here.

Case 1: $n = 0$

For this case, the delay is less than one sampling period, i.e.,

$$0 < a\tau \leq T$$

The characteristic equation given by Eq. (26.13) becomes

$$z^2 + [K(1 - d) - b]z - K(b - d) = 0 \tag{26.14}$$

For convenience, this may be written

$$z^2 + \gamma z - \alpha = 0 \tag{26.15}$$

where $\gamma = [K(1 - d) - b]$ and $\alpha = K(b - d)$.

Applying the bilinear transformation given by Eq. (24.4) to Eq.(26.15) gives

$$\frac{(w + 1)^2}{(w - 1)^2} + \frac{w + 1}{w - 1}\gamma - \alpha = 0 \tag{26.16}$$

After algebraic rearrangement, one obtains

$$w^2(1 + \gamma - \alpha) + w[2(1 + \alpha)] + 1 - \gamma - \alpha = 0$$

Replacing γ by $K(1 - d) - b$ and α by $K(b - d)$ gives

$$w^2[1 - b + K(1 - b)] + 2w[1 + K(b - d)] + 1 - [K(1 - d) - b + K(b - d)] = 0$$

or

$$w^2(1 - b)(1 + K) + 2w[1 + K(b - d)] + 1 - [K(1 - d) - b + K(b - d)] = 0$$

The coefficient of w^2 is positive since $b = e^{-T/\tau}$ is always positive and less than one. For stability, the Routh test requires that all coefficients be positive; therefore

$$1 + K(b - d) > 0$$

and

$$1 + b - K(1 + b - 2d) > 0$$

These inequalities may be rewritten

$$K < \frac{1}{d - b} \tag{26.17}$$

$$K < \frac{1 + b}{1 + b - 2d} \tag{26.18}$$

Both of these inequalities must be satisfied simultaneously. The best way to understand the result is to plot the stability boundaries as shown in Fig. 26.2 where the ultimate gain K_u is plotted as a function of a. Recall that d is a function of a as shown under Eq. (26.11). For stability, K must fall under the boundary as indicated in the figure. One can see that K_u reaches a maximum at a_{max} in Fig. 26.2. The value of a_{max} is determined by the intersection of the two constraints, which leads to

$$\exp(a_{max}) = \frac{(1 + b)^2}{b(3 + b)}$$

For the value of T/τ of 0.8 used in Fig. 26.2, one can compute that $a_{max} = 0.304$.

Figure 26.2 shows that up to a certain point, adding delay time to a system with $T/\tau = 0.8$ increases the ultimate gain. One suggested design criterion is to set $K/K_u = 0.5$. The assumption is made that the "relative stability" of the loop is constant, for constant K/K_u. This would imply for this particular example that adding delay time to the loop up to $a = 0.304$ will be beneficial, since increasing K gives less steady-state offset and also faster system response. However, the relative stability of the sampled-data system does not remain the same for constant K/K_u. Figure 26.3 illustrates this point. The response of the system to a unit-

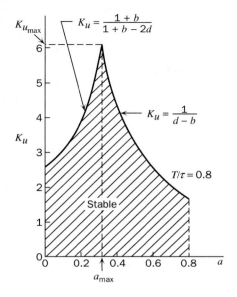

FIGURE 26-2
Stability boundary of a sampled-data system as a function of delay time.

step change in set point is shown for various amounts of delay time in the loop. Although K/K_u is maintained at 0.5, increasing the delay time toward a_{max} has definitely destabilized the system. Hence, use of constant K/K_u is not a good design rule for sampled-data systems and the conclusion that control can be improved by adding delay time is false.

Case 2: $n = 1$

For this case, $T < a\tau \le 2T$

One can see from Eq. (26.13) that the order of the characteristic equation is three. Using the same stability analysis as for the case for $n = 0$, one can show that stability requires that the following inequalities hold simultaneously

$$(2d + 1 - 3b)K < 3 - b \tag{26.19}$$

$$(2d - 1 - b)K < 1 + b \tag{26.20}$$

$$(d - b)^2 K^2 + (1 - b)(1 + b - d)K - 1 < 0 \tag{26.21}$$

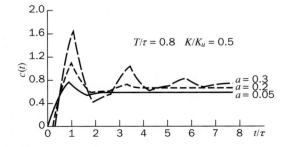

FIGURE 26-3
Closed-loop transient response of sampled-data system for a unit step in set point.

The stability boundaries represented by these three equations have been plotted by Mosler et al. (1966). The stability constraints for the case of $n = 2$, which require four inequalities, also may be found in Mosler (1966). This demonstrates that as the value of transport lag $(a\tau)$ increases relative to a fixed sampling period T, the order of the characteristic equation increases and the stability criteria become more and more complex.

TRANSIENT RESPONSE OF CLOSED-LOOP SAMPLED-DATA SYSTEMS

We shall consider the transient response for the system shown in Fig. 26.1 for a step change in set point. For this particular disturbance, the block diagram may be drawn as shown in Fig. 26.4. For the special case of a step change in set point, the sampling switch and the hold in Fig. 26.1 can be placed in the forward loop of Fig. 26.4. For a step change in R occurring just before a sampling instant, it does not matter whether the sampling occurs before the comparator or after the comparator. The reason for redrawing the block diagram is that the expression for $C(z)$ for Fig. 26.4 is simpler than the expression for Fig. 26.1.

Using the method described in Chap. 23, one can obtain for $C(z)$

$$C(z) = \frac{G(z)R(z)}{1 + G(z)} \tag{26.22}$$

where $G(z) = G_c G_p G_h(z)$

The expression for $C(z,m)$ is

$$C(z,m) = \frac{G(z,m)R(z)}{1 + G(z)} \tag{26.23}$$

Equations (26.22) and (26.23) can also be found from Table 23.1.

If one were to obtain expressions for $C(z)$ and $C(z,m)$ for the system in Fig. 26.1, the result would be as follows:

$$C(z) = \frac{G_c G_p R(z)}{1 + G(z)} \tag{26.24}$$

and

$$C(z,m) = RG_c G_p(z,m) - \frac{G(z,m)RG_c G_p(z)}{1 + G(z)} \tag{26.25}$$

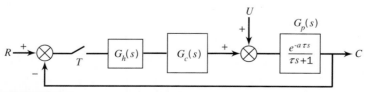

FIGURE 26-4
Rearranged block diagram.

If one were to use Eq. (26.24) or Eq. (26.25) for a *step change* in R, the result for $c(nT)$ and $c_\lambda(nT)$ would be, of course, the same as that obtained using Eqs. (26.22) and (26.23); these latter two are simpler than Eqs. (26.24) and (26.25).

The diagram in Fig. 26.4 may replace the diagram in Fig. 26.1 for the more general set-point function $r(t)$, which is piecewise constant and where changes in r occur just before sampling instants. This more general function is a stair-step function.

The transient response for the system shown in Fig. 26.4 will be considered in detail for two cases:

Case I: no transport delay, i.e., $a = 0$
Case II: $0 < a\tau \le T$, i.e., $n = 0$

The results can be used to establish design criteria.

Case I: No Transport Lag

For this case, one can show that Eq. (26.22) becomes

$$\frac{C(z)}{R(z)} = \frac{K(1 - b)}{z + K(1 - b) - b}$$

where $b = e^{-T/\tau}$

For a unit-step change in set point, $R = z/(z - 1)$ and the response is expressed by

$$C(z) = \frac{K(1 - b)z}{(z - 1)[z + K(1 - b) - b]}$$

Inverting this expression gives

$$c(nT) = \frac{K}{K + 1}\{1 - [b - K(1 - b)]^n\} \qquad n = 0, 1, 2, \ldots \qquad (26.26)$$

This result was derived in Chap. 23 [Eq. (23.36)] and is presented again for convenience in developing design criteria. The transient response for a specific set of parameters, shown in Fig. 26.5, consists of arcs of exponential functions that intersect at sampling instants.

It is possible to make the loop gain of the system (K) so small that the response is overdamped. In fact, the value of K, below which the response is overdamped, can be found by examining Eq. (26.26). When the expression in brackets, $b - K(1 - b)$, is greater than 0, the system no longer oscillates; therefore, we may write

$$b - K(1 - b) > 0$$

or

$$K < \frac{b}{1 - b} \qquad \text{for overdamped response} \qquad (26.27)$$

An overdamped response is also shown in Fig. 26.5.

An overdamped response for a closed-loop system is usually considered too sluggish; consequently, the design criteria to be developed will be based on the underdamped response, such as the one shown in Fig. 26.5.

Since the peaks (maximum and minimum) of the underdamped response occur at sampling instants, Eq. (26.26) may be used to compute overshoot and decay ratio. For a stable response, the ultimate value of $c(t)$ is

$$c(nT)|_{n\to\infty} = \frac{K}{K+1} \tag{26.28}$$

This result, which is the same as that for a continuous proportional control system, should not be surprising if one recalls that the system's steady state is determined by steady-state relationships that are the same for both sampled-data control and continuous control.

The first peak in the direction of set-point change occurs at $t = T$ and the second peak in the same direction occurs at $t = 3T$. From this information one may compute from Eq. (26.26) the fractional overshoot and the decay ratio; the results are as follows:

$$\text{Fractional overshoot} = \frac{c(T) - c(\infty)}{c(\infty)} = K(1 - b) - b \tag{26.29}$$

$$\text{Decay ratio} = \frac{c(3T) - c(\infty)}{c(T) - c(\infty)} = [K(1 - b) - b]^2 \tag{26.30}$$

It is interesting to note that the relationship between decay ratio and fractional overshoot for the first-order, underdamped sampled system is the same as that for the second-order, underdamped continuous system, namely:

$$\text{Decay ratio} = (\text{fractional overshoot})^2 \tag{26.31}$$

For a decay ratio α^2, where $0 < \alpha < 1$, the loop gain may be computed from Eq. (26.29) to be

$$K_{\alpha^2} = \frac{\alpha + b}{1 - b} \tag{26.32}$$

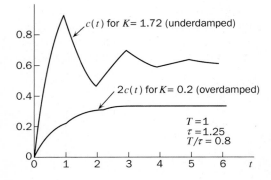

$c(t)$ for $K= 1.72$ (underdamped)

$2c(t)$ for $K= 0.2$ (overdamped)

$T=1$
$\tau =1.25$
$T/\tau = 0.8$

FIGURE 26-5
Transient response of a first-order, sampled-data system (no transport lag).

For any choice of sampling rate, one may determine from Eq.(26.32) the loop gain required for a desired decay ratio. Increasing the sampling rate (lower T) increases the speed of the response and decreases the period, which is $2T$. This increase in sampling rate will also provide a larger open-loop gain, which gives less steady-state offset. From Eqs. (23.37) and (26.32), we may write

$$\frac{K_{\alpha^2}}{K_u} = \frac{\alpha + b}{1 + b} \tag{26.33}$$

For quarter-decay ratio ($\alpha^2 = \frac{1}{4}$), one obtains from Eq. (26.33) for T ranging from 0 to ∞

$$\frac{1}{2} < \frac{K_{1/4}}{K_u} < \frac{3}{4} \tag{26.34}$$

The ratio varies from $\frac{1}{2}$ for $T = \infty$ to $\frac{3}{4}$ for $T = 0$.

One may contrast this result with the Ziegler-Nichols rule for proportional control of continuous systems (see Chap. 17). The Ziegler-Nichols rule requires that $K/K_u = \frac{1}{2}$; if this rule is used for a continuous system, one expects to obtain a "good" transient response, for which quarter-decay ratio is often considered good in industrial practice. We see from Eq. (26.34) that $K_{1/4}/K_u$ varies from $\frac{1}{2}$ to $\frac{3}{4}$ as the sampling period varies from infinity to zero. The Ziegler-Nichols rule and the rule provided by Eq. (26.34) are comparable.

Example 26.1. A simple example will help illustrate the use of the design equation, Eq. (26.32). For the sampled-data system shown in Fig. 26.6, determine the open-loop gain K for quarter-decay ratio for the following sampling periods: (a) $T = 1$, (b) $T = 0.5$, and (c) continuous control. Also find the period of oscillation and the offset for each case.

(a) $\tau = 1.25$ $T = 1.0$ $b = e^{-T/\tau} = e^{-0.8} = 0.449$

For quarter-decay ratio, $\alpha^2 = 0.25$ or $\alpha = 0.5$

Substituting α into Eq. (26.32) gives

$$K_{1/4} = \frac{\alpha + b}{1 - b} = \frac{0.5 + 0.449}{1 - 0.449} = 1.724$$

From Eq. (26.28), the ultimate value of c is

$$c(\infty) = \frac{K}{K + 1} = \frac{1.724}{2.724} = 0.633$$

$$\text{offset} = r(\infty) - c(\infty) = 1 - 0.633 = 0.334$$

$$\text{period} = 2T = 2$$

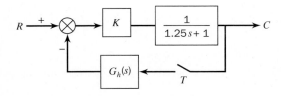

FIGURE 26-6
Sampled-data control system for Example 26.1.

(*b*) For this case, where $T/\tau = 0.4$ and $b = 0.670$, the answers are:

$$K_{1/4} = 3.550$$
$$\text{offset} = 0.219$$
$$\text{period} = 2T = 1$$

(*c*) For continuous proportional control of a first-order process, the transient response is never oscillatory; therefore, Eq. (26.32) does not apply.

This example illustrates that a faster sampling rate permits a higher proportional gain and less offset for a fixed decay ratio.

Case II: Transport Lag, $0 < a\tau \leq T$, $n = 0$

For this case, the response will be delayed by an amount $a\tau$. To determine the decay ratio and overshoot, we must be able to compute the peaks of the transient response. Because of the delay $a\tau$, the peaks will not occur until $a\tau$ after the sampling instants. This observation is based on an understanding of the behavior of a first-order system and the transport lag. The sketch shown in Fig. 26.7 illustrates the situation.

To determine the peak values, we must invert $C(z,m)$ with

$$\lambda T = T - a\tau \qquad \text{and} \qquad m = 1 - \lambda = a\tau/T$$

As shown in Eq.(26.23), we must obtain $G(z,m)$ in order to determine $C(z,m)$. The transfer function $G(s)$ is

$$G(s) = G_c G_p H(s) = \frac{Ke^{-a\tau s}}{\tau s + 1} \frac{(1 - e^{-Ts})}{s} \tag{26.35}$$

We cannot obtain $G(z,m)$ for $G(s)$ in Eq. (26.35) directly from the table of transforms because $G(s)$ contains $e^{-a\tau s}$ where $a\tau$ is a nonintegral number of sampling periods (i.e., $a\tau < T$). However, we can obtain $G(z,m)$ by using the approach taken in developing the modified Z-transform in Chap. 25. We express $G(s)$ as $G_\lambda(s)$ where $G_\lambda(s) = G(s)e^{-\lambda Ts}$. Recall that λT is the amount by which the response is to be shifted. We see from Fig. 26.7 that $\lambda T = T - a\tau$. Now $G_\lambda(s)$ can be written

$$G_\lambda(s) = \left\{ \frac{Ke^{-a\tau s}}{(\tau s + 1)s}(1 - e^{-Ts}) \right\} e^{-(T - a\tau)s} \tag{26.36}$$

This expression may be simplified to give

$$G_\lambda(s) = \frac{K(1 - e^{-Ts})e^{-Ts}}{s(\tau s + 1)} \tag{26.37}$$

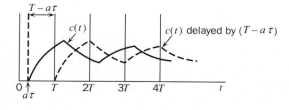

$c(t)$ delayed by $(T - a\tau)$

FIGURE 26-7
Response of first-order, sampled-data system with transport lag.

Note that the right side of Eq. (26.37) does not include a term involving a nonintegral power of T. Obtaining the Z-transform of $G_\lambda(s)$ gives an expression, which is $G(z, m)$. The details are shown in the following steps.

$$G_\lambda(z) = G(z,m) = K(1 - z^{-1})z^{-1}Z\left\{\frac{1}{s(\tau s + 1)}\right\}$$

$$G(z,m) = \frac{K(z - 1)}{z^2}\left[\frac{1}{z - 1} - \frac{e^{-T/\tau}}{z - e^{-T/\tau}}\right]$$

Simplifying this expression gives

$$G(z,m) = \frac{K(1 - b)}{z(z - b)} \tag{26.38}$$

The other terms needed to evaluate $C(z,m)$ in Eq. (26.23) are $G(z)$ and $R(z)$. $G(z)$ is given in Eq. (26.11). For a unit-step change in set point

$$R(z) = \frac{z}{z - 1} \tag{26.39}$$

Substituting Eqs. (26.11), (26.38), and (26.39) into Eq. (26.23) gives, after considerable algebraic manipulation

$$C(z, m) = \frac{K(1 - b)z}{(z - 1)\{z^2 + [K(1 - d) - b]z + K(d - b)\}} \tag{26.40}$$

Inversion of this expression will give the value of the delayed response $c[t - (T - a\tau)]$. These values are, of course, the peak values of $c(t)$ as illustrated in Fig. 26.7.

Mosler (1966) has inverted Eq. (26.40) by partial fraction expansion; the result is a rather complex expression. He used this result to obtain some design rules for determining the values of K and T that will produce a transient with quarter-decay ratio. The development of the rules is quite involved and beyond the scope of this book.

SUMMARY

In this chapter, the principles of sampled-data theory have been applied to the proportional control of a process, which represents a large class of systems in chemical processing, namely, a process that consists of a first-order process with transport lag $[e^{-a\tau s}/(\tau s + 1)]$. Since the transport lag parameter $(a\tau)$ may not be an integral number of sampling periods, the modified Z-transform was used to obtain the pulse transfer function of the system. As the order of the characteristic equation for the closed-loop system increases, the stability criteria become more and more complex and require that several inequalities be satisfied simultaneously for stability.

For the case of proportional control of a first-order system without transport lag, some simple design rules were developed for tuning the proportional controller to obtain a desired decay ratio.

PROBLEM

26.1. The stirred-tank control system shown in Fig. P26.1 blends a stream of concentrated solution with a process stream to maintain a desired concentration of solute in the outlet stream. The flow rates and concentrations are indicated in the diagram. The chemical analysis, which must be done manually by withdrawing a small sample from the tank, takes 1.0 min. At the end of each analysis, the chemist sets a dial immediately to a value corresponding to the concentration just determined. The dial, in turn, feeds a concentration signal to the controller. As soon as one sample is analyzed, a new one is withdrawn from the tank and analyzed.

The flow rate through the valve varies linearly from 0 to 0.02 liter/min as the valve-top pressure varies from 3 to 15 psig. Under normal conditions, the process stream is free of solute. However, from time to time, a load change may occur in the form of a change in concentration of solute in the process stream entering the tank.

(*a*) Show that the system is equivalent to a sampled-data control system and draw its block diagram.

(*b*) From the design rules developed by Mosler (1966, Eq. 65), one can show that the value of K_c required for quarter-decay ratio and fast sampling $(T = a\tau)$ is 10.3 psi/(g/l). Using this value of K_c, sketch the transient response for c and q for a step change in c_i of magnitude 0.5 g/l. Determine the extreme values of c, p, and q during the transient. Determine the value of $c(\infty)$.

(*c*) If the chemist uses a continuous analyzer having no lag, but still sets the dial manually as just described, every 60 sec, show how the block diagram changes and determine K_c to obtain quarter-decay ratio. Use the design rule given by Eq. (26.32) to determine this gain.

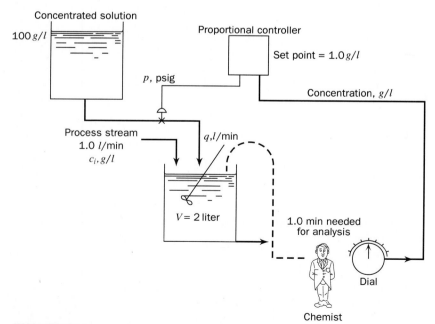

FIGURE P26-1

CHAPTER
27

DESIGN
OF SAMPLED-DATA
CONTROLLERS

In this chapter, sampled-data control theory will be applied to the design of direct-digital control algorithms. In the most general terms, direct-digital control is the automatic control of a process by means of a digital computer. Today the single-station, analog-type continuous controller (pneumatic or electronic) has been almost completely replaced by an instrument that is essentially a small, self-contained digital computer. Such instruments are described as microprocessor-based controllers. This change, of course, was brought about by the great decrease in the cost of computing components and the tremendous increase in the speed of computation. In this chapter, the design philosophy for designing special purpose controllers will be developed and illustrated with some examples.

The block diagram of the control system to be considered is shown in Fig. 27.1 The elements of the block that are implemented by the computer are enclosed by a dotted line and labeled "computer." These elements, which consist of two impulse-modulation switches, the Z-transform of the digital control algorithm $D(z)$, and the zero-order hold $G_h(s)$, will be described later. For the moment, it is necessary to understand only the general operating features of the control system.

To simplify the discussion, $G_p(s)$ in Fig. 27.1 contains the valve, the current-to-pressure converter, and the process. The transfer function for the measuring element has been taken as one; for this reason, no measurement block is shown in the feedback path of Fig. 27.1. The output from the hold is a current (or voltage) signal.

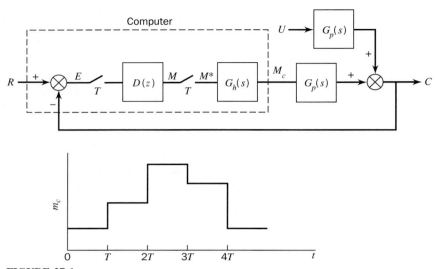

FIGURE 27-1
Block Diagram for a computer control system.

Every T units of time, the computer reads and stores the measured value of the process variable C. The computer operates on this signal, according to the algorithm $D(z)$ stored in it, to produce a signal to the valve M_c. It is assumed that the computation of M_c is instantaneous, relative to the sampling period of the process. For many chemical processes that are slow, this is a reasonable assumption. By means of the hold, the signal to the valve, M_c, is held constant (i.e., clamped) between sampling instants. Consequently, the valve response during transient operation of the process will resemble a stair-step function. The control algorithm is simply a mathematical description that tells the computer how to calculate the signal to the valve each sampling instant.

The digital computer implements an algorithm of the form

$$m(nT) = \sum_{i=0}^{k} g_i e[(n-i)T] - \sum_{j=1}^{p} h_j m[(n-j)T] \qquad (27.1)$$

This equation gives the value at which $m_c(t)$ is to be held constant during the following sampling period, that is,

$$m_c(t) = m(nT) \qquad \text{for} \qquad nT \le t < (n+1)T$$

The term T is the sampling period and g_i and h_j are constants. The set of constants (g_i, h_j) constitutes the control algorithm. In the following pages, methods will be developed for finding these constants for a specific design of a controller.

To understand Eq. (27.1) more readily, consider the case where $k = 2$ and $p = 2$. If we want to compute m at the one-hundredth sampling instant, Eq. (27.1) is written

$$m(100T) = g_0 e(100T) + g_1 e(99T) + g_2 e(98T) - h_1 m(99T) - h_2 m(98T)$$

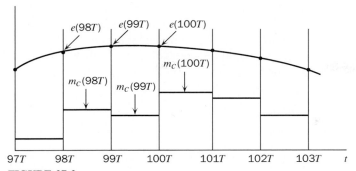

FIGURE 27-2
Typical relationship between $m_c(t)$ and $e(t)$ for a computer control system.

Figure 27.2 illustrates the nature of the signals used in this expression for $m(100T)$. Notice that $m(100T)$ is computed instantaneously at $t = 100T$. For this particular example, the computation requires the present value of error, two past values of error, and two past values of manipulated variable. The more constants (g_i, h_j) in the algorithm, the more complex it becomes in terms of computer storage and computer time needed to solve the algorithm.

To illustrate how an algorithm of the form of Eq. (27.1) is derived, several algorithms will be derived for a process consisting of a first-order process with transport lag.

ALGORITHMS FOR A FIRST-ORDER WITH TRANSPORT LAG MODEL

A variety of algorithms will be derived for a process with a transfer function that is first-order with transport lag, that is,

$$G_p(s) = \frac{e^{-a\tau s}}{\tau s + 1} \tag{27.2}$$

Figure 27.1 is redrawn as Fig. 27.3 with $G_p(s)$ expressed as a first-order process with transport lag and $G_h(s)$ expressed as a zero-order hold. In Fig. 27.3 the clamped signal M_c is obtained from the zero-order hold, which obtains its input

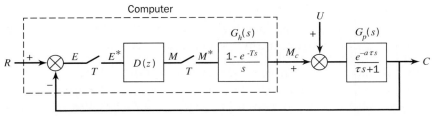

FIGURE 27-3
Computer control system.

signal from an impulse-modulated sampling switch. Since this signal from the switch is a pulsed signal, it is given the symbol M^* in the diagram. At the outset of this discussion, the reader should realize that the components enclosed in the dotted line do not represent any physical components or hardware; they are simply mathematical symbols that aid in deriving the control algorithm. From Fig. 27.3, we may write

$$M(z) = D(z)E(z) \qquad (27.3)$$

To obtain the algorithm in the form of Eq. (27.1), $D(z)$ may be written

$$D(z) = \frac{g_0 + g_1 z^{-1} + g_2 z^{-2} + \cdots}{1 + h_1 z^{-1} + h_2 z^{-2} + \cdots} = \frac{M(z)}{E(z)} \qquad (27.4)$$

Cross-multiplying this expression and solving for M(z) give

$$M(z) = g_0 E(z) + g_1 z^{-1} E(z) + \cdots - \{h_1 z^{-1} M(z) + h_2 z^{-2} M(z) + \cdots\} \quad (27.5)$$

Recognizing the term $z^{-i} E(z)$ to be equivalent to the Z-transform of the error delayed by i sampling periods, $e(nT - iT)$, we may write Eq. (27.5) in the time domain as

$$m(nT) = g_0 e(nT) + g_1 e[(n-1)T] + g_2 e[(n-2)T] + \cdots$$
$$-\{h_1 m[(n-1)T] + h_2 m[(n-2)T] + \cdots\} \qquad (27.6)$$

Since this expression matches Eq. (27.1), we see that Eq. (27.4) is a satisfactory expression for $D(z)$.

Performance Specifications

Before the details of the design method for digital control algorithms are presented, the performance specifications for the control system will be listed. The *minimal prototype* response of Bergen and Ragazzini (1954) considered the response of the system only at sampling instants. The requirements for the minimal prototype response are given in the following list.

1. The compensation algorithm must be physically realizable (i.e., no prediction is needed by the algorithm).
2. The output of the system must have zero steady-state error at sampling instants.
3. The output should equal the set point in a minimum number of sampling periods.

However, for the practical application of a digital control algorithm to a real system, several additional requirements are important. These are:

4. The digital control algorithm should be open-loop stable.
5. Unstable or nearly unstable pole-zero cancellations should be avoided, since exact cancellation in real processes is impossible, and the resulting closed-loop system may be unstable or excessively oscillatory.

6. The design should consider the entire response of the system in order to eliminate hidden oscillations (intersample ripple).
7. In addition to the system output responding in a desired manner to the input disturbance selected for the design of the control algorithm, the system output should be satisfactory for other possible disturbances.

These additional constraints are necessary to ensure that the proposed algorithm will perform satisfactorily on real systems. To meet these requirements, the resulting system may respond with a settling time longer than the minimal prototype settling time (item 3 in the preceding list). The concept of a minimum settling time is used only as a theoretical performance criterion. In real systems, where modeling error and noise are present, it is not possible to bring the state of the system completely to rest. This does not negate the value of the theoretical concept of minimum settling time, because systems designed to meet this requirement are likely to give satisfactory performance in tests on real processes. In addition, all digital control algorithms to be presented here contain the equivalent of an integrator, which ensures zero steady-state offset for step disturbances, regardless of modeling error and the location of the disturbance in the loop. Finally, the minimal prototype response concept provides the basis for a systematic approach to the design of digital control algorithms.

Analysis and Design of Sampled-Data Controllers

From Fig. 27.3, we may write directly from observation

$$E(s) = R(s) - G_p G_h(s) M^*(s) - U G_p(s) \qquad (27.7)$$

and

$$M^*(s) = D^*(s) E^*(s) \qquad (27.8)$$

Taking the Z-transform of each equation gives

$$E(z) = R(z) - G_p G_h(z) M(z) - U G_p(z) \qquad (27.9)$$

$$M(z) = D(z) E(z) \qquad (27.10)$$

Combining the last two equations gives

$$E(z) = \frac{R(z)}{1 + G(z) D(z)} - \frac{U G_p(z)}{1 + G(z) D(z)} \qquad (27.11)$$

where $G(z) = G_p G_h(z)$

We may also obtain from Fig. 27.3

$$C(z) = R(z) - E(z) \qquad (27.12)$$

Combining Eqs. (27.11) and (27.12) gives

$$C(z) = \frac{G(z) D(z) R(z)}{1 + G(z) D(z)} + \frac{U G_p(z)}{1 + G(z) D(z)} \qquad (27.13)$$

Note that $R(z)$ in Eqs. (27.11) and (27.13) is not bound to other transfer functions. This can be advantageous in design computations.

Design Methods

It is convenient to design $D(z)$ for a load change or a set-point change. For a given disturbance in load or set point, the designer proposes a desired response at sampling instants, which means that $C(nT)$ must be specified. This desired response will be written as C_d. From the desired response $C_d(nT)$, one obtains $C_d(z)$. Consider the case of only a load disturbance, i.e., $R(z) = 0$. Solving Eq. (27.13) for $D(z)$ and replacing C by C_d to indicate that the desired response of C has been selected by the designer give

$$D(z) = \frac{1}{G(z)}\left[\frac{UG_p(z)}{C_d(z)} - 1\right] \qquad \text{(Design equation for load change)} \quad (27.14)$$

Equation (27.14) is the equation to be used to design $D(z)$ for a load change.

In a similar manner, one can obtain from Eq. (27.13) a design equation for set point change, i.e. $U(z) = 0$. The result is

$$D(z) = \frac{C_d(z)}{G(z)[R(z) - C_d(z)]} \qquad \text{(Design equation for set-point change)}$$

$$(27.15)$$

It is necessary that the highest power of z in the numerator of $D(z)$ not exceed the highest power of z in the denominator. If this restriction is not satisfied, the algorithm will require knowledge of the future values of the error, i.e., prediction. An algorithm not satisfying this restriction is called unrealizable. Note that Eq. (27.4) is written in such a form that it does not admit the case where the highest power of z in the numerator exceeds the highest power of z in the denominator. If an unrealizable $D(z)$ is obtained, we obtain the expression given by Eq. (27.4) multiplied by some positive power of z.

To show how the control algorithms are obtained, several detailed examples that apply to Fig. 27.3 will be presented.

Example 27.1. ($T = a\tau$, fast sampling, load)
With regard to Fig. 27.3, consider the design of $D(z)$ for $T = a\tau$. This relation between the sampling period and the process transport lag will be referred to as *fast sampling*. Later, an example involving *slow sampling* will be considered, in which $T > a\tau$.

Consider a unit-step change in load to enter the system at an instant of sampling. The Z-transform of the output can be written in general form:

$$C_d(z) = \eta_0 + \eta_1 z^{-1} + \eta_2 z^{-2} + \cdots \qquad (27.16)$$

where the coefficients (η_i) correspond to the desired outputs of the system at sampling instants. It is the task of the designer to specify a desired output that leads to a realizable algorithm and fulfills the performance specifications listed earlier. In some cases, the nature of the physical process being controlled will aid in choosing a suitable $C_d(z)$ as expressed by Eq. (27.16).

The response diagram in Fig. 27.4 will help explain how the output $C_d(z)$ is selected by the designer. Because the transport lag in $G_p(s)$ is one sampling period T, the output will remain at 0 for the first two sampling instants, that is,

$$c(0) = 0, \ c(T) = 0$$

The response of c, which is the usual first-order exponential rise starting at $t = T$, will reach $1 - e^{-T/\tau}$ at $t = 2T$, that is,

$$c(2T) = 1 - e^{-T/\tau} = 1 - b \tag{27.17}$$

where $b = e^{-T/\tau}$.

The control algorithm in $D(z)$ cannot start to respond until $t = 2T$, at which time the error e has become $-(1-b)$. The value of the manipulated variable m_c generated by the algorithm at $t = 2T$ will depend on the actual algorithm used. However, regardless of the value of m_c at $t = 2T$, c will not change its course until $t = 3T$ because of the transport lag e^{-Ts}. The output of c from $t = 2T$ to $t = 3T$ will continue as an exponential rise and reach $1 - e^{-2T/\tau}$ at $t = 3T$, that is,

$$c(3T) = 1 - e^{-2T/\tau} = 1 - b^2$$

After $t = 3T$, the designer is free to choose any values of c at sampling instants. If c is chosen as zero at sampling instants beyond $t = 3T$, the resulting algorithm will be called a minimal prototype algorithm because the output returns to the set point in a minimum number of sampling periods. The response shown in Fig. 27.4 illustrates the minimal prototype response for a unit-step change in load.

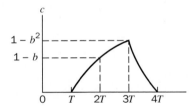

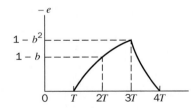

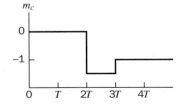

FIGURE 27-4
Minimal prototype response to a unit-step change in load for Example 27.1.

The important point to be emphasized is that c follows the response $1 - e^{-t/\tau}$ from $t = T$ to $t = 3T$, regardless of the algorithm $D(z)$. The values of η in Eq. (27.16) for this example are summarized below

$$\eta_0 = 0$$
$$\eta_1 = 0$$
$$\eta_2 = 1 - b$$
$$\eta_3 = 1 - b^2$$
$$\eta_4, \eta_5, \ldots, \eta_i = 0 \quad \text{(for minimal prototype)}$$

The designer, of course, may choose nonzero values of η_i for $i > 3$, for which case the response will be a nonminimal prototype response.

For the minimal prototype response, we obtain from Eq. (27.16)

$$C_d(z) = (1 - b)z^{-2} + \left(1 - b^2\right)z^{-3}$$

or

$$C_d(z) = \frac{(1 - b)z + 1 - b^2}{z^3} \tag{27.18}$$

For the system under consideration in Fig. 27.3, we have

$$G(s) = G_p G_h(s) = \frac{1 - e^{-Ts}}{s} \frac{e^{-Ts}}{\tau s + 1}$$

for which the corresponding $G(z)$ is

$$G(z) = \frac{1 - b}{z(z - b)} \tag{27.19}$$

For a unit-step change in U,

$$UG_P(s) = \frac{e^{-Ts}}{s(\tau s + 1)}$$

for which the corresponding Z-transform is

$$UG_p(z) = \frac{1 - b}{(z - 1)(z - b)} \tag{27.20}$$

Substituting Eqs. (27.18), (27.19), and (27.20) into Eq. (27.14) gives, after considerable algebraic manipulation

$$D(z) = \frac{1 + b + b^2}{1 - b} \frac{z\left[z - \dfrac{b(1 + b)}{1 + b + b^2}\right]}{(z - 1)[z + (1 + b)]} \tag{27.21}$$

It is instructive to find the expression for the manipulated variable in the form of Eq. (27.6). This is the form that one must use to write a computer program for control of the process. We may write Eq. (27.21) in the form

$$D(z) = \frac{M(z)}{E(z)} = \frac{\alpha z(z - \gamma)}{(z - 1)[z + (1 + b)]} \tag{27.22}$$

where $\alpha = (1 + b + b^2)/(1 - b)$
$\gamma = b(1 + b)/(1 + b + b^2)$

Cross-multiplying Eq. (27.22) gives

$$(z - 1)[z + (1 + b)]M(z) = \alpha z(z - \gamma)E(z)$$

Expanding the terms and rearranging give the result

$$M(z) = \alpha E(z) - \alpha\gamma E(z)z^{-1} - bM(z)z^{-1} + (1 + b)M(z)z^{-2} \qquad (27.23)$$

This form matches Eq. (27.1) or Eq. (27.5), and one can see that the algorithm is quite simple, with $k = 1$ and $p = 2$. The form corresponding to Eq. (27.6) is

$$m(nT) = \alpha e(nT) - \alpha\gamma e(nT - T) - bm(nT - T)$$
$$+ (1 + b)m(nT - 2T) \qquad (27.24)$$

To obtain the sequence of values of manipulated variable for a unit-step change in load, we proceed as will be shown. Before the load change occurs, assume that the system is at steady state for which case $e(t) = 0$ and $m(t) = 0$. Furthermore, assume that the load disturbance occurs at $t = 0$, i.e., at $n = 0$. These are the usual initial steady-state conditions that are used in testing the dynamic performance of a control system.

To see how Eq. (27.24) is used by a computer, the computation can be organized as shown in Table 27.1. At each sampling instant, $m(nT)$ is calculated from Eq. (27.24). For the example under consideration, we know that $m = 0$ and $e = 0$ for $t \leq 0$. The leftmost column in the table gives the time at sampling instants when the computation is made. For convenience in computation, the coefficients of the terms on the right side of Eq. (27.24) are placed in a row under the appropriate terms of this equation.

The calculation of $m(nT)$ for several values of n are now shown; these calculation steps are the same as those the computer would follow in implementing Eq. (27.24).

For m(0). Substituting $n = 0$ into Eq. (27.24) gives

$$m(0) = \alpha e(0) - \alpha\gamma e(-T) - bm(-T) + (1 + b)m(-2T) \qquad (27.25)$$

TABLE 27.1
Computation of $m(nT)$ from computer control algorithm

n	$e(nT)$	$e[(n - 1)T]$	$m(nT)$	$m[(n - 1)T]$	$m[(n - 2)T]$
	α	$-\alpha\gamma$		$-b$	$1 + b$
0	0	0	0	0	0
1	0	0	0	0	0
2	$-(1 - b)$	0	$-(1 + b + b^2)$	0	0
3	$-(1 - b^2)$	$-(1 - b)$	-1	$-(1 + b + b^2)$	0
4	0	$-(1 - b^2)$	-1	-1	$-(1 + b + b^2)$
5	0	0	-1	-1	-1

From the initial conditions stated earlier,

$$e(0) = e(-T) = 0$$

and

$$m(-T) = m(-2T) = 0$$

Introducing these values into Eq. (27.25) gives $m(0) = 0$.

In preparation for the next computation of $m(nT)$, the values of $e(nT)$, $m(nT)$, and $m[(n-1)T]$ are shifted to the next sampling instant as shown by the arrows in the table.

For m(T). To compute $m(T)$, we change n in Eq. (27.24) to 1 to obtain

$$m(T) = \alpha e(T) - \alpha\gamma e(0) - bm(0) + (1+b)m(-T) \qquad (27.26)$$

Substituting the appropriate values of e and m as given in the table into this expression gives $m(T) = 0$.

For m(2T). Letting $n = 2$ in Eq. (27.24) gives

$$m(2T) = \alpha e(2T) - \alpha\gamma e(T) - bm(T) + (1+b)m(0) \qquad (27.27)$$

At $t = 2T$, the disturbance has worked its way through the transport lag and the error now differs from zero. We have at $t = 2T$ as shown in the table or in Fig. 27.4

$$e(2T) = -(1-b)$$
$$e(T) = 0$$
$$m(T) = 0$$
$$m(0) = 0$$

Substituting these values into Eq. (27.27) gives $m(2T) = \alpha[-(1-b)]$.
Introducing the expression for α from Eq. (27.22) gives $m(2T) = -(1+b+b^2)$

For m(3T). Letting $n = 3$ in Eq. (27.24) gives

$$m(3T) = \alpha e(3T) - \alpha\gamma e(2T) - bm(2T) + (1+b)m(T) \qquad (27.28)$$

At $t = 3T$,

$$e(3T) = -(1-b^2)$$
$$e(2T) = -(1-b)$$
$$m(2T) = -(1+b+b^2)$$
$$m(T) = 0$$

Substituting these values into Eq. (27.28) gives

$$m(3T) = \alpha[-(1-b^2)] - \alpha\gamma[-(1-b)] - b[-(1+b+b^2)]$$

Reducing this expression algebraically gives $m(3T) = -1$.

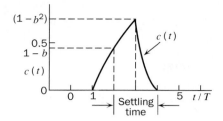

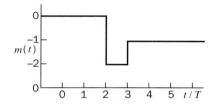

FIGURE 27-5
Response under fast sampling, load, minimal proto-type algorithm to a unit-step change in load (a = 0.5).

If one continues in this sequential manner, which is how the computer handles the computation, one can show that

$$m(4T) = m(5T) = m(nT) = \cdots = -1$$

In other words, the manipulated variable reaches -1 at $t = 3T$, and remains at this value thereafter. A graph showing the response and the manipulated variable is shown in Fig. 27.5 for the case where $a = 0.5$. For this case

$$b = e^{-T/\tau} = e^{-0.5} = 0.606$$

and

$$m(2T) = -1.974$$

The manipulated variable $m(nT)$ that results for this case is not surprising when one considers the nature of the first-order system with transport lag. In fact, for this simple process, one can calculate the values of manipulated variable directly, without use of Eq. (27.24). However, for other disturbances and for more complex algorithms, the calculation becomes very involved without a systematic approach such as that given by Eq. (27.24).

Settling Time

A useful parameter for describing a transient response of a control system is *settling time*. For a load change, the settling time is defined as the time required to reduce the error to zero; this time is measured from the sampling instant for which nonzero error is recorded to the sampling instant for which the output returns to the set point and remains at the set point at future sampling instants. For the example under consideration, the settling time t_s is $2T$. This can be seen most easily from Fig. 27.5.

For a set-point disturbance, the settling time is defined as the time required to reduce the error to zero; this time is measured from the sampling instant for which the set-point change is first detected to the sampling instant for which the response returns to the set point and remains there at future sampling instants.

OBTAINING $M(z)$ DIRECTLY FROM KNOWLEDGE OF $C(z)$. If one wishes to compute $M(z)$ without using the sequential method just discussed and shown in Table 27.1, the following direct procedure can be used.

For the servo problem, where $U(s) = 0$, one can write directly from Fig. 27.3

$$C(z) = M(z)G(z) \tag{27.29}$$

where $G(z) = G_h G_p(z)$

Solving for $M(z)$ gives

$$M(z) = C(z)/G(z) \tag{27.30}$$

For the regulator problem, where $R(s) = 0$, one can obtain directly from Fig. 27.3

$$M(z) = -D(z)C(z) \tag{27.31}$$

For Example 27.1, one can use Eq. (27.31) to obtain $M(z)$ for a unit-step change in load; C in Eq. (27.31) is replaced by C_d of Eq. (27.18). The reader should try this approach to see that it leads to the same results as those obtained in Table 27.1.

Example 27.2. ($T > a\tau$, slow sampling, load)

In this example, the minimal prototype $D(z)$ will be designed for the following conditions:

$T > a\tau$ (slow sampling)

$U = 1/s$ (load disturbance)

$G_p = e^{-a\tau s}/(\tau s + 1)$, $a\tau < T$

The fact that $a\tau$ is not an integral number of sampling periods will make the derivation of this algorithm more complicated. Based on the response of a first-order system with transport lag, the minimal prototype response shown in Fig. 27.6 can be written:

$c(0) = 0$

$c(T) = 1 - e^{-(T-a\tau)/\tau}$

$c(2T) = c(3T) = \cdots c(nT) = 0$

The desired response C_d is therefore

$$C_d(z) = 0 + [1 - e^{-(T-a\tau)/\tau}]z^{-1} + 0z^{-2} + \cdots$$

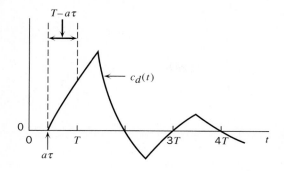

FIGURE 27-6
Minimal prototype response for Example 27.2.

or

$$C_d(z) = (1 - d)z^{-1} \qquad (27.32)$$

where $d = e^{(a - T/\tau)}$

For a load algorithm, we use Eq. (27.14) to obtain $D(z)$. Notice that the peaks in Fig. 27.6 occur at time $a\tau$ into each sampling interval. For this problem, we may use Eq. (26.11) with $K = 1$ and $n = 0$ to obtain, after simplification

$$G(z) = \frac{1 - d}{z(z - b)}\left[z + \frac{d - b}{1 - d}\right] \qquad (27.33)$$

For this example,

$$UG_p(s) = \frac{e^{-a\tau s}}{s(\tau s + 1)} = e^{-a\tau s}\frac{1/\tau}{s + (1/\tau)}$$

To obtain $UG_p(z)$, we shall make use of the modified Z-transform as was done in Chap 25. With this in mind,

let $e^{-\lambda T s} = e^{-a\tau s}$

or $\lambda T = a\tau$

or $\lambda = a\tau/T$

We can now write the m parameter in the modified Z-transforms as

$$m = 1 - \lambda = 1 - (a\tau/T)$$

The Z-transform of $UG_p(s)$ becomes

$$UG_p(z) = Z\left\{\frac{e^{-a\tau s}}{s(\tau s + 1)}\right\} = Z_m\left\{\frac{1}{s(\tau s + 1)}\right\}$$

with $m = 1 - (a\tau/T)$.

From the table of transforms (Table 22.1), we obtain

$$UG_p(z) = \frac{1}{z - 1} - \frac{e^{-(1 - a\tau/T)T/\tau}}{z - b}$$

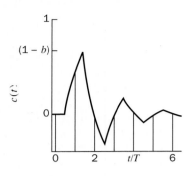

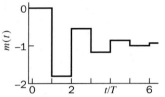

FIGURE 27-7
Response under slow sampling, load, minimal prototype algorithm to unit-step in load. $a = 0.5$, $T/\tau = 1$.

This may be simplified to

$$UG_p(z) = \frac{(1 - d)\left[z + \dfrac{d - b}{1 - d}\right]}{(z - 1)(z - b)} \qquad (27.34)$$

where $b = e^{-T/\tau}$ and $d = e^{(a - T/\tau)}$.

Introducing Eqs. (27.32), (27.33), and (27.34) into Eq. (27.14) gives

$$D(z) = \frac{z(1 - db)\left[z - \dfrac{b(1 - d)}{1 - db}\right]}{(1 - d)^2(z - 1)\left[z - \dfrac{d - b}{1 - d}\right]} \qquad (27.35)$$

It is instructive to examine the continuous response for this example as shown in Fig. 27.7. Although the response is zero at sampling instants after $t = T$, there is intersample ripple. Furthermore, one can show that the manipulated variable does not settle down, as was the case in Example 27.1, but continues to oscillate with decreasing amplitude as shown in Fig. 27.7. The reason for this unsatisfactory behavior is that the minimal prototype response is too demanding in returning the process variable to the set point. If the designer selects a nonminimal prototype response, which permits the response to return to the set point at $3T$ or later, the intersample ripple will be eliminated. A possible nonminimal prototype response is shown in Fig. 27.8.

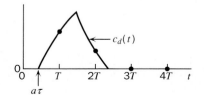

FIGURE 27-8
Possible nonminimal prototype response for Example 27.2.

Response of Sampled-Data System to Other Disturbances

As stated in item 7 of the design specifications in this chapter, it is important to test the control algorithm for inputs that differ from the input for which the algorithm was designed. The computation for other inputs is straightforward, but it can be quite tedious. One uses Eq. (27.13) with the appropriate $R(z)$ or $UG_p(z)$. For the algorithm developed in Example 27.1, the response to two different inputs will be considered: a step change in set point, and a ramp change in load.

STEP CHANGE IN SET POINT. For a set-point change, Eq. (27.13) becomes:

$$C(z) = \frac{G(z)D(z)R(z)}{1 + G(z)D(z)} \tag{27.36}$$

For a unit-step change in R

$$R(z) = \frac{z}{z - 1}$$

Substituting this expression for $R(z)$ and those for $G(z)$ and $D(z)$ from Eqs. (27.19) and (27.21), respectively, into Eq. (27.36) gives

$$C(z) = \frac{\dfrac{1 - b}{z(z - b)} \dfrac{\alpha z(z - \gamma)}{(z - 1)[z + (1 + b)]} \dfrac{z}{z - 1}}{1 + \dfrac{1 - b}{z(z - b)} \dfrac{\alpha z(z - \gamma)}{(z - 1)[z + (1 + b)]}} \tag{27.37}$$

The inversion of this expression gives a result, shown graphically in Fig. 27.9.

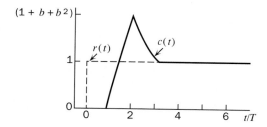

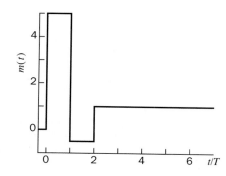

FIGURE 27-9
Response under fast sampling, load, minimal prototype algorithm to a unit-step change in set point ($a = 0.5$).

Notice that the response gives an overshoot with a settling time of $3T$ and the response is free of intersample ripple. From this result, one concludes that the response is satisfactory for a step change in set point.

RAMP CHANGE IN LOAD. For the algorithm developed in Example 27.1, consider a ramp change in load, for which $u(t) = t$ or $U(s) = 1/s^2$. For a load change, Eq. (27.13) becomes

$$C(z) = \frac{UG_p(z)}{1 + G(z)D(z)} \tag{27.38}$$

For this case, one can show that

$$UG_p(z) = \frac{1}{z}\left[\frac{Tz}{(z-1)^2} - \frac{z(1-b)}{(1/\tau)(z-1)(z-b)}\right] \tag{27.39}$$

Introducing this expression for $UG_p(z)$ and $G(z)$ and $D(z)$ from Eqs. (27.19) and (27.21), respectively, into Eq. (27.38) gives, after considerable algebraic manipulation

$$C(z) = \frac{z^{-3}}{z-1}[z + (1+b)]\{z[T - \tau(1-b)] - [bT - (1-b)\tau]\}$$

Inverting this expression by the method of long division gives

$$c(0) = 0$$
$$c(T) = 0$$
$$c(2T) = T - (1-b)\tau$$
$$c(3T) = 2T - (1-b^2)\tau$$
$$c(nT) = \tau a(1-b)(2+b) \qquad \text{for } n \geq 4$$

The response to the ramp input is shown in Fig. 27.10. The response is stable and shows offset that is typical for proportional control of continuous systems when subject to a ramp input in load.

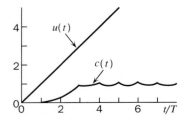

FIGURE 27-10
Response under fast sampling, load, minimal prototype algorithm for ramp change in load ($a = 0.5$).

The two examples given here should be sufficient to show how the control algorithm $D(z)$ can be obtained. The interested reader will find a number of algorithms developed by Mosler et al. (1967), two of which have been presented here as Examples 27.1 and 27.2.

$D(z)$ FOR CONVENTIONAL CONTROLLERS

The conventional continuous controllers discussed in Chap. 10 have their equivalent forms in sampled-data control. The control algorithms in terms of $D(z)$ for PI and PID controllers will be developed in this section.

PI Control

The PI control law can be written as

$$m(t) = K_c e(t) + \frac{K_c}{\tau_I} \int_0^t e(t)\, dt \tag{27.40}$$

To develop $D(z)$ we first write Eq. (27.40) for $m(nT)$ and $m[(n-1)T]$ as follows:

$$m(nT) = K_c e(nT) + \frac{K_c}{\tau_I} \int_0^{nT} e(t)\, dt \tag{27.41}$$

$$m[(n-1)T] = K_c e[(n-1)T] + \frac{K_c}{\tau_I} \int_0^{(n-1)T} e(t)\, dt \tag{27.42}$$

Subtracting Eq. (27.42) from Eq. (27.41) gives

$$m(nT) - m[(n-1)T] = K_c\{e(nT) - e[(n-1)T]\} + \frac{K_c}{\tau_I} \int_{(n-1)T}^{nT} e(t)\, dt \tag{27.43}$$

To convert this equation into a form that involves past values of m and present and past values of e, as is required by Eq. (27.1), we must approximate the definite integral. Many possible approximations can be used, but the one used here will consider $e(t)$ to remain at $e(nT)$ during the time interval $(n-1)T$ to nT. The nature of this approximation is shown in Fig. 27.11. The approximation may be written

$$\int_{(n-1)T}^{nT} e(t)\, dt \cong Te(nT) \tag{27.44}$$

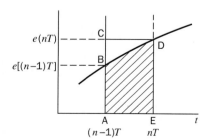

FIGURE 27-11
Approximation of a definite integral: ABDEA = exact value, ACDEA = approximate value.

Using this approximation in Eq. (27.43) and solving for $m(nT)$ gives

$$m(nT) = m[(n-1)T] + K_c\{e(nT) - e[(n-1)T]\} + \frac{K_c T}{\tau_I}e(nT)$$

In terms of Z-transforms, this equation becomes

$$M(z) = M(z)z^{-1} + K_c E(z) - K_c E(z)z^{-1} + (K_c T/\tau_I)E(z)$$

Solving for $M(z)/E(z)$, which is $D(z)$, gives the following expression for $D(z)$ for the PI controller:

$$D(z) = \frac{M(z)}{E(z)} = \frac{K_c}{\alpha}\frac{z-\alpha}{z-1} \tag{27.45}$$

where $\alpha = \dfrac{\tau_I}{\tau_I + T}$

Before developing $D(z)$ for the PID controller, it will be instructive to study the nature of the response for $D(z)$ in Eq. (27.45) to a unit-step change in E. The block diagram for this case is shown in Fig. 27.12. For this block diagram, we write directly

$$M(z) = D(z)E(z) \tag{27.46}$$

where $E(z) = z/(z-1)$

For this example, let $K_c = \tau_I = T = 1$, then $\alpha = 0.5$ and $D(z)$ from Eq. (27.45) becomes

$$D(z) = \frac{1}{0.5}\frac{z-0.5}{z-1} = \frac{2z-1}{z-1}$$

Using this expression for $D(z)$ in Eq. (27.46) gives

$$M(z) = \frac{z}{z-1}\frac{2z-1}{z-1} = \frac{2z^2-z}{z^2-2z+1}$$

By long division, one obtains

$$M(z) = 2 + 3z^{-1} + 4z^{-2} + 5z^{-3} + \cdots$$

In terms of the time domain, this equation gives

$$m(0) = 2$$
$$m(1) = 3$$
$$m(2) = 4$$
$$m(3) = 5$$

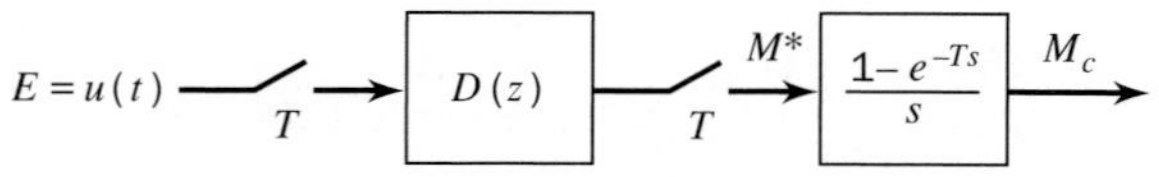

$E = u(t) \longrightarrow T \longrightarrow \boxed{D(z)} \longrightarrow T \xrightarrow{M^*} \boxed{\dfrac{1-e^{-Ts}}{s}} \xrightarrow{M_c}$

FIGURE 27-12
Open-loop step response of a sampled-data PI controller.

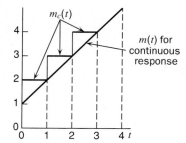

FIGURE 27-13
Comparison of sampled-data response and continuous response of a PI controller subjected to a step change in input.

A graph of the sampled-data response and the response for a continuous controller for the same parameters ($K_c = 1$, $\tau_I = 1$) is shown in Fig. 27.13.

Notice that the sampled-data response equals the continuous response at sampling instants for $n \geq 1$. As the sampling period is reduced, the sampled-data response approaches the continuous response. Based on this observation, we can see that the sampled-data controller is a reasonable approximation of the continuous controller.

PID Control

In a similar manner to the development of $D(z)$ for PI control, we shall develop $D(z)$ for PID control. The PID control law may be written

$$G_c(s) = K_c e + \frac{K_c}{\tau_I} \int_0^t e\, dt + K_c \tau_D \frac{de}{dt}$$

Writing $m(t)$ at nT and $(n-1)T$ gives

$$m(nT) = K_c \left\{ e(nT) + \frac{1}{\tau_I} \int_0^{nT} e(t)\, dt + \tau_D \frac{de}{dt}\bigg|_{t=nT} \right\} \qquad (27.47)$$

and

$$m[(n-1)T] = K_c \left\{ e[(n-1)T] + \frac{1}{\tau_I} \int_0^{(n-1)T} e(t)\, dt + \tau_D \frac{de}{dt}\bigg|_{t=(n-1)T} \right\}$$

$$(27.48)$$

For this case, the approximation to the integral will be the same as used for PI control [Eq. (27.44)]. The approximation for the derivative will be taken as a simple backward difference approximation; thus

$$\frac{de}{dt}\bigg|_{t=nT} \cong \frac{e(nT) - e[(n-1)T]}{T} \qquad (27.49)$$

This simply states that the derivative is approximately equal to the change in e over one sampling period divided by the sampling period.

If we introduce these approximations for the integral and the derivative into Eqs. (27.47) and (27.48), we obtain after subtracting Eq. (27.48) from Eq. (27.47)

$$m(nT) - m[(n-1)T] = K_c \left\{ e(nT) - e[(n-1)T] + \frac{T}{\tau_I} e(nT) \right.$$

$$\left. + \frac{\tau_D}{T} \left[e(nT) - 2e[(n-1)T] + e[(n-2)T] \right] \right\} \qquad (27.50)$$

Converting this equation to the z-domain and solving for $M(z)/E(z)$, which is $D(z)$, finally gives the algorithm:

$$D(z) = \frac{K_c[z^2 - \beta z + \gamma]}{\mu z(z-1)} \qquad (27.51)$$

where $\beta = \dfrac{(T + 2\tau_D)\tau_I}{T^2 + T\tau_I + \tau_I\tau_D}$

$\gamma = \dfrac{\tau_I + \tau_D}{T^2 + T\tau_I + \tau_I\tau_D}$

$\mu = \dfrac{T\tau_I}{T^2 + T\tau_I + \tau_I\tau_D}$

The nature of the response for a unit-step change in input for $K_c = \tau_I = T = 1$ and $\tau_D = 2$ is shown in Fig. 27.14. The details of obtaining this result are left as an exercise for the reader. The response of the continuous PID controller to a unit-step change in input for the same parameters (K_c, τ_I, τ_D) is also shown in Fig. 27.14. Notice that the impulse at $t = 0$ for the continuous response is replaced by a pulse during the first sampling period that reaches a value of 4.0 instead of infinity. After $t = 1$, the sampled-data response is the same as for the PI sampled-data response shown in Fig. 27.13. As τ_D is increased, the pulse during the first sampling period will become larger, thereby approximating more closely the jump to infinity for the continuous response.

The simple backward difference formula used to approximate the derivative term in the PID control algorithm can be replaced by a higher order difference

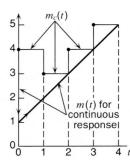

FIGURE 27-14
Comparison of sampled-data response and continuous response for a PID controller subjected to a step change in input.

approximation to give an alternate version of $D(z)$. In fact, many alternate difference approximations for the integral term and the derivative term can be used to give a variety of forms for $D(z)$. As the sampling period T is reduced, the response of the control system using different forms of $D(z)$ for PI or PID control should approach the response for continuous versions of the algorithms. One of the problems at the end of this chapter involves the calculation of the response of a system that uses the $D(z)$ for a PI controller given by Eq. (27.45). In general, the replacement of a continuous controller by its equivalent sampled-data version will give a less stable response for the same set of controller parameters (K_c, τ_I, τ_D).

SUMMARY

In this chapter a systematic procedure for the design of direct-digital control algorithms was described. The procedure requires that a model for the process be known and that the location of the disturbance (set point or load) and the type of disturbance (step, ramp, etc.) be specified. These requirements are similar to those for designing a controller by the internal model control procedure discussed in Chap. 18. The design procedure presented here gives the designer a wide choice of the desired response of the control system; this choice is usually based on knowledge of the response of the process model. The minimal prototype response is an ideal response that reduces the error (at sampling instants) to zero in the least time. The control algorithm $D(z)$ obtained by the design procedure can be written in a form that can be used by a digital computer to control the process. The need to test a proposed algorithm for a disturbance other than the one used to design the algorithm was emphasized and illustrated by examples.

The equivalent sampled-data control algorithms for conventional (PI and PID) control were derived and the open-loop response for each algorithm was compared to the response for the corresponding continuous algorithms. As the sampling period T decreases, the response of the digital algorithm approaches that of the continuous algorithm.

PROBLEMS

27.1. Derive $D(z)$ for the control system shown in Fig. P27.1 for a unit-step change in R and for a response in which C is returned to the set point in one sampling period and remains there at sampling instants. Notice that C_d is $1/(z-1)$ for this problem.

Express the manipulated variable m in terms of present and past values of e and m. Plot $m(t)$ and $c(t)$ during the first few periods.

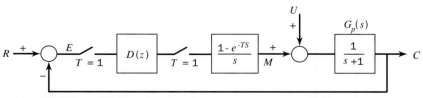

FIGURE P27-1

27.2. (*a*) Determine the pulse transfer function, $D(z)$, for the system shown in Fig. P27.2 if the input disturbance is a unit-step function and if the output is to reach the set point one sampling period after the disturbance occurs. Plot the manipulated variable. Notice that C_d is $1/(z - 1)$ for this problem.

(*b*) If the input is $r(t) = tu(t)$, plot the output if the $D(z)$ of part (*a*) is used.

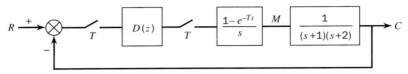

FIGURE P27-2

27.3. The sampled-data system shown in Fig. P27.3 uses the following algorithm

$$D(z) = \frac{1}{(1 - b)} \frac{z(z - b)}{(z + 1)(z - 1)}$$

where $b = e^{-T/\tau} = e^{-1}$

For the process $\tau = 1$, $a = 1$, $T = a\tau = 1$. If a unit-step enters as a load change (i.e., $U(s) = 1/s$), determine $C(z)$. Plot the continuous response $c(t)$. Determine values of $c(t)$ at $t = 1, 1.5, 2, 3,$ and 4. Determine $m_c(t)$ for $t = 0, 1, 2, 3,$ and 4.

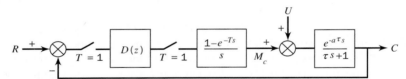

FIGURE P27-3

27.4. Determine the minimal prototype response for a unit-step change in load for the control system shown in Fig. P27.4 for the following plant transfer functions:

(*a*) $G_p = e^{-Ts}/(s + 1)$

(*b*) $G_p = 1/(2s + 1)$

(*c*) $G_p = e^{-2Ts}/s$

(*d*) $G_p = 1/(s^2 + 0.4s + 1)$

Express your result as $C_d = \eta_0 + \eta_1 z^{-1} + \eta_2 z^{-2} + \cdots$

Give numerical values of $\eta_0, \eta_1, \eta_2, \ldots$

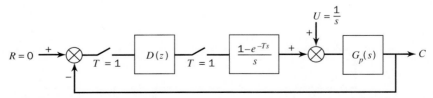

FIGURE P27-4

27.5. (a) For the sampled-data process shown in Fig. P27.5a show that

$$G(z) = \frac{K(z - a)}{z(z - 1)(z - b)}$$

where $b = e^{-T/\tau}$

$\qquad a = 2 - 1/b$

$\qquad K$ is proportional to K_c

For $\tau = 2$, determine the value of K_c for which the system becomes unstable. Use the root locus method. For $K_c = 1$, determine $c(nT)$.

(b) For the process in Fig. P27.5b, use for $D(z)$ the following PI equivalent:

$$D(z) = \frac{K_c}{\alpha} \frac{z - \alpha}{z - 1}$$

where $\alpha = \dfrac{\tau_I}{\tau_I + T}$

For $K_c = 1$, $\tau_I = 1$, and $\tau = 2$, determine $c(nT)$ and compare with $c(nT)$ of part (a).

(c) For the continuous control process shown in Fig. P27.5c, determine the ultimate value of K_c. Compare this value of K_c with that of part (a) to see the effect of sampling.

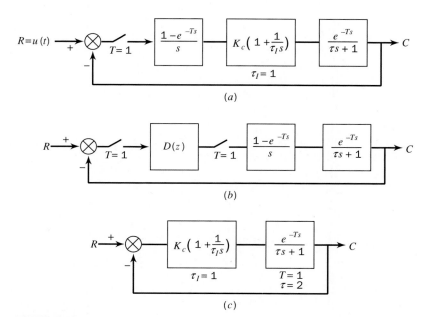

(a)

(b)

(c)

FIGURE P27-5

PART
VII

STATE-SPACE
METHODS

CHAPTER
28

STATE-SPACE REPRESENTATION OF PHYSICAL SYSTEMS

Up to this point, we have described dynamic physical systems by means of differential equations and transfer functions. Another method of description, which is widely used in all branches of control theory, is the state-space method. In fact, other disciplines of engineering (e.g., electrical engineering) introduce the state-space description before the transfer function description. The reader who plans to go beyond an introductory course in control or read from other engineering disciplines should be familiar with state-space methods. In the chapters of this part of the book, the state-space method will be developed and compared with the transfer function method. It is much easier to start with the transfer function method and then develop the state-space method. The mathematical background needed for the transfer function approach involves differential equations and Laplace transforms. The additional mathematical background needed for the state-space method involves matrix algebra. Nearly all students today receive information on matrices in their mathematics courses. For those who are rusty in this topic, it is recommended that they review some of the fundamental matrix operations. A brief review of matrix algebra is given in Appendix 28A.

The transfer function approach is sufficient to calculate the response of linear control systems. The state-space approach is especially valuable in the field of optimal control of linear or nonlinear systems. The concepts developed in this part of the book will be used in the next part on nonlinear control.

STATE VARIABLES

A linear physical system can be described mathematically by:

- an nth order differential equation
- a transfer function
- n first-order differential equations
- a matrix differential equation

So far, we have used the first two mathematical representations for describing physical systems. The third and fourth representations are referred to as state variable descriptions.

To illustrate these four methods of description, consider the familiar second-order process relating an output y to an input u. The four expressions for this process are listed below.

1. nth order differential equation ($n = 2$)

$$\tau^2 \frac{d^2y}{dt^2} + 2\zeta\tau \frac{dy}{dt} + y = u \qquad (28.1)$$

2. Transfer function. The transfer function corresponding to Eq. (28.1) is

$$\frac{Y(s)}{U(s)} = \frac{1}{\tau^2 s^2 + 2\zeta\tau s + 1} \qquad (28.2)$$

3. n first-order differential equations ($n = 2$). Equation (28.1) can be expressed by the following differential equations:

$$\dot{x}_1 = x_2 \qquad (28.3a)$$

$$\dot{x}_2 = -\frac{1}{\tau^2} x_1 - \frac{2\zeta}{\tau} x_2 + \frac{1}{\tau^2} u \qquad (28.3b)$$

where $x_1 = y$ and $x_2 = \dot{y}$

In Eqs. (28.3a) and (28.3b), x_1 and x_2 are the state variables.

To see that Eqs. (28.3) are the equivalent to Eq. (28.1), differentiate both sides of Eq. (28.3a); the result is

$$\ddot{x}_1 = \dot{x}_2 \qquad (28.4)$$

In Eq. (28.3b), we may now replace $\dot{x}_2$ by $\ddot{x}_1$ and x_2 by $\dot{x}_1$; the result is

$$\ddot{x}_1 = -\frac{1}{\tau^2} x_1 - \frac{2\zeta}{\tau} \dot{x}_1 + \frac{1}{\tau^2} u \qquad (28.5)$$

Since $x_1 = y$, we may write

$$\dot{x}_1 = \dot{y} \qquad \text{and} \qquad \ddot{x}_1 = \ddot{y}$$

Using these expressions in Eq. (28.5) gives

$$\ddot{y} = -\frac{1}{\tau^2} y - \frac{2\zeta}{\tau} \dot{y} + \frac{1}{\tau^2} u \qquad (28.6)$$

Equation (28.6) is, of course, the same as Eq. (28.1). We shall see later that other choices for x_1 and x_2 are possible; at this point, the reader is asked to accept Eqs. (28.3a) and (28.3b) as a valid description of the second-order system under consideration.

4. Matrix differential equation. Equations (28.3a) and (28.3b) can be written as one matrix differential equation as follows:

$$\dot{\mathbf{x}} = \mathbf{A}\mathbf{x} + \mathbf{b}u \tag{28.7}$$

where

$$\dot{\mathbf{x}} = \begin{bmatrix} \dot{x}_1 \\ \dot{x}_2 \end{bmatrix} \qquad \mathbf{x} = \begin{bmatrix} x_1 \\ x_2 \end{bmatrix}$$

$$\mathbf{A} = \begin{bmatrix} 0 & 1 \\ \dfrac{-1}{\tau^2} & \dfrac{-2\zeta}{\tau} \end{bmatrix} \qquad \mathbf{b} = \begin{bmatrix} 0 \\ \dfrac{1}{\tau^2} \end{bmatrix} \qquad u \text{ is a scalar}$$

The representation given by Eqs. (28.3) and the representation given by Eq. (28.7) are exactly the same; Equation (28.7) is in a more compact form. The state variables x_1 and x_2 are represented by the column vector $\mathbf{x}$. The coefficients of the state variables on the right sides of Eqs. (28.3a) and (28.3b) are the elements of the matrix $\mathbf{A}$. In this example, there is only one input or forcing term, u, which is a scalar. Each term on the right side of Eq. (28.7) must be a vector containing two elements (i.e., a 2×1 matrix). In order for the expression given by Eq. (28.7) to agree with Eqs. (28.3a) and (28.3b), the coefficient of u must be a vector with the upper element zero. With some practice, the reader will be able to look at a matrix expression such as Eq. (28.7) and quickly see the equivalent set of differential equations.

The output y in representations 1 or 2 often represents a physical variable of interest, such as the temperature of a process or the position of a mechanical system. The alternate state variable representation given by Eqs. (28.3) or Eq. (28.7) contains two state variables, one of which is y and the other the derivative of y (i.e., $\dot{y}$). In this case only y may be of interest to the control engineer; $\dot{y}$ is available, but may not be of interest since it cannot always be measured easily. (For example, there is no easy way to measure the rate of change of temperature if y represents temperature.)

State-Space Description

In general, a physical system can be described by state variables as follows

$$\begin{aligned} \dot{x}_1 &= f_1(x_1, x_2, \ldots, x_n, u_1, u_2, \ldots, u_m) \\ \dot{x}_2 &= f_2(x_1, x_2, \ldots, x_n, u_1, u_2, \ldots, u_m) \end{aligned} \tag{28.8}$$

$$\vdots$$

$$\dot{x}_n = f_n(x_1, x_2, \ldots, x_n, u_1, u_2, \ldots, u_m)$$

where $x_1, x_2, \ldots, x_n$ are n state variables and $u_1, u_2, \ldots, u_m$ are m inputs or forcing terms. The above set of equations may be written as a matrix expression as follows

$$\dot{\mathbf{x}} = f(\mathbf{x}, \mathbf{u})$$

If the system parameters vary with time, the vector f will contain explicit functions of time. An example for an element of f might be the expression on the right side of the following equation:

$$\dot{x}_1 = 2tx_1 + x_2 + u_1 + u_2$$

In this chapter, we shall be concerned with time-invariant systems for which $\dot{x}_i$ is a linear combination of state variables and the coefficients are constant. For the time-invariant case, we may write the general term $\dot{x}_i$ in Eq. (28.8) as follows:

$$\dot{x}_i = a_{i1}x_1 + a_{i2}x_2 + \cdots + a_{in}x_n + b_{i1}u_1 + \cdots + b_{im}u_m \tag{28.9}$$

for $i = 1, 2, 3, \ldots, n$

The equivalent matrix expression for Eq. (28.9) is

$$
\begin{bmatrix} \dot{x}_1 \\ \dot{x}_2 \\ \vdots \\ \dot{x}_n \end{bmatrix} =
\begin{bmatrix}
a_{11} & a_{12} & \cdots & a_{1n} \\
a_{21} & a_{22} & \cdots & a_{2n} \\
\vdots & \vdots & \vdots & \vdots \\
a_{n1} & a_{n2} & \cdots & a_{nn}
\end{bmatrix}
\begin{bmatrix} x_1 \\ x_2 \\ \vdots \\ x_n \end{bmatrix}
$$

$$
+
\begin{bmatrix}
b_{11} & b_{12} & \cdots & b_{1m} \\
b_{21} & b_{22} & \cdots & b_{2m} \\
\vdots & \vdots & \cdots & \vdots \\
b_{n1} & b_{n2} & \cdots & b_{nm}
\end{bmatrix}
\begin{bmatrix} u_1 \\ u_2 \\ \vdots \\ u_m \end{bmatrix}
\tag{28.10}
$$

Writing this in the more compact matrix form, we have

$$\dot{\mathbf{x}} = \mathbf{Ax} + \mathbf{Bu} \tag{28.11}$$

In this expression, there are m different inputs where $m \leq n$. The nature of the linear physical system expressed by Eq. (28.11) is completely stated by the matrices $\mathbf{A}$ and $\mathbf{B}$. For the time-invariant system, the elements of $\mathbf{A}$ and $\mathbf{B}$ are constants.

The outputs of interest to the control engineer may differ from the state variables (x_i). The most general statement for relating the output to the state variables is

$$\mathbf{y} = \mathbf{Cx} \tag{28.12}$$

where $\mathbf{y}$ is the vector of outputs $(y_1, y_2, \ldots, y_p)$ chosen by the control engineer for some practical reason. The matrix $\mathbf{C}$ is a $p \times n$ matrix containing constant elements. The way in which the matrix $\mathbf{C}$ is selected will be clarified in the example to follow. In summary, the state-space description for a linear time-invariant system is given by Eqs. (28.11) and (28.12).

Example 28.1. For the two-tank noninteracting liquid level system shown in Fig. 28.1, obtain the state-space description as expressed by Eqs. (28.11) and (28.12). The output y of interest is the level in tank 2. Notice that streams enter both tanks.

For this example, let the state variables be the physical variables h_1 and h_2, which are the levels in tanks 1 and 2. These state variables are called physical variables because they can be easily measured or observed. (In another example, we shall consider a different set of state variables.)

For the liquid-level system shown in Fig. 28.1 we may write

$$A_1 \frac{dh_1}{dt} = u_1 - \frac{h_1}{R_1} \tag{28.13}$$

$$A_2 \frac{dh_2}{dt} = u_2 + \frac{h_1}{R_1} - \frac{h_2}{R_2} \tag{28.14}$$

or

$$\frac{dh_1}{dt} = -\frac{1}{R_1 A_1} h_1 + \frac{1}{A_1} u_1 \tag{28.15}$$

$$\frac{dh_2}{dt} = \frac{1}{R_1 A_2} h_1 - \frac{1}{A_2 R_2} h_2 + \frac{1}{A_2} u_2 \tag{28.16}$$

These equations can be written as follows

$$\dot{\mathbf{h}} = \mathbf{A}\mathbf{h} + \mathbf{B}\mathbf{u} \tag{28.17}$$

where

$$\mathbf{h} = \begin{bmatrix} h_1 \\ h_2 \end{bmatrix} \qquad \mathbf{A} = \begin{bmatrix} \dfrac{-1}{R_1 A_1} & 0 \\ \dfrac{1}{R_1 A_2} & \dfrac{-1}{R_2 A_2} \end{bmatrix} \qquad \mathbf{B} = \begin{bmatrix} \dfrac{1}{A_1} & 0 \\ 0 & \dfrac{1}{A_2} \end{bmatrix}$$

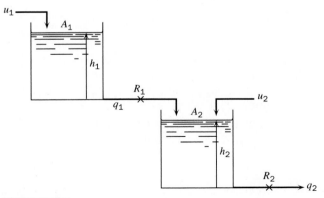

FIGURE 28-1
Liquid-level system for Examples 28.1 and 28.2:

$$A_1 = 1, A_2 = 0.5, R_1 = 0.5, R_2 = 2/3$$

If the output is to be the level in tank 2 (h_2), we have

$$y = \mathbf{Ch}$$

where $\mathbf{y} = y_1 = h_2$ $\mathbf{C} = [0 \quad 1]$

In this case $\mathbf{y}$ is a scalar (i.e., a 1×1 matrix).

The choice of output can be stated in many ways. Regardless of the choice, the output is related to the state variables by Eq. (28.12). To see how the matrix $\mathbf{C}$ depends on the choice of output, consider the following examples:

If $\mathbf{y}$ is to be a scalar that is the arithmetic average of the levels in the two tanks one can show that, $\mathbf{C} = [0.5 \quad 0.5]$.

If the output is to be h_1 and h_2, one can show that

$$\mathbf{C} = \begin{bmatrix} 1 & 0 \\ 0 & 1 \end{bmatrix}$$

Selection of State Variables

To the beginner, the selection of state variables may seem mysterious. The state variables of a system are the smallest set of variables that contain sufficient information to permit all future states to be determined. Although the number of state variables is fixed, the actual selection of these state variables is not unique. If possible, it is convenient to choose state variables that are directly related to physical variables which can be measured or observed (e.g., temperature, level, composition, position, velocity, etc.) For mechanical systems, transducers are available for measuring velocity; for this reason, velocity is considered a physical variable. On the other hand, since the measurement of rate of change of composition is not easily made, this variable is not usually considered a physical variable.

If one solves a dynamic problem by means of an analog computer or by means of a simulation language such as TUTSIM or ACSL,* which involves simulated integrator blocks, one legitimate set of state variables is the output from each integrator.

In the control literature, the types of state variables have been classified as follows.

1. *Physical variables*   State variables are called physical variables when they are readily measured and observed (level, temperature, composition, etc.). Physical variables were discussed at the beginning of this chapter and illustrated for a liquid-level system in Example 28.1 where $x_1 = h_1$ and $x_2 = h_2$.
2. *Phase variables*   State variables that are chosen to be the dependent variable and its successive derivatives are called phase variables. Phase variables were

---

*Although the analog computer will not be discussed in this book, simulation software, such as TUTSIM or ACSL (Advanced Computer Simulation Language), will be discussed in a later chapter. These simulation languages contain integrator blocks that are equivalent to the response of an integrator in an analog computer

selected at the beginning of this chapter in Eqs. (28.3a) and (28.3b) where $x_1 = y$ and $x_2 = \dot{y}$.

3. *Canonical variables*   If the state variables are selected to be canonical variables, the result is that the matrix **A** is diagonal. At this point, it is sufficient to say that canonical variables are selected as state variables for ease in matrix computation. In general, the canonical variables are not readily identified with physical variables.

   In addition to the types of state variables listed above, any other legitimate set of variables can be selected. In Example 28.1, we used physical variables, namely the levels in the tanks of the liquid-level system. In the next two examples, the method for selecting state variables will be shown.

   Example 28.2.   For the two-tank liquid-level system of Example 28.1, shown in Fig. 28.1, obtain the state-space description as expressed by Eqs. (28.11) and (28.12) when phase variables are selected for the state variables. To simplify the problem, let $u_2 = 0$, i.e., there is only one input u_1.

   For the system shown in Fig. 28.1, one can show that

   $$\frac{H_2(s)}{U_1(s)} = \frac{R_2}{(\tau_1 s + 1)(\tau_2 s + 1)} \tag{28.18}$$

   where $\tau_1 = A_1 R_1$ and $\tau_2 = A_2 R_2$

   Introducing the parameters in Fig. 28.1 into Eq. (28.18) gives

   $$\frac{H_2(s)}{U_1(s)} = \frac{2/3}{(\frac{1}{2}s + 1)(\frac{1}{3}s + 1)} \tag{28.19}$$

   or

   $$\frac{H_2(s)}{U_1(s)} = \frac{4}{(s + 2)(s + 3)} \tag{28.20}$$

   To obtain the differential equation corresponding to Eq. (28.20), we cross-multiply to obtain

   $$(s + 2)(s + 3)H_2 = 4U_1$$

   or

   $$(s^2 + 5s + 6)H_2 = 4U_1$$

   This may be expressed as the following differential equation:

   $$\ddot{h}_2 + 5\dot{h}_2 + 6h_2 = 4u_1 \tag{28.21}$$

   Let the state variables be the following phase variables:

   $$x_1 = h_2 \tag{28.22}$$

   $$x_2 = \dot{h}_2 \tag{28.23}$$

We may now write

$$\dot{x}_1 = x_2 [= \dot{h}_2] \qquad (28.24)$$

$$\dot{x}_2 = \ddot{h}_2 \qquad (28.25)$$

Equation (28.21) becomes

$$\dot{x}_2 + 5x_2 + 6x_1 = 4u_1 \qquad (28.26)$$

The system can be described by Eqs. (28.24) and (28.26):

$$\dot{x}_1 = x_2 \qquad (28.27a)$$

$$\dot{x}_2 = -6x_1 - 5x_2 + 4u_1 \qquad (28.27b)$$

In terms of a matrix expression, Eqs. (28.27) may be written:

$$\dot{\mathbf{x}} = \mathbf{Ax} + \mathbf{b}u_1$$

where $\mathbf{A} = \begin{bmatrix} 0 & 1 \\ -6 & -5 \end{bmatrix}$ $\mathbf{b} = \begin{bmatrix} 0 \\ 4 \end{bmatrix}$

If the output $\mathbf{y}$ is to be the level in tank 2,

$$\mathbf{y} = \mathbf{Cx}$$

where

$$\mathbf{C} = [1 \quad 0]$$

Example 28.3. For the PI control system shown in Fig. 28.2, obtain a state-space representation in the form of Eq. (28.7); thus

$$\dot{\mathbf{x}} = \mathbf{Ax} + \mathbf{b}r$$

where r is a scalar. Let

$$x_1 = c \qquad (28.28)$$

$$x_2 = \dot{c} = \dot{x}_1 \qquad (28.29)$$

With this choice of state variables, we have selected phase variables.

From Fig. 28.2, we may write

$$\frac{C(s)}{M(s)} = \frac{K_p}{(\tau_1 s + 1)(\tau_2 s + 1)} = \frac{K_p / \tau_1 \tau_2}{\left(s + \frac{1}{\tau_1}\right)\left(s + \frac{1}{\tau_2}\right)}$$

or

$$\frac{C(s)}{M(s)} = \frac{A}{(s + a)(s + b)} \qquad (28.30)$$

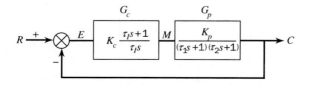

FIGURE 28-2
PI Control System for Example 28.3.

where $A = K_p/\tau_1\tau_2$

$\qquad a = 1/\tau_1$

$\qquad b = 1/\tau_2$

Cross-multiplying Eq. (28.30) gives

$$[s^2 + (a + b)s + ab]C(s) = AM(s)$$

or

$$\ddot{c} + (a + b)\dot{c} + abc = Am \qquad (28.31)$$

From Eqs. (28.28), (28.29), and (28.31) we obtain

$$\dot{x}_1 = x_2 \qquad (28.32)$$

$$\dot{x}_2 = -abx_1 - (a + b)x_2 + Am \qquad (28.33)$$

We must now obtain the state variables associated with the PI controller. From Fig. 28.2, we obtain

$$\frac{M(s)}{E(s)} = K_c\frac{(\tau_I s + 1)}{\tau_I s}$$

or

$$\tau_I sM(s) = K_c\tau_I sE(s) + K_cE(s)$$

In terms of the time domain, this expression becomes

$$\dot{m} = K_c\dot{e} + (K_c/\tau_I)e \qquad (28.34)$$

From the signals entering and leaving the comparator, we may write

$$e = r - c$$

or, since $x_1 = c$, we may write

$$e = r - x_1 \qquad (28.35a)$$

and

$$\dot{e} = \dot{r} - \dot{x}_1 \qquad (28.35b)$$

Combining Eqs. (28.34) and (28.35) gives

$$\dot{m} = K_c(\dot{r} - \dot{x}_1) + (K_c/\tau_I)(r - x_1)$$

or

$$\dot{m} = K_c\dot{r} - K_cx_2 + (K_c/\tau_I)r - (K_c/\tau_I)x_1 \qquad (28.36)$$

At this stage, we are faced with the difficulty of having a derivative term on the right side of Eq. (28.36). In state-space representation, all variables on the right side must be state variables, not derivatives of state variables. One way to handle the present difficulty is to define a new state variable x_3; let

$$x_3 = m - K_cr \qquad (28.37)$$

or

$$\dot{x}_3 = \dot{m} - K_c\dot{r} \qquad (28.38)$$

Combining Eqs. (28.38) and (28.36) leads to

$$\dot{x}_3 = -(K_c/\tau_I)x_1 - K_c x_2 + (K_c/\tau_I)r \tag{28.39}$$

or

$$\dot{x}_3 = -\alpha x_1 - K_c x_2 + \alpha r \tag{28.40}$$

where $\alpha = K_c/\tau_I$

Summarizing the state variable equations given by Eqs. (28.32), (28.33), and (28.40) and using the definition of x_3 in Eq. (28.37) give

$$\dot{x}_1 = x_2$$
$$\dot{x}_2 = -abx_1 - (a + b)x_2 + Ax_3 + AK_c r$$
$$\dot{x}_3 = -\alpha x_1 - K_c x_2 + \alpha r$$

where $A = K_p/\tau_1\tau_2$

$$a = 1/\tau_1 \qquad b = 1/\tau_2 \qquad \alpha = K_c/\tau_I$$

The $\mathbf{A}$ and $\mathbf{b}$ terms in $\dot{\mathbf{x}} = \mathbf{A}\mathbf{x} + \mathbf{b}r$ are

$$\mathbf{A} = \begin{bmatrix} 0 & 1 & 0 \\ -ab & -(a+b) & A \\ -\alpha & -K_c & 0 \end{bmatrix} \qquad \mathbf{b} = \begin{bmatrix} 0 \\ AK_c \\ \alpha \end{bmatrix}$$

If m is required as a function of t, it can always be found by solving Eq. (28.37) for m; thus

$$m = x_3 + K_c r$$

SUMMARY

State-space representation is an alternative to the transfer function representation of a physical system that we have used up to this point. A transfer function that relates an output variable to an input variable represents an nth-order differential equation. In the state-space representation, the nth-order differential equation is written as n first-order differential equations in terms of n state variables. These n differential equations can also be written in a more compact form as a matrix differential equation:

$$\dot{\mathbf{x}} = \mathbf{A}\mathbf{x} + \mathbf{B}\mathbf{u}$$

For an nth-order dynamic system, the number of state variables is fixed at n, but the selection of the variables is not unique. Of the many sets of state variables that one can choose, we discussed three sets that are useful in control theory; namely, physical variables, phase variables, and canonical variables. The state-space representation gives all of the dynamic detail of a system (e.g., the dependent variable and its successive derivatives for the case of phase variables). Whether or not this detail is needed depends on the problem being solved. We shall see the value of state-space representation in multivariable control and in nonlinear control in later chapters.

APPENDIX 28A
ELEMENTARY MATRIX ALGEBRA

The purpose of this section is to provide in a convenient location a review of some of the elementary operations of matrix algebra for use in state-space methods. It is expected that the reader has had a course in linear algebra discussing the concepts of a vector and a matrix and the operations performed on them.

VECTORS. An n-dimensional column vector is an ordered series of elements (numbers): $x_1, x_2, \ldots, x_n$ and is written as

$$\mathbf{x} = \begin{bmatrix} x_1 \\ x_2 \\ \vdots \\ x_n \end{bmatrix}$$

Multiplication of a vector by a scalar ($\lambda \mathbf{x}$) results in a vector for which each element is multiplied by λ.

MATRICES. A matrix is a rectangular array of elements (numbers) that takes the form:

$$\mathbf{A} = \begin{bmatrix} a_{11} & a_{12} & \cdots & a_{1m} \\ a_{21} & a_{22} & \cdots & a_{2m} \\ \vdots & \vdots & \vdots & \vdots \\ a_{n1} & a_{n2} & \cdots & a_{nm} \end{bmatrix}$$

in which the elements are written a_{ij}. The subscript i refers to the ith row and j to the jth column. $\mathbf{A}$ is called an $n \times m$ matrix where n is the number of rows and m is the number of columns. If $n = m$, the matrix is called a square matrix. If $m = 1$, the matrix is a column vector ($n \times 1$). If $n = 1$, the matrix is a row vector ($1 \times m$).

The transpose of a matrix, $\mathbf{A}^T$, is a matrix for which the rows and columns of the matrix $\mathbf{A}$ are interchanged. If the diagonal elements (a_{ij}) of a square matrix are unity and all off-diagonal elements are zero, then the matrix is called a unit matrix and is given the symbol $\mathbf{I}$.

If $\mathbf{A} = \mathbf{A}^T$ for a square matrix, the matrix $\mathbf{A}$ is said to be symmetrical.

When two matrices are added (or subtracted), the corresponding elements are added (or subtracted), thus

$$\mathbf{A} + \mathbf{B} = \begin{bmatrix} a_{11} + b_{11} & a_{12} + b_{12} & \cdots & a_{1m} + b_{1m} \\ a_{21} + b_{21} & a_{22} + b_{22} & \cdots & a_{2m} + b_{2m} \\ \vdots & \vdots & \vdots & \vdots \\ a_{n1} + b_{n1} & a_{n2} + b_{n2} & \cdots & a_{nm} + b_{nm} \end{bmatrix}$$

The product of two matrices $\mathbf{C} = \mathbf{AB}$ is a matrix whose elements are obtained by the expression

$$c_{ij} = \sum_{k=1}^{m} a_{ik}b_{kj} \qquad \text{for } i = 1 \ldots n$$

$$\text{and } j = 1 \ldots p$$

where $\mathbf{A}$ is an $n \times m$ matrix and $\mathbf{B}$ is an $m \times p$ matrix. The matrix $\mathbf{C}$ is an $n \times p$ matrix.

INVERSE OF A MATRIX. The inverse of a matrix is related to the concept of division for numbers. The inverse of a number x is written $1/x$ or x^{-1}. The product of a number x and its inverse is equal to unity. The inverse of a matrix $\mathbf{A}$ is written $\mathbf{A}^{-1}$ and the product of a matrix and its inverse is equal to the unit matrix; thus

$$\mathbf{A}^{-1}\mathbf{A} = \mathbf{I}$$

The expression used for matrix inversion for the examples used in this chapter takes the form:

$$\mathbf{A}^{-1} = \frac{\text{adj } \mathbf{A}}{|\mathbf{A}|} \tag{28A.1}$$

where $|\mathbf{A}|$ is the determinant of $\mathbf{A}$ and adj $\mathbf{A}$ is the adjoint of $\mathbf{A}$. These two terms will now be described .

The determinant of a matrix $|\mathbf{A}|$ is a scalar which is computed from the elements of the matrix as follows:

$$|\mathbf{A}| = a_{i1}A_{i1} + a_{i2}A_{i2} + \cdots + a_{in}A_{in}$$

or

$$|\mathbf{A}| = \sum_{j=1}^{n} a_{ij}A_{ij} \qquad \text{for any } i \tag{28A.2}$$

where A_{ij}, the cofactor of the element a_{ij}, is computed as

$$A_{ij} = (-1)^{i+j}M_{ij}$$

The determinant M_{ij} is the minor of the element a_{ij} and is defined as follows. If the row and column containing the element a_{ij} are deleted from a square matrix $\mathbf{A}$, the determinant of the resulting matrix, which is an $(n-1) \times (n-1)$ matrix, is the minor M_{ij}. An alternate expression for the calculation of a determinant which uses the elements of a specific column and its cofactors is as follows:

$$|\mathbf{A}| = \sum_{i=1}^{n} a_{ij}A_{ij} \qquad \text{for any } j \tag{28A.3}$$

A determinant of a matrix with two equal rows or columns is zero.

We now define the adjoint of a matrix. Let the matrix $\mathbf{B}$ be an $n \times n$ matrix whose elements b_{ij} are the cofactors A_{ji} of $\mathbf{A}$, i.e., the transpose of the cofactor matrix. $\mathbf{B}$ is the adjoint of $\mathbf{A}$; thus $\mathbf{B} = \text{adj } \mathbf{A} = $ transpose of cofactor matrix or

$$\text{adj } \mathbf{A} = \begin{bmatrix} A_{11} & A_{21} & \cdots & A_{n1} \\ A_{12} & A_{22} & \cdots & A_{n2} \\ \vdots & \vdots & \vdots & \vdots \\ A_{1n} & A_{2n} & \cdots & A_{nn} \end{bmatrix}$$

Some useful properties of the inverse are

$$(\mathbf{AB})^{-1} = \mathbf{B}^{-1}\mathbf{A}^{-1}$$
$$(\mathbf{A}^{-1})^{\mathrm{T}} = (\mathbf{A}^{\mathrm{T}})^{-1}$$
$$(\mathbf{A}^{-1})^{-1} = \mathbf{A}$$

The derivations of relationships presented here, as well as other properties of matrices, can be found in textbooks on linear algebra (see Anton, 1984).

EXAMPLES

1. Evaluate the determinant of $\mathbf{A}$ for the following matrix

$$\mathbf{A} = \begin{bmatrix} 2 & 3 & 5 \\ 1 & 0 & 1 \\ 2 & 1 & 0 \end{bmatrix}$$

For this problem, we use Eq. (28A.2) with $i = 1$ (i.e., use row 1)

$$|\mathbf{A}| = 2\begin{bmatrix} 0 & 1 \\ 1 & 0 \end{bmatrix} - 3\begin{bmatrix} 1 & 1 \\ 2 & 0 \end{bmatrix} + 5\begin{bmatrix} 1 & 0 \\ 2 & 1 \end{bmatrix}$$

$$|\mathbf{A}| = 2[(0)(0) - (1)(1)] - 3[(1)(0) - (1)(2)] + 5[(1)(1) - (0)(2)]$$
$$|\mathbf{A}| = 2(-1) - 3(-2) + 5(1) = 9$$

2. Find the inverse of the matrix

$$\mathbf{A} = \begin{bmatrix} 2 & 3 \\ 1 & 4 \end{bmatrix}$$

$$\mathbf{A}^{-1} = \frac{\text{adj } \mathbf{A}}{|\mathbf{A}|}$$

The determinant of $\mathbf{A}$ is

$$|\mathbf{A}| = (2)(4) - (3)(1) = 5$$

The matrix of minors is

$$\begin{bmatrix} 4 & 1 \\ 3 & 2 \end{bmatrix}$$

The cofactor matrix is

$$\begin{bmatrix} 4 & -1 \\ -3 & 2 \end{bmatrix}$$

The adjoint of the matrix, which is the transpose of the cofactor matrix, is

$$\text{adj } \mathbf{A} = \begin{bmatrix} 4 & -3 \\ -1 & 2 \end{bmatrix}$$

therefore $\mathbf{A}^{-1} = \dfrac{1}{5}\begin{bmatrix} 4 & -3 \\ -1 & 2 \end{bmatrix} = \begin{bmatrix} \frac{4}{5} & -\frac{3}{5} \\ -\frac{1}{5} & \frac{2}{5} \end{bmatrix}$

3. Obtain the inverse of the matrix

$$\mathbf{A} = \begin{bmatrix} 2 & 3 & 1 \\ 1 & 2 & 3 \\ 3 & 1 & 2 \end{bmatrix}$$

One can show that

$$|\mathbf{A}| = 18$$

The cofactor matrix is

$$\begin{bmatrix} 1 & 7 & -5 \\ -5 & 1 & 7 \\ 7 & -5 & 1 \end{bmatrix}$$

The adjoint matrix is

$$\begin{bmatrix} 1 & -5 & 7 \\ 7 & 1 & -5 \\ -5 & 7 & 1 \end{bmatrix}$$

$$\mathbf{A}^{-1} = \frac{1}{18}\begin{bmatrix} 1 & -5 & 7 \\ 7 & 1 & -5 \\ -5 & 7 & 1 \end{bmatrix}$$

PROBLEMS

28.1. In the liquid level process shown in Fig. P28.1, the three tanks are interacting. The process may be described by:

$$\dot{x} = \mathbf{A}x + \mathbf{B}u$$

where $\mathbf{x} = \begin{bmatrix} x_1 \\ x_2 \\ x_3 \end{bmatrix}$ and $\mathbf{u} = \begin{bmatrix} u_1 \\ u_2 \end{bmatrix}$

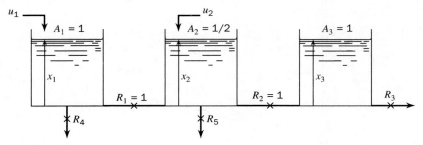

FIGURE P28-1

(*a*) If

$$
\mathbf{A} = \begin{bmatrix} -3 & 1 & 0 \\ 2 & -3 & 2 \\ 0 & 1 & -3 \end{bmatrix}
$$

determine values of R_3, R_4, and R_5. If one of these values of R is negative, what is your interpretation?

(*b*) Determine **B**.

28.2. For the system shown in Fig. P28.2, find **A** and **b** in

$$\dot{\mathbf{x}} = \mathbf{Ax} + \mathbf{bu}$$

The tanks are interacting. The following data apply:

$$A_1 = 1, \ A_2 = \frac{1}{2}, \ R_1 = \frac{1}{2}, \ R_2 = 2, \ R_3 = 1$$

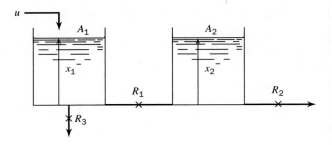

FIGURE P28-2

CHAPTER
29

TRANSFER FUNCTION MATRIX

In the previous chapter, we have seen that a linear dynamic system can be expressed in terms of the following equations

$$\dot{\mathbf{x}} = \mathbf{Ax} + \mathbf{Bu} \tag{29.1}$$

$$\mathbf{y} = \mathbf{Cx} \tag{29.2}$$

where $\mathbf{x}$ = column vector of n state variables $(x_1, x_2, \ldots, x_n)$

$\mathbf{u}$ = column vector of m inputs or forcing terms $(u_1, u_2, \ldots, u_m)$

$\mathbf{y}$ = column vector of p outputs $(y_1, y_2, \ldots, y_p)$

$\mathbf{A}$ = $n \times n$ matrix of coefficients

$\mathbf{B}$ = $n \times m$ matrix of coefficients

$\mathbf{C}$ = $p \times n$ matrix of coefficients

One of the objectives of this chapter is to show how one solves Eqs. (29.1) and (29.2) in a systematic manner.

Before discussing the solution of the matrix differential equation of Eq. (29.1), consider the scalar differential equation

$$dx/dt = Ax + Bu \tag{29.3}$$

In this equation all of the terms are scalars. The solution to Eq. (29.3) can be written as the sum of the complementary function and the particular integral as follows:

$$x(t) = e^{At}x(0) + \int_0^t e^{A(t-\tau)}Bu(\tau)d\tau \tag{29.4}$$

Equation (29.4) is a well known result that has been derived in many books on the solution of ordinary differential equations.

TRANSITION MATRIX

Let us now turn our attention to the solution of the matrix differential equation

$$\dot{\mathbf{x}} = \mathbf{A}\mathbf{x} \tag{29.5}$$

This is Eq. (29.1) for the case of no inputs (i.e., $\mathbf{u} = 0$). The initial conditions for Eq. (29.5) may be expressed as $\mathbf{x}(0)$. One can show that the solution to Eq. (29.5) with initial conditions $\mathbf{x}(0)$ is given by

$$\mathbf{x}(t) = \left\{ \mathbf{I} + \mathbf{A}t + \frac{\mathbf{A}^2}{2!}t^2 + \dots + \frac{\mathbf{A}^k}{k!}t^k \right\} \mathbf{x}(0) \tag{29.6}$$

The infinite series of matrix terms within the braces is given the symbol $e^{\mathbf{A}t}$. This symbol is chosen to recall that the infinite series of the scalar term e^{at} is

$$1 + at + \frac{a^2}{2!}t^2 + \dots + \frac{a^k}{k!}t^k$$

Using the symbol $e^{\mathbf{A}t}$, we may write Eq. (29.6) as

$$\mathbf{x}(t) = e^{\mathbf{A}t}\mathbf{x}(0) \tag{29.7}$$

The symbol $e^{\mathbf{A}t}$ is an $n \times n$ matrix in which each element contains a power series of t. The solution to Eq. (29.1) can be shown to be

$$\mathbf{x}(t) = e^{\mathbf{A}t}\mathbf{x}(0) + \int_0^t e^{\mathbf{A}(t-\tau)}\mathbf{B}\mathbf{u}(\tau)d\tau \tag{29.8}$$

Notice that Eq. (29.8) resembles Eq. (29.4), which is the solution for the scalar differential equation. Since $e^{\mathbf{A}t}$ is awkward and perhaps misleading as to its nature, $e^{\mathbf{A}t}$ is sometimes replaced by $\boldsymbol{\phi}(t)$; thus

$$\boldsymbol{\phi}(t) = e^{\mathbf{A}t} \quad \text{(transition matrix)} \tag{29.9}$$

Either of the terms $\boldsymbol{\phi}(t)$ and $e^{\mathbf{A}t}$ can be used for the transition matrix. In this book, we shall use $e^{\mathbf{A}t}$.

Example 29.1.   Solution of a matrix differential equation.   Solve the following matrix differential equation

$$\dot{\mathbf{x}} = \begin{bmatrix} -1 & 1 \\ 0 & -2 \end{bmatrix} \mathbf{x} + \begin{bmatrix} 0 \\ 1 \end{bmatrix} u(t)$$

where $u(t)$ is a unit-step function and

$$\mathbf{x}(0) = \begin{bmatrix} -1 \\ 0 \end{bmatrix}$$

One can show that

$$e^{\mathbf{A}t} = \begin{bmatrix} e^{-t} & e^{-t} - e^{-2t} \\ 0 & e^{-2t} \end{bmatrix}$$

In the next section, the method used to obtain the elements of this matrix will be developed. Applying Eq. (29.8) gives

$$\mathbf{x}(t) = \begin{bmatrix} e^{-t} & e^{-t} - e^{-2t} \\ 0 & e^{-2t} \end{bmatrix}\begin{bmatrix} -1 \\ 0 \end{bmatrix} + \int_0^t \begin{bmatrix} e^{-(t-\tau)} & e^{-(t-\tau)} - e^{-2(t-\tau)} \\ 0 & e^{-2(t-\tau)} \end{bmatrix}\begin{bmatrix} 0 \\ 1 \end{bmatrix} d\tau$$

or

$$\mathbf{x}(t) = \begin{bmatrix} -e^{-t} \\ 0 \end{bmatrix} + \begin{bmatrix} e^{-(t-\tau)} - 0.5e^{-2(t-\tau)} \\ 0.5e^{-2(t-\tau)} \end{bmatrix}\Bigg|\begin{matrix} \tau = t \\ \tau = 0 \end{matrix}$$

or

$$\mathbf{x}(t) = \begin{bmatrix} 0.5 - 2e^{-t} + 0.5e^{-2t} \\ 0.5 - 0.5e^{-2t} \end{bmatrix}$$

Determining $e^{\mathbf{A}t}$

One method for determining the elements of the transition matrix $e^{\mathbf{A}t}$ is to use Laplace transforms. Consider the matrix differential equation of Eq. (29.1).

$$\dot{\mathbf{x}} = \mathbf{A}\mathbf{x} + \mathbf{B}\mathbf{u}$$

If we take the Laplace transform of each side, we obtain

$$s\mathbf{X}(s) - \mathbf{x}(0) = \mathbf{A}\mathbf{X}(s) + \mathbf{B}\mathbf{U}(s)$$

or

$$s\mathbf{X}(s) - \mathbf{A}\mathbf{X}(s) = \mathbf{x}(0) + \mathbf{B}\mathbf{U}(s)$$

Solving for $\mathbf{X}(s)$ gives

$$(s\mathbf{I} - \mathbf{A})\mathbf{X}(s) = \mathbf{x}(0) + \mathbf{B}\mathbf{U}(s) \qquad (29.10)$$

To obtain an expression for $\mathbf{X}(s)$, pre-multiply both sides of Eq. (29.10) by $(s\mathbf{I} - \mathbf{A})^{-1}$; thus

$$(s\mathbf{I} - \mathbf{A})^{-1}(s\mathbf{I} - \mathbf{A})\mathbf{X}(s) = (s\mathbf{I} - \mathbf{A})^{-1}\mathbf{x}(0) + (s\mathbf{I} - \mathbf{A})^{-1}\mathbf{B}\mathbf{U}(s)$$

This equation becomes

$$\mathbf{X}(s) = (s\mathbf{I} - \mathbf{A})^{-1}\mathbf{x}(0) + (s\mathbf{I} - \mathbf{A})^{-1}\mathbf{B}\mathbf{U}(s) \qquad (29.11)$$

To obtain $\mathbf{x}(t)$ from Eq. (29.11), we may take the inverse transform; thus

$$\mathbf{x}(t) = L^{-1}\{(s\mathbf{I} - \mathbf{A})^{-1}\mathbf{x}(0)\} + L^{-1}\{(s\mathbf{I} - \mathbf{A})^{-1}\mathbf{B}\mathbf{U}(s)\} \qquad (29.12)$$

By comparing Eqs. (29.8) and (29.12), we see that

$$e^{\mathbf{A}t} = L^{-1}\{(s\mathbf{I} - \mathbf{A})^{-1}\} \qquad (29.13)$$

and

$$\int_0^t e^{\mathbf{A}(t-\tau)}\mathbf{B}\mathbf{u}(\tau)d\tau = L^{-1}\{(s\mathbf{I} - \mathbf{A})^{-1}\mathbf{B}\mathbf{U}(s)\} \qquad (29.14)$$

TRANSFER FUNCTION MATRIX

When $\mathbf{x}(0) = 0$, a case frequently used in control applications, we obtain from Eq. (29.11)

$$\mathbf{X}(s) = (s\mathbf{I} - \mathbf{A})^{-1}\mathbf{B}\mathbf{U}(s) \qquad (29.15)$$

This may be written

$$\mathbf{X}(s) = \mathbf{G}(s)\mathbf{U}(s) \qquad (29.16)$$

where

$$\mathbf{G}(s) = (s\mathbf{I} - \mathbf{A})^{-1}\mathbf{B} \qquad \text{(transfer function matrix)} \qquad (29.17)$$

The term $\mathbf{G}(s)$ is called the *transfer function matrix* and serves the same purpose as the transfer function for the scalar case; namely, it relates a set of state variables $\mathbf{X}(s)$ to a set of inputs $\mathbf{U}(s)$.

If we prefer to relate the output to the input as expressed by Eq. (29.2), we may proceed as follows.
Taking the Laplace transform of both sides of Eq. (29.2) gives

$$\mathbf{Y}(s) = \mathbf{C}\mathbf{X}(s) \qquad (29.18)$$

Combining Eqs. (29.15) and (29.18) gives

$$\mathbf{Y}(s) = \mathbf{C}(s\mathbf{I} - \mathbf{A})^{-1}\mathbf{B}\mathbf{U}(s)$$

We may now write

$$\mathbf{Y}(s) = \mathbf{G}_1(s)\mathbf{U}(s) \qquad (29.19)$$

where

$$\mathbf{G}_1(s) = \mathbf{C}(s\mathbf{I} - \mathbf{A})^{-1}\mathbf{B} \qquad (29.20)$$

The term $\mathbf{G}_1(s)$ in Eq. (29.20) is also a transfer function matrix that relates the output vector $\mathbf{Y}$ to the input vector $\mathbf{U}$.

Example 29.2. Determine the transfer function matrix for the 2-tank liquid-level system shown in Fig. 29.1. As developed in Example 28.1 [Eq. (28.17)] of the previous chapter, this system is described by

$$\dot{\mathbf{h}} = \mathbf{A}\mathbf{h} + \mathbf{B}\mathbf{u}$$

where

$$\mathbf{A} = \begin{bmatrix} -2 & 0 \\ 4 & -3 \end{bmatrix} \qquad \mathbf{B} = \begin{bmatrix} 1 & 0 \\ 0 & 2 \end{bmatrix}$$

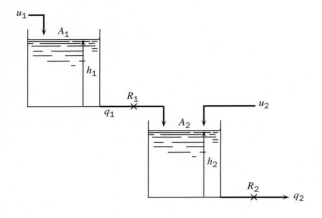

FIGURE 29-1
Liquid-level system for Example
29.2: $A_1 = 1$, $A_2 = 0.5$, $R_1 = 0.5$, $R_2 = 2/3$.

From the definition of the transfer function matrix of Eq. (29.17), we write

$$\mathbf{G}(s) = (s\mathbf{I} - \mathbf{A})^{-1}\mathbf{B}$$

The inverse of $(s\mathbf{I} - \mathbf{A})$ is obtained as follows (see Appendix 28A for details on the inversion of a matrix):

$$(s\mathbf{I} - \mathbf{A})^{-1} = \frac{\text{adj}(s\mathbf{I} - \mathbf{A})}{|s\mathbf{I} - \mathbf{A}|}$$

$$(s\mathbf{I} - \mathbf{A}) = \begin{bmatrix} s+2 & 0 \\ -4 & s+3 \end{bmatrix}$$

$$\text{cofactor of } (s\mathbf{I} - \mathbf{A}) = \begin{bmatrix} s+3 & 4 \\ 0 & s+2 \end{bmatrix}$$

We can now find the adjoint:

$$\text{adj}(s\mathbf{I} - \mathbf{A}) = \begin{bmatrix} s+3 & 0 \\ 4 & s+2 \end{bmatrix}$$

The determinant of $(s\mathbf{I} - \mathbf{A})$ is

$$\begin{vmatrix} s+2 & 0 \\ -4 & s+3 \end{vmatrix} = (s+2)(s+3)$$

We can now determine the inverse of $(s\mathbf{I} - \mathbf{A})$.

$$(s\mathbf{I} - \mathbf{A})^{-1} = \frac{\begin{bmatrix} s+3 & 0 \\ 4 & s+2 \end{bmatrix}}{(s+2)(s+3)} \tag{29.21}$$

$$\mathbf{G}(s) = \frac{\begin{bmatrix} s+3 & 0 \\ 4 & s+2 \end{bmatrix}}{(s+2)(s+3)} \begin{bmatrix} 1 & 0 \\ 0 & 2 \end{bmatrix} = \frac{\begin{bmatrix} s+3 & 0 \\ 4 & 2(s+2) \end{bmatrix}}{(s+2)(s+3)}$$

Simplifying this expression gives

$$\mathbf{G}(s) = \begin{bmatrix} \dfrac{1}{s+2} & 0 \\ \dfrac{4}{(s+2)(s+3)} & \dfrac{2}{s+3} \end{bmatrix}$$

From Eq. (29.16) we write

$$\mathbf{H}(s) = \mathbf{G}(s)\mathbf{U}(s)$$

therefore

$$\begin{bmatrix} H_1(s) \\ H_2(s) \end{bmatrix} = \begin{bmatrix} \dfrac{1}{s+2} & 0 \\ \dfrac{4}{(s+2)(s+3)} & \dfrac{2}{s+3} \end{bmatrix} \begin{bmatrix} U_1(s) \\ U_2(s) \end{bmatrix} \qquad (29.22)$$

From Eq. (29.22), we obtain

$$H_1(s) = \frac{1}{s+2}U_1(s)$$

and

$$H_2(s) = \frac{4}{(s+2)(s+3)}U_1(s) + \frac{2}{s+3}U_2(s)$$

For given inputs, the above equations may be inverted to obtain $h_1(t)$ and $h_2(t)$. For the case of $U_1(s) = 1/s$ and $U_2(s) = 0$, we get

$$H_1(s) = \frac{1}{s(s+2)} = \frac{0.5}{s(0.5s+1)}$$

and

$$H_2(s) = \frac{4}{s(s+2)(s+3)}$$

Inversion of $H_1(s)$ and $H_2(s)$ gives

$$h_1(t) = 0.5(1 - e^{-2t})$$
$$h_2(t) = (2/3)\left[1 - 0.5\left(3e^{-2t} - 2e^{-3t}\right)\right]$$

The results given above can be obtained, of course, by the methods presented earlier in this book.

The transition matrix can be obtained by applying Eq. (29.13) to Eq. (29.21):

$$e^{\mathbf{A}t} = L^{-1}\left\{ \begin{bmatrix} \dfrac{1}{s+2} & 0 \\ \dfrac{4}{(s+2)(s+3)} & \dfrac{1}{s+3} \end{bmatrix} \right\}$$

Inverting each term in the matrix gives

$$e^{\mathbf{A}t} = \begin{bmatrix} e^{-2t} & 0 \\ 4(e^{-2t} - e^{-3t}) & e^{-3t} \end{bmatrix}$$

This matrix can be used in Eq. (29.8) to calculate $h_1(t)$ and $h_2(t)$. The result will be the same as obtained by inversion of Eq. (29.22).

SUMMARY

The matrix differential equation

$$\dot{\mathbf{x}} = \mathbf{Ax} + \mathbf{Bu}$$

used to describe a control system by the state-space method can be solved for the vector of state variables ($\mathbf{x}$) by use of the transfer function matrix. It consists of a matrix of transfer functions that relate the state variables to the inputs. The transfer function matrix serves the same purpose in a multiple-input multiple-output system as the transfer function does for a single-input single-output system. The transfer function matrix is obtained from the matrix differential equation by application of Laplace transforms.

PROBLEMS

29.1. Determine $\mathbf{x}(t)$ for the system

$$\dot{\mathbf{x}} = \mathbf{Ax} + \mathbf{Bu}$$

where $e^{\mathbf{A}t} = \begin{bmatrix} e^{-5t} & -e^{-2t} + e^{-5t} \\ 0 & e^{-2t} \end{bmatrix}$

$$\mathbf{x}(0) = \begin{bmatrix} 0 \\ 1 \end{bmatrix} \qquad \mathbf{u}(t) = \begin{bmatrix} 3 \\ 1 \end{bmatrix} \qquad \mathbf{B} = \begin{bmatrix} 1 & 2 \\ 0 & -4 \end{bmatrix}$$

CHAPTER
30

MULTIVARIABLE
CONTROL

Up to this point, the fundamentals of process dynamics and control have been illustrated by single-input single-output (SISO) systems. The processes encountered in the real world are usually multiple-input multiple-output systems (MIMO). To explore these concepts, consider the interacting, two-tank liquid-level system in Fig. 30.1 where there is one input, the flow to tank 1 (m_1) and one output, the level in tank 2 (h_2). In this figure, h_2 is related to m_1 by a second-order transfer function. From the point of view of a SISO system, the relation between h_2 and m_1 may be represented by the block diagram in Fig. 30.1b. One may place a feedback control system around the open-loop system of Fig. 30.1b to maintain control of H_2.

Now consider the same process of Fig. 30.1 in which there are two inputs (m_1 and m_2) and two outputs (h_1 and h_2). This system is shown in Fig. 30.2a. A change in m_1 alone will affect both outputs (h_1 and h_2). A change in m_2 alone will also change both outputs. (Remember that this is an interacting process for which the level in tank 1 is affected by the level in tank 2.) The interaction between inputs and outputs can be seen more clearly by the block diagram of Fig. 30.2b. In this diagram, the transfer functions show how the change in one of the inputs affects both of the outputs. For example, if a change occurs in only M_1, the responses of H_1 and H_2 are

$$H_1(s) = G_{11}(s)M_1(s)$$
$$H_2(s) = G_{21}(s)M_1(s)$$

The transfer functions in Fig. 30.2b will be worked out for a specific set of process parameters in Example 30.1. (If the tanks were noninteracting, $G_{12} = 0$, with

453

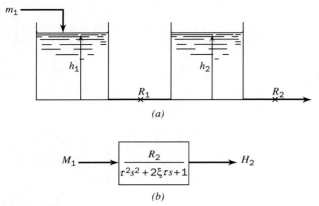

FIGURE 30-1
Single-input single-output system (SISO): (*a*) two-tank interacting level system, (*b*) block diagram for SISO system.

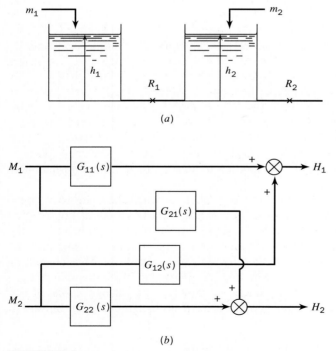

FIGURE 30-2
Multiple-input multiple-output system (MIMO): (*a*) level process, (*b*) block diagram.

the result that a change in flow to tank 2 would not affect H_1.) If both H_1 and H_2 are to be controlled, a single control loop will not be sufficient; in this case two control loops are needed. The addition of control loops to the interacting system will be considered in the next section.

CONTROL OF INTERACTING SYSTEMS

The problem of controlling the outputs of an MIMO system will be discussed by means of a 2×2 system shown in Fig. 30.3. The problem can be extended to the case of more than two pairs of inputs and outputs by the same procedure described here. The control objective is to control C_1 and C_2 independently, in spite of changes in M_1 and M_2 or other load variables not shown. Two control loops are added to the diagram of Fig. 30.3 as shown in Fig. 30.4. Each loop has a block for the controller, the valve, and the measuring element. In principle, the multiloop control system of Fig. 30.4 will maintain control of C_1 and C_2. However, because of the interaction present in the system, a change in R_1 will also cause C_2 to vary because a disturbance enters the lower loop through the transfer function G_{21}. Because of interaction, both outputs (C_1 and C_2) will change if a change is made in either input alone. If G_{21} and G_{12} provide weak interaction, the two-controller scheme of Fig. 30.4 will give satisfactory control. In the extreme, if $G_{12} = G_{21} = 0$, we have no interaction and the two control loops are isolated from each other.

 To completely eliminate the interaction between outputs and set points, two more controllers (cross-controllers) are added to the diagram of Fig. 30.4 to give the diagram shown in Fig. 30.5. In principle, these cross-controllers can be designed to eliminate interaction. The following analysis, which is expressed in matrix form, will lead to the method of design for cross-controllers that will eliminate interaction.

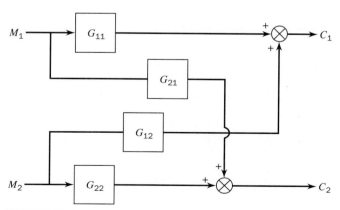

FIGURE 30-3
MIMO system for two pairs of inputs and outputs.

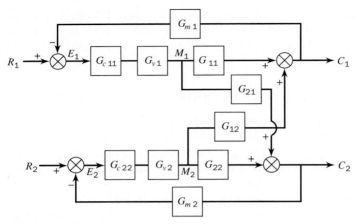

FIGURE 30-4
Multiloop control system with two controllers.

Response of Multiloop Control System

From Fig. 30.5, we may write by direct observation the following relationships in the form of the matrix expression

$$\mathbf{C} = \mathbf{G}_p\mathbf{M} \tag{30.1}$$

where $\mathbf{G}_p = \begin{bmatrix} G_{11} & G_{12} \\ G_{21} & G_{22} \end{bmatrix}$ $\quad \mathbf{C} = \begin{bmatrix} C_1 \\ C_2 \end{bmatrix}$ $\quad \mathbf{M} = \begin{bmatrix} M_1 \\ M_2 \end{bmatrix}$

We also may write from Fig. 30.5

$$M_1 = G_{v1}G_{c11}E_1 + G_{v1}G_{c12}E_2 \tag{30.2}$$
$$M_2 = G_{v2}G_{c21}E_1 + G_{v2}G_{c22}E_2 \tag{30.3}$$

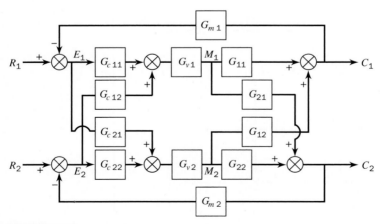

FIGURE 30-5
Multiloop control system with two primary controllers and two cross-controllers.

where G_{v1} and G_{v2} are the transfer functions for the valves. Equations (30.2) and (30.3) may be written in matrix form as

$$\mathbf{M} = \mathbf{G}_v \mathbf{G}_c \mathbf{E} \tag{30.4}$$

where $\quad \mathbf{G}_v = \begin{bmatrix} G_{v1} & 0 \\ 0 & G_{v2} \end{bmatrix} \qquad$ (valve matrix)

$$\mathbf{G}_c = \begin{bmatrix} G_{c11} & G_{c12} \\ G_{c21} & G_{c22} \end{bmatrix} \qquad \text{(controller matrix)}$$

$$\mathbf{E} = \begin{bmatrix} E_1 \\ E_2 \end{bmatrix}$$

From Fig. 30.5, we write directly

$$E_1 = R_1 - G_{m1} C_1 \tag{30.5}$$
$$E_2 = R_2 - G_{m2} C_2 \tag{30.6}$$

where E_1 and E_2 are the error signals from the comparators. Equations (30.5) and (30.6) can be written in the matrix form

$$\mathbf{E} = \mathbf{R} - \mathbf{G}_m \mathbf{C} \tag{30.7}$$

where $\quad \mathbf{G}_m = \begin{bmatrix} G_{m1} & 0 \\ 0 & G_{m2} \end{bmatrix} \qquad$ (measuring element matrix)

$$\mathbf{R} = \begin{bmatrix} R_1 \\ R_2 \end{bmatrix}$$

From Eqs. (30.1) and (30.4), we obtain

$$\mathbf{C} = \mathbf{G}_p \mathbf{G}_v \mathbf{G}_c \mathbf{E} \tag{30.8}$$

If we let $\mathbf{G}_o = \mathbf{G}_p \mathbf{G}_v \mathbf{G}_c$, Eq. (30.8) becomes

$$\mathbf{C} = \mathbf{G}_o \mathbf{E} \tag{30.9}$$

Combining Eqs. (30.7) and (30.9) gives

$$\mathbf{C} = \mathbf{G}_o \mathbf{R} - \mathbf{G}_o \mathbf{G}_m \mathbf{C} \tag{30.10}$$

We may now solve Eq. (30.10) for $\mathbf{C}$ to obtain

$$\mathbf{C} = [\mathbf{I} + \mathbf{G}_o \mathbf{G}_m]^{-1} \mathbf{G}_o \mathbf{R} \tag{30.11}$$

Notice that the closed-loop behavior expressed by this matrix equation is analogous to the closed-loop response of a SISO system, which may be written

$$C(s) = \frac{G_o(s)}{1 + G_o(s) G_m(s)} R(s) \tag{30.12}$$

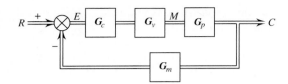

FIGURE 30-6
Block diagram for MIMO control system in terms of matrix blocks.

The matrix term $[\mathbf{I}+\mathbf{G}_o\mathbf{G}_m]^{-1}$ is equivalent to the scalar term $1/[1+G_o(s)G_m(s)]$.

A block diagram equivalent to the diagram for the MIMO control system in Fig. 30.5 is shown in Fig. 30.6. In this diagram, the blocks are filled with the matrices in Eqs. (30.1), (30.4), and (30.7). The double line indicates that more than one variable is being transmitted. Each block contains a matrix of transfer functions that relates an output vector to an input vector. The diagram can be simplified by multiplying the three matrices in the forward loop together and calling the result $\mathbf{G}_o$, as was done to obtain Eq. (30.9). The simplified diagram is shown in Fig. 30.7.

Noninteracting Control

In order for no interaction to occur between $\mathbf{C}$ and $\mathbf{R}$ in Fig. 30.5 (i.e., R_1 affects only C_1 and R_2 affects only C_2), the off-diagonal elements of $[\mathbf{I} + \mathbf{G}_o\mathbf{G}_m]^{-1}\mathbf{G}_o$ in Eq. (30.11) must be zero. Since $\mathbf{I}$ and $\mathbf{G}_m$ are diagonal, $[\mathbf{I} + \mathbf{G}_o\mathbf{G}_m]^{-1}\mathbf{G}_o$ will be diagonal if $\mathbf{G}_o$ is diagonal. Multiplication of the matrices in the expression for $\mathbf{G}_o$ is now shown:

$$\mathbf{G}_o = \mathbf{G}_p\mathbf{G}_v\mathbf{G}_c$$

$$\mathbf{G}_o = \begin{bmatrix} G_{11} & G_{12} \\ G_{21} & G_{22} \end{bmatrix}\begin{bmatrix} G_{v1} & 0 \\ 0 & G_{v2} \end{bmatrix}\begin{bmatrix} G_{c11} & G_{c12} \\ G_{c21} & G_{c22} \end{bmatrix}$$

The result of multiplying these matrices gives

$$\mathbf{G}_o = \begin{bmatrix} G_{11}G_{v1}G_{c11} + G_{12}G_{v2}G_{c21} & G_{11}G_{v1}G_{c12} + G_{12}G_{v2}G_{c22} \\ G_{21}G_{v1}G_{c11} + G_{22}G_{v2}G_{c21} & G_{21}G_{v1}G_{c12} + G_{22}G_{v2}G_{c22} \end{bmatrix} \quad (30.13)$$

Setting the off-diagonal elements to zero and solving for G_{c12} and G_{c21} give

$$G_{c12} = -\frac{G_{12}G_{v2}G_{c22}}{G_{11}G_{v1}} \quad (30.14)$$

$$G_{c21} = -\frac{G_{21}G_{v1}G_{c11}}{G_{22}G_{v2}} \quad (30.15)$$

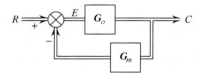

FIGURE 30-7
Reduced block diagram for MIMO control system where $\mathbf{G}_o = \mathbf{G}_p\mathbf{G}_v\mathbf{G}_c$.

The following example will give some experience with the computations involved in applying the theory developed so far in this chapter.

Example 30.1. For the two-tank, interacting liquid-level system shown in Fig. 30.8, develop the block diagram for an MIMO system corresponding to Fig. 30.3.
Material balances around tank 1 and tank 2 give the following differential equations:

$$A_1\dot{c}_1 = m_1 - \frac{c_1 - c_2}{R_1} - \frac{c_1}{R_3} \tag{30.16}$$

$$A_2\dot{c}_2 = m_2 + \frac{c_1 - c_2}{R_1} - \frac{c_2}{R_2} \tag{30.17}$$

Introducing the parameters given in Fig. 30.8 into Eqs. (30.16) and (30.17) gives

$$\dot{c}_1 = m_1 - 3c_1 + 2c_2 \tag{30.18}$$

$$\dot{c}_2 = 2m_2 + 4c_1 - 5c_2 \tag{30.19}$$

These equations may be written in matrix form as

$$\dot{\mathbf{c}} = \mathbf{Ac} + \mathbf{Bm}$$

where

$$\mathbf{A} = \begin{bmatrix} -3 & 2 \\ 4 & -5 \end{bmatrix} \qquad \mathbf{B} = \begin{bmatrix} 1 & 0 \\ 0 & 2 \end{bmatrix}$$

We use Eq. (29.15) to obtain

$$\mathbf{C}(s) = (s\mathbf{I} - \mathbf{A})^{-1}\mathbf{BM}(s) \tag{30.20}$$

Writing Eq. (30.20) in the form of Eq. (30.1) gives

$$\mathbf{C} = \mathbf{G}_p\mathbf{M}$$

where $\mathbf{G}_p = (s\mathbf{I} - \mathbf{A})^{-1}\mathbf{B}$

After several steps involving the inversion of $(s\mathbf{I} - \mathbf{A})$ and multiplying the result of inversion by $\mathbf{B}$, one gets

$$G_p = \frac{\begin{bmatrix} s + 5 & 4 \\ 4 & 2(s + 3) \end{bmatrix}}{(s + 1)(s + 7)} \tag{30.21}$$

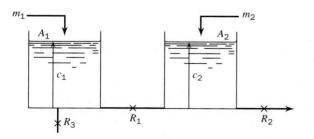

FIGURE 30-8
Process for Example 30.1: $A_1 = 1$, $A_2 = 1/2$, $R_1 = 1/2$, $R_2 = 2$, $R_3 = 1$.

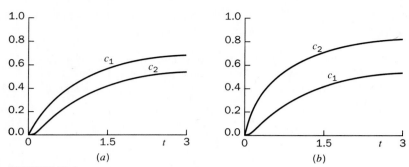

FIGURE 30-9
Open-loop response for Example 30.1. (a) $M_1 = 1/s$, $M_2 = 0$, (b) $M_2 = 1/s$, $M_1 = 0$.

The block diagram can now be drawn as shown in Fig. 30.3 with

$$G_{11} = \frac{s+5}{(s+1)(s+7)} \qquad G_{12} = \frac{4}{(s+1)(s+7)}$$

$$G_{21} = \frac{4}{(s+1)(s+7)} \qquad G_{22} = \frac{2(s+3)}{(s+1)(s+7)}$$

Notice that the diagonal elements of $\mathbf{G}_p(s)$ are of the form

$$\frac{\alpha(s+\beta)}{(s+1)(s+7)}$$

These elements, which relate c_1 to m_1 and c_2 to m_2, will produce a second-order response to a step change in input that has a finite slope at the origin because of the numerator term $s + \beta$. In contrast, the off-diagonal elements have second-order transfer functions without numerator dynamics, for which case the step response will be second-order with zero slope at the origin. The responses of c_1 and c_2 for unit-step changes in m_1 and m_2 taken separately are shown in Fig. 30.9.

Example 30.2. For the two-tank liquid-level system of Example 30.1, determine the controller transfer function matrix $\mathbf{G}_c$ needed to eliminate interaction. The primary controllers are to be proportional, i.e., $G_{c11} = K_1$, $G_{c22} = K_2$. The diagram of the control system is shown in Fig. 30.10. The block labeled controller contains the

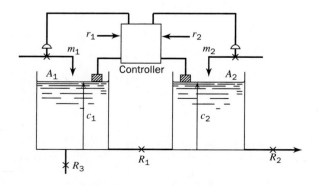

FIGURE 30-10
Process for Example 30.2. $A_1 = 1$, $A_2 = 1/2$, $R_1 = 1/2$, $R_2 = 2$, $R_3 = 1$, $G_{c11} = K_1$, $G_{c22} = K_2$.

four transfer functions that are the elements of $\mathbf{G}_c$. In this problem, $\mathbf{G}_v$ is a unit diagonal matrix i.e., $G_{v1} = G_{v2} = 1$.

From Eqs. (30.14) and (30.15) we obtain

$$G_{c12} = -\frac{G_{12}G_{c22}}{G_{11}} = -\frac{4}{(s+1)(s+7)}K_2\frac{(s+1)(s+7)}{s+5}$$

or

$$G_{c12} = \frac{-4K_2}{s+5} \tag{30.22}$$

$$G_{c21} = -\frac{G_{21}G_{c11}}{G_{22}} = -\frac{4}{(s+1)(s+7)}K_1\frac{(s+1)(s+7)}{2(s+3)}$$

or

$$G_{c21} = \frac{-2K_1}{s+3} \tag{30.23}$$

Having found the transfer functions for the cross-controllers, we can now determine the nature of the uncoupled response of c_1 to a change in r_1 and of c_2 to a change in r_2.

Inserting $G_{v1} = G_{v2} = 1$ and the expressions for G_{c12} and G_{c21} from Eqs. (30.14) and (30.15) into Eq. (30.13) gives for $\mathbf{G}_o$

$$\mathbf{G}_o = \begin{bmatrix} G_{11}G_{c11} + G_{12}G_{c21} & 0 \\ 0 & G_{21}G_{c12} + G_{22}G_{c22} \end{bmatrix} \tag{30.24}$$

Inserting the appropriate elements of the $\mathbf{G}_p$ matrix [Eq. (30.21)] and the $\mathbf{G}_c$ matrix in Eq. (30.24) gives after considerable simplification

$$\mathbf{G}_o = \begin{bmatrix} \dfrac{K_1}{s+3} & 0 \\ 0 & \dfrac{2K_2}{s+5} \end{bmatrix} \qquad \text{(decoupled system)} \tag{30.25}$$

The block diagram for this decoupled MIMO system is shown in Fig. 30.11. Assuming that the measurement matrix $\mathbf{G}_m$ is a unit diagonal matrix, the diagram

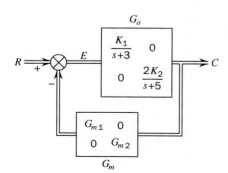

FIGURE 30-11
Block diagram for decoupled system in Example 30.2.

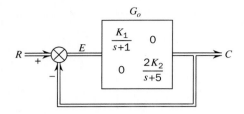

FIGURE 30-12
Simplified block diagram for Example 30.2.

in Fig. 30.11 can be simplified to the unity feedback diagram of Fig. 30.12. From Fig. 30.12, we may write directly

$$\mathbf{C} = \mathbf{G}_o\mathbf{E}$$

$$\mathbf{E} = \mathbf{R} - \mathbf{C}$$

therefore $\mathbf{C} = \mathbf{G}_o\mathbf{R} - \mathbf{G}_o\mathbf{C}$

or

$$\begin{bmatrix} C_1 \\ C_2 \end{bmatrix} = \begin{bmatrix} G_{o11} & 0 \\ 0 & G_{o22} \end{bmatrix}\begin{bmatrix} R_1 \\ R_2 \end{bmatrix} - \begin{bmatrix} G_{o11} & 0 \\ 0 & G_{o22} \end{bmatrix}\begin{bmatrix} C_1 \\ C_2 \end{bmatrix}$$

From this expression, we may write

$$C_1 = G_{o11}R_1 - G_{o11}C_1$$

$$C_2 = G_{o22}R_2 - G_{o22}C_2$$

Solving for $C_1(s)$ gives

$$C_1(s) = \frac{G_{o11}}{1 + G_{o11}}R_1(s)$$

Inserting G_{o11} from Eq. (30.25) gives

$$C_1(s) = \frac{\dfrac{K_1}{s+3}}{1 + \dfrac{K_1}{s+3}}R_1(s) \qquad (30.26)$$

In a similar way, one can show that

$$C_2(s) = \frac{\dfrac{2K_2}{s+5}}{1 + \dfrac{2K_2}{s+5}}R_2(s) \qquad (30.27)$$

The result shows that the cross-controllers of Eqs. (30.22) and (30.23) give two separate noninteracting control loops as shown in Fig. 30.13.

The response of the control system of Fig. 30.10 is shown in Fig. 30.14 for a unit-step change in R_1. In Fig. 30.14a, no cross-controllers are present in the matrix $\mathbf{G}_c$. In Fig. 30.14b, cross-controllers having the transfer functions given by Eq. (30.22) and (30.23) are present. As expected, for the case of no cross-controllers, one sees from Fig. 30.14a that a request for a unit-step change in r_1

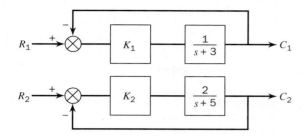

FIGURE 30-13
Decoupled control system for Example 30.2 where primary controllers are proportional.

causes both c_1 and c_2 to change. For the case where cross-controllers are present, one sees from Fig. 30.14*b* that a change in r_1 does not affect c_2 as demanded by a decoupled system.

To avoid the offset associated with proportional control, we can use PI controllers for the primary controllers for the decoupled system. To study the effect of PI controllers for the decoupled system, let

$$G_{c11} = K_1\left(1 + \frac{1}{s}\right) \qquad \text{and} \qquad G_{c22} = K_2\left(1 + \frac{1}{s}\right)$$

For this case, the cross-controller transfer functions may be obtained from, Eqs. (30.14) and (30.15); the results are

$$G_{c12} = \frac{-4K_2(s + 1)}{s(s + 5)} \qquad \text{and} \qquad G_{c21} = \frac{-2K_1(s + 1)}{s(s + 3)}$$

A simulation using these four controller transfer functions with $K_1 = K_2 = 4$ is shown in Fig. 30.15. From the transient response, we see that c_1 moves toward the set point of 1.0 and that c_2 does not change, as is expected for a decoupled system.

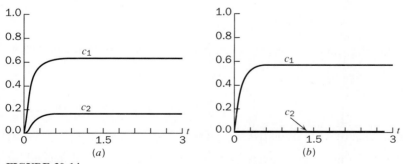

FIGURE 30-14
Response for control system in Example 30.2 for $R_1 = 1/s$, $R_2 = 0$, $G_{c11} = K_1 = 4$, $G_{c22} = K_2 = 4$. (*a*) no cross-controllers, (*b*) cross-controllers present.

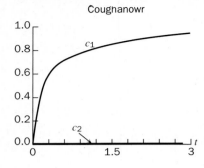

FIGURE 30-15
Response of decoupled control system in Example 30.2 for PI primary controllers: $G_{c11} = G_{c22} = 4(1 + 1/s)$, $R_1 = 1/s$, $R_2 = 0$.

STABILITY OF MULTIVARIABLE SYSTEMS

Determining the stability for a multivariable control system, such as the one in Fig. 30.4 or Fig. 30.5, can be much more complicated than for an SISO system. The transfer function for the closed-loop response of an MIMO system is given by Eq. (30.11):

$$\mathbf{C} = [\mathbf{I} + \mathbf{G}_o\mathbf{G}_m]^{-1}\mathbf{G}_o\mathbf{R}$$

To invert this expression, we write

$$\mathbf{C} = \frac{\text{adj } [\mathbf{I} + \mathbf{G}_o\mathbf{G}_m]\mathbf{G}_o\mathbf{R}}{|\mathbf{I} + \mathbf{G}_o\mathbf{G}_m|} \tag{30.28}$$

The numerator of this expression is an $n \times n$ matrix; the denominator is a nth order polynomial. To simplify the following argument, let the matrix in Eq. (30.28) be 2×2. Let the elements of the numerator, after expansion, be written as follows:

$$\text{adj}[\mathbf{I} + \mathbf{G}_o\mathbf{G}_m]\mathbf{G}_o\mathbf{R} = \begin{bmatrix} \beta_{11}(s) & \beta_{12}(s) \\ \beta_{21}(s) & \beta_{22}(s) \end{bmatrix} \tag{30.29}$$

Let the elements of $\mathbf{G}_o\mathbf{G}_m$ be written as follows:

$$\mathbf{G}_o\mathbf{G}_m = \begin{bmatrix} \alpha_{11}(s) & \alpha_{12}(s) \\ \alpha_{21}(s) & \alpha_{22}(s) \end{bmatrix} \tag{30.30}$$

Expansion of the determinant in Eq. (30.28), using Eq. (30.30), is shown below

$$|\mathbf{I} + \mathbf{G}_o\mathbf{G}_m| = \begin{vmatrix} 1 + \alpha_{11}(s) & \alpha_{12}(s) \\ \alpha_{21}(s) & 1 + \alpha_{22}(s) \end{vmatrix}$$

or

$$|\mathbf{I} + \mathbf{G}_o\mathbf{G}_m| = [1 + \alpha_{11}(s)][1 + \alpha_{22}(s)] - \alpha_{12}(s)\alpha_{21}(s) \tag{30.31}$$

Equation (30.31) is a polynomial expression, for which the order will depend on the order of the transfer functions in $\mathbf{G}_o$ and $\mathbf{G}_m$. Equation (30.28) can now be written in terms of the expansions shown in Eqs. (30.29) and (30.31) as follows:

$$
\mathbf{C} = \begin{bmatrix} \dfrac{\beta_{11}(s)}{|\mathbf{I} + \mathbf{G}_o\mathbf{G}_m|} & \dfrac{\beta_{12}(s)}{|\mathbf{I} + \mathbf{G}_o\mathbf{G}_m|} \\[4mm] \dfrac{\beta_{21}(s)}{|\mathbf{I} + \mathbf{G}_o\mathbf{G}_m|} & \dfrac{\beta_{22}(s)}{|\mathbf{I} + \mathbf{G}_o\mathbf{G}_m|} \end{bmatrix}
$$

Since each term contains the polynomial $|\mathbf{I} + \mathbf{G}_o\mathbf{G}_m|$ in the denominator, the stability of the multivariable system will depend on the roots of the polynomial equation

$$
|\mathbf{I} + \mathbf{G}_o\mathbf{G}_m| = 0 \qquad \text{(characteristic equation)} \qquad (30.32)
$$

Equation (30.32) is the characteristic equation of the multivariable system. Although Eq. (30.32) has been derived here for the case where $\mathbf{G}_o\mathbf{G}_m$ is a 2 × 2 matrix, one can show that Eq. (30.32) applies to the general MIMO system of Fig. 30.7 in which $\mathbf{G}_o\mathbf{G}_m$ is a matrix of any size ($n \times n$). If the roots of the characteristic equation are in the left half of the complex plane, we know that the system is stable. One method to be used for examining the stability of a multivariable system is to apply the Routh test to the characteristic equation of Eq. (30.32). In practice, the characteristic equation can be of high order for a simple 2 × 2 multivariable control system. Example 30.3 illustrates the determination of stability for a multivariable control system.

Example 30.3. For the control system of Example 30.2, which is shown in Fig. 30.10, determine stability for the case where $G_{c11} = K_1$, $G_{c22} = K_2$, and there are no cross-controllers present (i.e., $G_{c12} = G_{c21} = 0$) also let $\mathbf{G}_m$ and $\mathbf{G}_v$ be unit matrices. From Example 30.1, we have for the elements of $\mathbf{G}_p$

$$
G_{11} = \frac{s + 5}{(s + 1)(s + 7)} \qquad G_{12} = \frac{4}{(s + 1)(s + 7)}
$$

$$
G_{21} = \frac{4}{(s + 1)(s + 7)} \qquad G_{22} = \frac{2(s + 3)}{(s + 1)(s + 7)}
$$

Since $\mathbf{G}_v = \mathbf{I}$, $\mathbf{G}_o = \mathbf{G}_p\mathbf{G}_c$. Since $\mathbf{G}_m = \mathbf{I}$, the characteristic equation of Eq. (30.32) can now be written as

$$
|\mathbf{I} + \mathbf{G}_p\mathbf{G}_c| = 0 \qquad (30.33)
$$

Introducing the elements of the matrices $\mathbf{G}_p$ and $\mathbf{G}_c$ into Eq. (30.33) gives, after expansion of the determinant

$$
[(s + 1)(s + 7) + K_1(s + 5)][(s + 1)(s + 7) + 2K_2(s + 3)] - 16K_1K_2 = 0
$$

For given values of K_1 and K_2, this expression can be expanded into a fourth order polynomial equation of the form

$$
s^4 + \alpha s^3 + \beta s^2 + \gamma s + \Delta = 0 \qquad (30.34)
$$

where α, β, γ, and Δ will include the gains K_1 and K_2.

The Routh test can be applied to Eq. (30.34) to determine whether or not the system is stable. From this simple example, the reader can appreciate the algebraic tedium that may be needed to determine the stability of a multivariable system.

One way to express the stability of this system is to plot the stability boundaries on a graph of K_1 versus K_2. The region within the boundaries gives the combinations of values of K_1 and K_2 for which the system is stable. Since the details of stability boundaries is beyond the scope of this chapter, the reader may consult Seborg, Edgar, and Mellichamp (1989) for examples of stability boundaries for multivariable systems.

SUMMARY

Most of the systems encountered are multiple-input multiple-output (MIMO) systems. Such systems have several inputs and several outputs that are often interacting, meaning that a disturbance at any input causes a response in some or all of the outputs. This interaction in an MIMO system makes control and stability analysis of the system very complicated compared to a single-input single-output (SISO) system. A convenient way to describe an MIMO system is by means of a block diagram in which each block contains a matrix of transfer functions that relates an input vector to an output vector.

It is often desirable to have a control system decoupled so that certain outputs can be controlled independently of other outputs. A systematic procedure was described for decoupling a control system by including cross-controllers along with the principal controllers. This approach to decoupling requires an accurate model of the system; the number of controllers (principal controllers and cross-controllers) increases rapidly with the number of inputs and outputs. A system represented by two inputs and two outputs requires as many as four controllers; a system of three inputs and three outputs requires as many as nine controllers, and so on.

The characteristic equation for a multivariable control system, from which one can determine stability by examining its roots, can be of high order for a relatively simple system. Expressing stability boundaries in terms of controller parameters becomes complex because of the large number of controller parameters that can be adjusted.

PROBLEMS

30.1. For the liquid-level system shown in Fig. P30.1 determine the cross-controller transfer functions that will decouple the system. Fill in each block of the diagram shown in Fig. 30.5 with a transfer function obtained from an analysis of the control system. The transfer function for each feedback measuring element is unity. The following data apply:

$$A_1 = 1, \ A_2 = 0.5, \ \text{Res}_1 = 0.5, \ \text{Res}_2 = 2/3, \ G_{c11} = K_1, \ G_{c22} = K_2$$

The resistance on the outlet of a tank has been denoted by Res to avoid confusion with the symbol for set point (R).

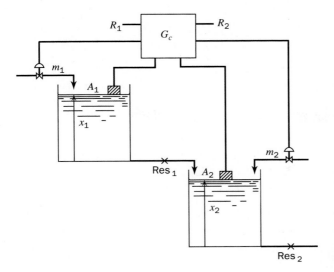

FIGURE P30-1

30.2. (*a*) For the interacting liquid-level system shown in Fig. P30.2, draw *very neatly* a block diagram that corresponds to Fig. 30.4. Each block should contain a transfer function obtained from an analysis of the liquid-level system. There are no cross-controllers in this system. The transfer function for each feedback element is unity. The following data apply:

$$A_1 = 1, A_2 = 1/2, \text{Res}_1 = 1/2, \text{Res}_2 = 2, \text{Res}_3 = 1$$

(*b*) Obtain the characteristic equation of this system in the form

$$s^n + \alpha s^{n-1} + \beta s^{n-2} + \cdots = 0$$

Obtain expressions for α, β, etc. in terms of K_2, $(K_1 = 1)$

(*c*) How would you determine stability limits for this interacting control system?

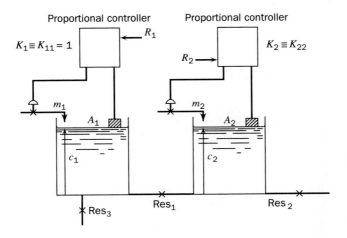

FIGURE P30-2

PART
VIII

NONLINEAR CONTROL

CHAPTER
31

EXAMPLES
OF NONLINEAR
SYSTEMS

In the previous chapters, we have confined our attention to the behavior of linear systems or to the analysis of linearized equations representative of nonlinear systems in the vicinity of the steady-state condition. While much useful information can be obtained from such analysis, it frequently is desirable or necessary to consider nonlinearities in control system design.

No real physical system is truly linear, particularly over a wide range of operating variables. Hence, to be complete, a control system design should allow for the possibility of a large deviation from steady-state behavior and resulting nonlinear behavior. The purpose of the next three chapters is to introduce some of the tools that can be used for this purpose and to indicate some of the complications that arise when nonlinear systems are considered.

DEFINITION OF A NONLINEAR SYSTEM

A nonlinear system is one for which the principle of superposition does not apply. Thus, by superposition, the response of a linear system to the sum of two inputs is the same as the sum of the responses to the individual inputs. This behavior, which allows us to characterize completely a linear system by a transfer function, is not true of nonlinear systems.

As an example, consider a liquid-level system. If the outflow is proportional to the square root of the tank level, superposition does not hold and the system is nonlinear. If the tank will always operate near the steady-state condition, the square-root behavior may be adequately represented by a straight line and super-

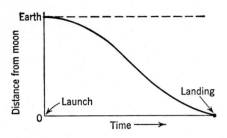

FIGURE 31-1
Distance-time plot for moon rocket.

position applied, as we have done before. On the other hand, if the tank level should fall to half the steady-state value, we would no longer expect the transfer function derived on the linearized basis to apply. The analysis becomes more complicated, as we shall see in our introduction to the study of nonlinear systems.

THE PHASE PLANE

The analysis of nonlinear dynamic systems may often be conceptually simplified by changing to a coordinate system known as *phase space*. In this coordinate system, time no longer appears explicitly, it being replaced by some other property of the system. For example, consider the flight of a rocket to the moon. In a grossly oversimplified manner, we may describe this motion by a plot of the distance of the rocket from the moon versus time. If all goes well, we would like such a plot to resemble Fig. 31.1. Note the initial acceleration during launch and the final deceleration at landing. We may, however, also represent this motion by a plot of rocket velocity versus distance from moon. This plot is shown in Fig. 31.2, where velocity is defined as d(distance from moon)$/dt$. Figure 31.2 is called a phase diagram of the rocket motion. Time now appears merely as a parameter along the curve of the rocket motion. It has been replaced as a coordinate by the rocket velocity. Although in the present example Fig. 31.2 may not be of significant advantage over Fig. 31.1, we shall find phase diagrams very helpful in the analysis of certain nonlinear control systems.

To begin our study of phase diagrams, we convert a linear motion studied previously in Chap. 8 to the phase plane. The linear motion will be that of the spring-mass-damper system.

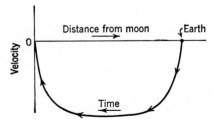

FIGURE 31-2
Velocity-distance plot for moon rocket.

PHASE-PLANE ANALYSIS
OF DAMPED OSCILLATOR

The differential equation describing the motion of the system of Fig. 8.1 in response to a unit-step function is

$$\tau^2 \frac{d^2Y}{dt^2} + 2\zeta\tau\frac{dY}{dt} + Y = 1 \tag{31.1}$$

Equation (31.1) has previously been solved to yield the motion in the form of $Y(t)$ versus t as shown in Fig. 8.2. For phase analysis, however, we want the motion in terms of velocity versus position, $\dot{Y}$ versus Y, where the dot notation is used to indicate differentiation with respect to t. Hence, we rewrite Eq. (31.1) as

$$\frac{dY}{dt} = \dot{Y}$$

$$\frac{d\dot{Y}}{dt} = \frac{-Y - 2\zeta\tau\dot{Y} + 1}{\tau^2} \tag{31.2}$$

It is usually convenient in phase-plane analysis to write the variables in terms of deviation about the *final condition*. In this case, the spring will ultimately come to rest at $Y = 1$. Hence we define

$$X = Y - 1$$

$$\dot{X} = \dot{Y}$$

Then, Eq. (31.2) becomes

$$\frac{dX}{dt} = \dot{X}$$

$$\frac{d\dot{X}}{dt} = \frac{-X - 2\zeta\tau\dot{X}}{\tau^2} \tag{31.3}$$

These are now viewed as two simultaneous, first-order differential equations in the variables X and $\dot{X}$.

To solve Eqs. (31.3), we may use the methods presented in Chaps. 28 and 29. For this purpose, let $X_1 = X$ and $X_2 = \dot{X}$. Eqs. (31.3) may be written in the form

$$\dot{\mathbf{X}} = \mathbf{AX} \tag{31.4}$$

where $\mathbf{X} = \begin{bmatrix} X_1 \\ X_2 \end{bmatrix}$ $\mathbf{A} = \begin{bmatrix} 0 & 1 \\ \dfrac{-1}{\tau^2} & \dfrac{-2\zeta}{\tau} \end{bmatrix}$

Equation (31.4) is in the standard form of a matrix differential equation [Eq. (28.7)]. Notice that the term $\mathbf{b}u$ of Eq. (28.7) is not present because no forcing term is present in Eqs. (31.3). Equation (31.4) may be solved by use of Eq. (29.7):

$$\mathbf{X}(t) = e^{\mathbf{A}t}\mathbf{X}(0) \qquad (29.7)$$

where
$$e^{\mathbf{A}t} = L^{-1}\{(s\mathbf{I} - \mathbf{A})^{-1}\} \qquad (29.13)$$

Following the usual steps required to solve these equations gives the result

$$\begin{aligned} X_1 &= X = C_1 e^{s_1 t} + C_2 e^{s_2 t} \\ X_2 &= \dot{X} = s_1 C_1 e^{s_1 t} + s_2 C_2 e^{s_2 t} \end{aligned} \qquad (31.5)$$

where
$$C_1 = \frac{s_2 X_0 - \dot{X}_0}{s_2 - s_1}$$

$$C_2 = \frac{\dot{X}_0 - s_1 X_0}{s_2 - s_1}$$

and X_0 and $\dot{X}_0$ are the initial conditions; thus $X_0 = X(0)$ and $\dot{X}_0 = \dot{X}(0)$. The terms s_1 and s_2 are the roots of the characteristic equation

$$|s\mathbf{I} - \mathbf{A}| = 0 \qquad (31.6)$$

Expanding this equation gives

$$\tau^2 s^2 + 2\zeta\tau s + 1 = 0$$

This quadratic equation has two roots:

$$s_{1,2} = \frac{-\zeta \pm \sqrt{\zeta^2 - 1}}{\tau}$$

If we take s_2 as the root with the positive sign

$$s_2 = \frac{-\zeta + \sqrt{\zeta^2 - 1}}{\tau}$$

the constants take the form

$$\begin{aligned} C_1 &= \frac{\tau}{2\sqrt{\zeta^2 - 1}}(s_2 X_0 - \dot{X}_0) \\ C_2 &= \frac{\tau}{2\sqrt{\zeta^2 - 1}}(\dot{X}_0 - s_1 X_0) \end{aligned} \qquad (31.7)$$

Equations (31.5) and (31.7) together give $X(t)$ and $\dot{X}(t)$ for all possible initial conditions X_0 and $\dot{X}_0$. For a given set of initial conditions, we compute C_1 and C_2 from (31.7), and then each value of t in Eq. (31.5) yields a pair of values for X and $\dot{X}$. These may be plotted as a point on an $\dot{X}X$ diagram (i.e., a phase plane). the locus of these points as t varies from zero to infinity will be a curve in the $\dot{X}X$ plane. As an example, consider the case $X_0 = -1$, $\dot{X}_0 = 0$, $\zeta < 1$. The solution is already known to us in the form of X versus t (Chap. 8) and is replotted in Fig. 31.3 for convenience, together with a plot of $\dot{X}$

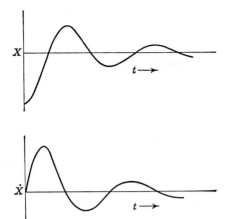

FIGURE 31-3
Typical motion of second-order system.

versus t. If these curves are replotted as $\dot{X}$ versus X, with t as a parameter, the result is as shown in Fig. 31.4. The reader should carefully compare Figs. 31.3 and 31.4 to be satisfied that they are indeed equivalent. The relationship between the two may be expressed by the statement that Fig. 31.3 is a parametric representation of Fig. 31.4. Having only the curve X versus t of Fig. 31.3, one can construct Fig. 31.4.

To explore the phase-diagram concept further, note that division of the second of Eqs. (31.3) by the first yields

$$\frac{d\dot{X}}{dX} = \frac{-X - 2\zeta\tau\dot{X}}{\tau^2\dot{X}} \tag{31.8}$$

in which the variable t has been eliminated. Equation (31.8) may be recognized as a homogeneous first-order differential equation. Hence, the substitution $\dot{X} = VX$ yields

$$\frac{X\,dV}{dX} = \frac{-1 - 2\zeta\tau V}{\tau^2 V} - V = \frac{-(1 + 2\zeta\tau V + \tau^2 V^2)}{\tau^2 V}$$

an equation which is separable in X and V. This can then be easily solved for V in terms of X. Finally, replacing $V = \dot{X}/X$ gives the solution for $\dot{X}$ versus X, or

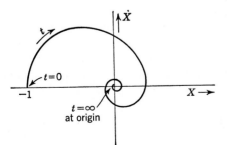

FIGURE 31-4
Phase plane corresponding to motion of Fig. 31.3.

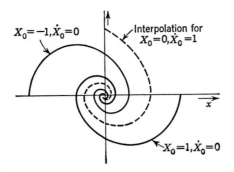

FIGURE 31-5
Interpolation on the phase plane.

the equation for the curve of Fig. 31.4. The algebraic details of this rather tedious process are omitted. (See Graham and McRuer, 1961, pp. 287–289.) The point of the discussion is to emphasize further the equivalence between the description of the motion as X versus t or $\dot{X}$ versus X.

A convenient feature of the phase diagram is that several motions, corresponding to different initial conditions, can be readily plotted on the same diagram. Thus, if we add to Fig. 31.4 a curve for the motion under the initial condition $X_0 = 1$, $\dot{X}_0 = 0$, we obtain Fig. 31.5. This new trajectory represents the motion of the system after it is stretched 2 units and released from rest. (This follows from the definition $X = Y - 1$.) Furthermore, we have also interpolated in Fig. 31.5 to obtain the motion corresponding to $X_0 = 0$, $\dot{X}_0 = 1$. As we shall see later, this interpolation is justified. Hence, it is evident that the phase diagram gives us the "big picture" of the motion of the underdamped spring-mass-damper system. No matter where the system starts, it spirals to the condition $X_0 = \dot{X}_0 = 0$, the steady-state position. This spiral motion in the phase plane corresponds to the oscillatory nature of the X versus t curve of Fig. 31.3.

Before beginning a more detailed study of the mechanics of phase analysis, it may be worthwhile to see how situations amenable to such analysis arise naturally in the physical world.

MOTION OF A PENDULUM

Consider the pendulum of Fig. 31.6. As the pendulum is moving in the direction shown, there are two forces acting to oppose its motion. These forces, which act tangentially to the circle of motion, are (1) the gravitational force $mg \sin \theta$ and (2) the friction in the pivot, which we suppose to be proportional to the tangential velocity of the mass, $BR(d\theta/dt)$. We shall assume the air resistance to be negligible and the rod to be of negligible mass. Application of Newton's second law gives

$$-mR\frac{d^2\theta}{dt^2} = mg \sin \theta + BR\frac{d\theta}{dt}$$

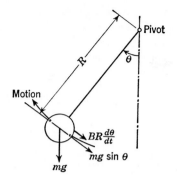

FIGURE 31-6
Forces acting on pendulum.

Rearrangement leads to

$$\frac{d^2\theta}{dt^2} + D\frac{d\theta}{dt} + \omega_n^2 \sin\theta = 0 \qquad (31.9)$$

where $D = \dfrac{B}{m}$

$$\omega_n^2 = \frac{g}{R}$$

This equation resembles the equation for the motion of the spring-mass-damper system. However, the presence of the term involving $\sin\theta$ makes the equation nonlinear.

Equation (31.9) has the following form in phase coordinates:

$$\frac{d\theta}{dt} = \dot{\theta}$$

$$\frac{d\dot{\theta}}{dt} = -\omega_n^2 \sin\theta - D\dot{\theta} \qquad (31.10)$$

and a phase diagram would be a plot of angular velocity $\dot{\theta}$ versus position θ. At this point, we can gain some insight by simple analysis of Eq. (31.10) without actually obtaining a solution.

Referring for the moment back to the spring-mass-damper system, we saw that the system ceased to oscillate when the point $X = \dot{X} = 0$ was reached. That is, all curves stopped at the origin of Fig. 31.5. The reason for this is quite clear; when $X = \dot{X} = 0$ is substituted into Eqs. (31.3), there is obtained

$$\frac{dX}{dt} = \frac{d\dot{X}}{dt} = 0$$

Since neither X nor $\dot{X}$ is changing with time, the motion ceases. Further examination of Eqs. (31.3) shows that $X = \dot{X} = 0$ is the *only* point at which both dX/dt and $d\dot{X}/dt$ are zero. Thus, we see that the mass will come to rest *only* when the situation of zero displacement and zero velocity is reached.

Now we perform a similar analysis on Eqs. (31.10). We are asking the following question: At what point or points in the phase plane ($\dot\theta$ versus θ diagram) do both $d\theta/dt$ and $d\dot\theta/dt$ become zero? From the first of these equations, we see that this can happen* only when $\dot\theta = 0$. Using this result in the second equation, it can be seen that it is also necessary that

$$\sin\theta = 0 \qquad\qquad (31.11)$$

Equation (31.11) is satisfied at any of the points

$$\theta = n\pi$$

where n is a positive or negative integer or zero. However, from a physical standpoint, we can really distinguish between only two of these points, which we take as $\theta = 0$ and $\theta = \pi$. Thus, the positions $\theta = 0, 2\pi, 4\pi, -2\pi$, etc., all look the same to us; i.e., the pendulum is hanging straight down. Similarly, the points $\theta = \pi, 3\pi$, etc., all correspond to the pendulum standing straight up.

Thus, the analysis leads to the conclusion that the motion will cease when the pendulum comes to rest in either of the positions shown in Fig. 31.7. In addition, it is clear from Eqs. (31.10) that, if the pendulum stops at any other point, the motion continues. Of course, this analysis agrees with our physical intuition. However, we expect to find a distinction between the stability characterics of the two equilibrium points, since the position at π is likely to be hard to attain and maintain. This distinction will be explored in more detail in Chap. 32.

---

*The reader should not be lulled into a false sense of security at this point. It would be wise to disregard the fact that $d\theta/dt$ and $\dot\theta$ are, in fact, the same quantity; $\dot\theta$ should be thought of as a coordinate in the phase plane, and $d\theta/dt$ as the rate of change with time of the other coordinate. The virtue of making this distinction will become clear in the next example, a chemical reactor.

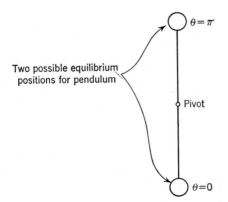

Two possible equilibrium positions for pendulum

$\theta = \pi$

Pivot

$\theta = 0$

FIGURE 31-7
Equilibrium positions for pendulum.

A CHEMICAL REACTOR

Consider the stirred-tank chemical reactor* of Fig. 31.8. The contents of the reactor are assumed to be perfectly mixed, and the reaction taking place is

$$A \rightarrow B \tag{31.12}$$

which occurs at a rate

$$R_A = kC_A e^{-E/RT} \tag{31.13}$$

where R_A = moles of A decomposing per hour per cubic foot of reacting mixture
   k = reaction velocity constant, hr^{-1}
   C_A = concentration of A in reacting mixture, moles/ft^3
   E = activation energy, a constant, Btu/mole
   R = universal gas law constant
   T = absolute temperature of reacting mixture

The reaction is exothermic; ΔH Btu of heat are generated for each mole of A that reacts. Hence, in order to control the reactor, cooling water is supplied to a cooling coil. The actual reactor temperature is compared with a set point, and the rate of cooling-water flow adjusted accordingly. To indicate this control mathematically, we write that $Q(T)$ Btu/hr of heat are removed through the cooling coil. In Chap. 32 we shall make a more detailed analysis of the dynamic behavior of the reactor. For the present preliminary analysis, it is not necessary to look carefully at $Q(T)$, and hence it is merely assumed that, as T rises, more heat is removed in the coil.

Let x_{A_0} = mole fraction of A in feed stream
   x_{B_0} = mole fraction of B in feed stream

---

*This example is based on the work of R. Aris and N. R. Amundson (1958).

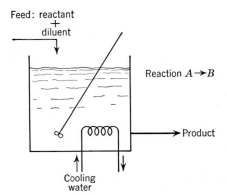

Feed: reactant
+
diluent

Reaction $A \rightarrow B$

Product

Cooling
water

FIGURE 31-8
Schematic of exothermic chemical reactor.

Then $(1 - x_{A_0} - x_{B_0})$ is the fraction of inerts in the feed stream. A mass balance on A,

(A in feed) $-$ (A in product) $-$ (A reacting) $=$ (A accumulating in reactor)

takes the form

$$F \rho x_{A_0} - F \rho x_A - k \rho V e^{-E/RT} x_A = \rho V \frac{dx_A}{dt} \qquad (31.14)$$

where F = feed rate, ft^3/hr

$\quad x_A$ = mole fraction of A in reactor

$\quad \rho$ = density of reacting mixture, moles/ft^3

$\quad V$ = volume of reacting mixture, ft^3

To arrive at Eq. (31.14) we have used Eq. (31.13) and made the following assumptions:

1. The density of the reacting mixture is constant, unaffected by the conversion of A to B.
2. The feed and product rates F are equal and constant.
3. Together, 1 and 2 imply that V, the volume of reacting mixture, is constant.
4. Perfect mixing occurs, so that x_A is the same in the reactor and product stream.

A similar mass balance may be derived for substance B. However, Eq. (31.12) shows that one mole of B appears for every mole of A destroyed. Hence

$$x_B - x_{B_0} = x_{A_0} - x_A \qquad (31.15)$$

Equation (31.15) permits us to circumvent the mass balance for x_B, since knowing x_A we can calculate x_B directly.

The energy balance on the reactor

(Sensible heat in feed) $\quad -$ (sensible heat in product)

$+$ (heat generated by reaction) $-$ (heat removed in cooling coil)

$\qquad\qquad\qquad\qquad = $ (energy accumulating in reactor)

can be written as

$$F \rho C_p (T_0 - T) + k \rho V (\Delta H) e^{-E/RT} x_A - Q(T) = \rho V C_p \frac{dT}{dt} \qquad (31.16)$$

where T_0 = temperature of feed stream

$\quad T$ = temperature in reactor

$\quad C_p$ = specific heat of reacting mixture

In writing Eq. (31.16), it is assumed that

1. The specific heat of the reacting mixture is constant, unaffected by the conversion of A to B.

2. The perfect mixing means that the temperatures of the reacting mixture and product stream are the same.

3. The heat of reaction ΔH is constant, independent of temperature and composition.

We remark here that these assumptions, as well as those made in Eq. (31.14), may be relaxed without affecting the conceptual aspects of the phase analysis. They are made only to keep the example as uncluttered as possible, without being trivial.

Equations (31.14) and (31.16) may be rearranged to the system

$$\frac{dx_A}{dt} = \frac{F}{V}(x_{A_0} - x_A) - ke^{-E/RT}x_A$$

$$\frac{dT}{dt} = \frac{F}{V}(T_0 - T) + \frac{k(\Delta H)}{C_p}e^{-E/RT}x_A - \frac{Q(T)}{\rho V C_p} \tag{31.17}$$

As a typical application of this system of equations, we might consider starting up the reactor, initially filled with a mixture at composition $x_A(0)$ and temperature $T(0)$. Suppose the feed rate, feed composition, feed temperature, and flow rate of cooling water are held constant and the reactor is operated in this manner until steady state is reached. To describe the transient behavior of the chemical reactor, one can solve Eqs. (31.17) by integrating them numerically, using a typical stepwise procedure such as the Euler or Runge-Kutta method. This will result in functions $x_A(t)$ and $T(t)$ for values of t from zero to some value (if one exists) at which, for practical purposes, $x_A(t)$ and $T(t)$ cease to change with t.

Alternatively, we may consider a phase-plane analysis of Eqs. (31.17) and seek solutions in the form of x_A versus T curves. Note that division of the first of Eqs. (31.17) by the second gives

$$\frac{dx_A}{dT} = \frac{(F/V)(x_{A_0} - x_A) - ke^{-E/RT}x_A}{(F/V)(T_0 - T) + \dfrac{k(\Delta H)}{C_p}e^{-E/RT}x_A - \dfrac{Q(T)}{\rho V C_p}} \tag{31.18}$$

The parameter t has been eliminated in Eq. (31.18), which is simply a differential equation relating x_A and T. As we shall see in Chap. 32, this phase-plane analysis of the chemical reactor offers significant advantages over the ordinary analysis.

In the chemical reactor, we no longer have the special relationship among the phase variables that we had in both previous cases. For both the spring and pendulum problems, we more or less artificially changed a second-order differential equation to two first-order equations by introducing the phase variable $\dot{X}$ (or $\dot{\theta}$). This phase variable was directly related to the other phase variable X (or θ) by the equation

$$\dot{X} = \frac{dX}{dt}$$

For the chemical reactor, there is no such simple relation between x_A and T.

We can study the steady-state solutions to Eqs. (31.17) without solving the equations, much as was done in the case of the damped pendulum of the previous example. As before, we note that steady state requires that x_A and T simultaneously cease to change with time,

$$\frac{dx_A}{dt} = \frac{dT}{dt} = 0$$

From Eqs. (31.17), this implies that

$$\frac{F}{V}(x_{A_0} - x_{A_s}) - ke^{-E/RT_s}x_{A_s} = 0$$

$$\frac{F}{V}(T_0 - T_s) + \frac{k(\Delta H)}{C_p}e^{-E/RT_s}x_{A_s} - \frac{Q(T_s)}{\rho VC_p} = 0 \tag{31.19}$$

where x_{A_s} and T_s are the steady-state values of x_A and T.

The first of Eqs. (31.19) can be solved for x_{A_s}, yielding

$$x_{A_s} = x_{A_0}\frac{1}{1 + (kV/F)e^{-E/RT_s}} \tag{31.20}$$

Substitution of (31.20) into the second of Eqs. (31.19) yields

$$\frac{k(\Delta H)x_{A_0}/C_p}{e^{E/RT_s} + kV/F} = \frac{Q(T_s)}{\rho VC_p} + \frac{F}{V}(T_s - T_0) \tag{31.21}$$

Equation (31.21) is implicit in T_s, the steady-state temperature. In physical terms, it expresses an equality between the heat generated by the reaction and the heat removed in the cooling coil and product stream. To emphasize this, we have arranged it so that the left side is the heat generation and the right side is the heat removal.

Solution of Eq. (31.21) for T_s requires numerical values for the various parameters. Without going into this much detail at present, we may obtain some qualitative information. To do this, we sketch the right and left sides of this equation as functions of T_s. A typical shape for the left side is given by the sigmoidal curve of Fig. 31.9. (See Aris and Amundsen, 1958, p. 121.) The unusual curvature, of course, is caused by the e^{E/RT_s} term in the denominator. To plot the right side, we must know $Q(T)$. While we have avoided specifying the form of $Q(T)$, we know it increases with T. If there were no control action, i.e., if the flow rate of cooling water were maintained constant regardless of T, then $Q(T)$ would increase almost linearly with T. This is because at constant water rate, the heat transfer in the coil is approximately proportional to the difference between T and the mean temperature of the cooling water. This latter temperature would not vary so rapidly as T at practical flow rates. However, since we expect to have control action, we know that the cooling-water flow rate will be increased with increasing T. Therefore, $Q(T)$ may be expected to increase faster than linearly with T, which means that the right side of (31.21) increases faster than linearly. Several typical curves of this right side are shown in Fig. 31.9.

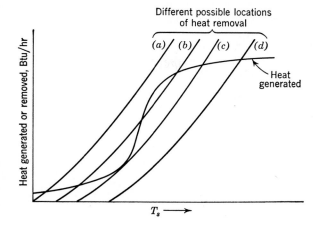

FIGURE 31-9
Steady-state generation and removal functions for exothermic chemical reactor.

A solution of Eq. (31.21) requires that the graphs of the right and left sides intersect. As shown in Fig. 31.9 there may be one, two, or three such intersections, depending on the relative locations of the heat generation (left side) and heat removal (right side). This means that there may be one, two, or three possible steady states for the reactor.

As we shall see in Chap. 32, the steady state actually attained by the reactor depends on initial conditions $x_A(0)$ and $T(0)$. The steady-state temperature T_s is then the temperature at the pertinent intersection, and the steady-state composition can be determined from Eq. (31.20). We shall also see that some of the steady states are unstable. In fact, the low-temperature steady state for curve (c) of Fig. 31.9, occurring as a point of tangency, is to be regarded with suspicion. Practically speaking, a perfect tangency would not occur. Minor variations in operating conditions (i.e., noise), which occur continually in actual process operation, may shift the curve (c) slightly to the left or right, resulting in two or zero low-temperature intersections, respectively.

SUMMARY

In this chapter, we have introduced the concept of a phase analysis and some of its basic elements. We have seen how physical situations give rise naturally to phase solutions. Furthermore, we have had our first look at true nonlinear behavior. In so doing, we have come to at least one interesting conclusion: a nonlinear motion or control-system response may have more than one steady-state solution. This was true for the chemical reactor and for the pendulum. In contrast, the linear motions and control-system responses we studied in the previous chapters had only one steady-state solution. In the next chapter, we shall discover still more differences which render nonlinear analysis more difficult than linear analysis.

CHAPTER
32

METHODS OF
PHASE-PLANE
ANALYSIS

The advantages of the phase analysis introduced in Chap. 31 can be more fully appreciated after some acquaintance with the tools available for such analysis. To give a detailed exposition of all, or even most, of the aspects of this subject is not intended. Instead, this chapter strives to indicate its flavor and to stimulate further study.

PHASE SPACE

In Chapter 31, we considered three examples for which the dynamic response can be described by two state variables. For the cases of the damped oscillator and the pendulum, the state variables were phase variables in which the dependent variable and its derivative (X, $\dot{X}$ or θ, $\dot{\theta}$) were chosen as the state variables. For the exothermic chemical reactor, the state variables selected were temperature and composition (T, x_A); these variables, which arose naturally in the analysis of the chemical reactor, were called physical variables in Chap. 28.

In general, an nth-order dynamic system can be described by n state variables. The state variables (x_1, x_2, . . . ,x_n) can be located in a coordinate system called phase space. Each value of t, say t_1, defines a point in this space: $x_1(t_1)$, $x_2(t_1)$, . . . ,$x_n(t_1)$. The solution curve is a locus of these points for all values of t. It is called a *trajectory* and connects successive states of the system. For

484

the damped oscillator presented in Chap. 31, the coordinate system was a plane with an axis for each state variable; we shall refer to this coordinate system as a phase plane. Figure 31.5 is a typical phase-plane representation of a dynamic system. When the physical system is third-order, the coordinate system consists of three axes, one for each state variable. Of course, systems of fourth- or higher-order require treatment in space that is of too many dimensions to be visualized. The graphic aspects of phase-space representation are advantageous primarily in the case of two dimensions (the phase plane) and to a limited extent for three dimensions. The bulk of practical use of phase-space analysis has been made in the two-dimensional autonomous (time invariant) case:

$$\frac{dx_1}{dt} = f_1(x_1, x_2)$$
$$\frac{dx_2}{dt} = f_2(x_1, x_2)$$

(32.1)

For this reason, we largely confine our attention in the remainder of this study to systems that may be written in the form of Eqs. (32.1). As we have seen, there is no loss in *conceptual generality,* but we cannot expect the *graphical aspects* of the material we shall develop to generalize to higher-dimensional phase space. The solution of the system (32.1) may be presented as a family of trajectories in the $x_2 x_1$ plane. If we are given the initial conditions

$$x_1(t_0) = x_{10}$$
$$x_2(t_0) = x_{20}$$

the initial state of the system is the point (x_{10}, x_{20}) in the $x_2 x_1$ plane and the trajectory may, in principle, be traced from this point.

By dividing the second of Eqs. (32.1) by the first, we obtain

$$\frac{dx_2}{dx_1} = \frac{f_2(x_1, x_2)}{f_1(x_1, x_2)}$$

(32.2)

Now dx_2/dx_1 is merely the slope of a trajectory, since a trajectory is a plot of x_2 versus x_1 for the system. Hence, at each point in the phase plane (x_1, x_2), Eq. (32.2) yields a unique value for the slope of a trajectory through the point, namely, $f_2(x_1, x_2)/f_1(x_1, x_2)$. This last statement should be amended to exclude any point (x_1, x_2) at which $f_1(x_1, x_2)$ and $f_2(x_1, x_2)$ are *both* zero. These important points are called *critical points* and will be examined in more detail below. Since the slope of the trajectory at a point, say (x_1, x_2), is by Eq. (32.2) unique, it is clear that trajectories cannot intersect except at a critical point, where the slope is indeterminate.

THE METHOD OF ISOCLINES

Let us now utilize this information about the trajectory slope to approximate the trajectory. We shall illustrate the technique with an example.

Example 32.1. Find the trajectory of the system

$$\frac{dx_1}{dt} = x_2$$

$$\frac{dx_2}{dt} = -5x_1 - 2x_2$$

(32.3)

which passes through the point

$$x_1 = 1 \qquad x_2 = 0$$

The slope of any trajectory is given by

$$\frac{dx_2}{dx_1} = -\frac{5x_1 + 2x_2}{x_2}$$

We search for all points through which the trajectories must have the same slope. If this slope is called S, then

$$-\frac{5x_1 + 2x_2}{x_2} = S$$

is the equation that must be satisfied by all points at which the slope is to be S. This may be rearranged to

$$x_2 = \frac{-5x_1}{S + 2}$$

which is the equation of a line through the origin in the $x_2 x_1$ plane. Thus, for example,

$$x_2 = -x_1$$

is the locus of all points at which the trajectories have slope 3. Similarly, the x_1 axis is the locus of points at which the slope of the trajectory is infinite. Such loci, which in this special case are straight lines, are called *isoclines*. Several isoclines, with the slopes indicated, are plotted on Fig. 32.1.

To sketch the desired trajectory, we first note that it starts at the point (1,0). At this point, according to Eqs. (32.3),

$$\frac{dx_1}{dt} = 0$$

$$\frac{dx_2}{dt} = -5$$

Hence, the trajectory starts out vertically downward. Between the isoclines $S = \infty$ and $S = 10$, the slope of the trajectory must vary between infinity and 10. The points on the $S = 10$ isocline, which would be reached if the trajectory had a constant slope of infinity or 10, are labeled a and b, respectively, in Fig. 32.1. The actual point at which the trajectory reaches the $S = 10$ isocline is taken as midway between a and b, which is equivalent to an averaging of the slopes. The construction is continued in this manner, and the trajectory sketched so as to connect the indicated points and to have the correct slope as it passes through each isocline. The short

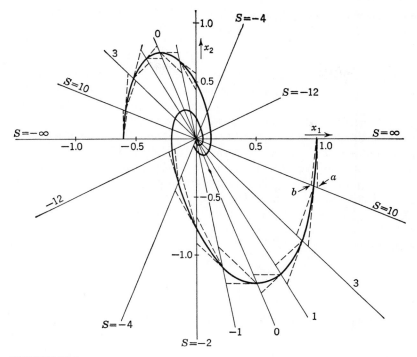

FIGURE 32-1
Isocline construction of phase plane for Eqs. (32.3).

dashed marks between isoclines, indicating the correct slope, are also helpful in satisfying this latter condition.

Another trajectory, starting from the point $(-0.6, 0)$, is shown on Fig. 32.1. This serves to emphasize that, once the isoclines have been located, interpolation is possible on the phase plane, and many trajectories representing various initial conditions are easily visualized or sketched.

There are other graphical techniques for construction of phase portraits. These are discussed in, for example, Thaler and Pastel (1962). The method of isoclines is usually suitable when the isocline equation

$$\frac{f_2(x_1, x_2)}{f_1(x_1, x_2)} = S$$

is not overly complicated and where a good overall knowledge of the phase plane is required. In practice, for more complex systems such as the chemical reactor of Chap. 31, the phase plane is often obtained by use of a computer.

Analysis of Critical Points

In the situations of most interest to us, Eq. (32.2) will represent the behavior of a (nonlinear) control system, as in Eq. (31.18). Therefore, we shall be interested in maintaining the system at or near a steady state. Since, from Eq. (32.1), a steady-state point is defined by

$$f_1(x_1, x_2) = f_2(x_1, x_2) = 0$$

it is clear that the steady states are critical points. At the critical points, the slope of the trajectory is undefined; hence, many trajectories may intersect at these points. In Fig. 32.1 the origin is a critical point. It can be seen from the isoclines that, in this case, all trajectories spiral into the origin. Hence, this particular system is such that, no matter what the initial state (i.e., for any disturbance which is applied), the system returns to steady state at the critical point.

The critical point of Fig. 32.1 is called a *focus*, because the trajectories spiral into it. This spiral motion of the trajectories corresponds to the oscillatory approach of the system to steady state. The oscillatory motion occurs because the system of Eqs. (32.3) is underdamped, as indicated by the characteristic equation

$$|s\mathbf{1} - \mathbf{A}| = 0$$

or

$$\begin{vmatrix} s & -1 \\ 5 & s+2 \end{vmatrix} = s^2 + 2s + 5 = 0$$

When put into standard form, this characteristic equation has parameters

$$\tau = \frac{1}{\sqrt{5}} \qquad \zeta = \frac{1}{\sqrt{5}}$$

Since $\zeta < 1$, the system is underdamped.

An overdamped system, such as that generated by the system

$$\frac{dx_1}{dt} = x_2$$

$$\frac{dx_2}{dt} = -5x_1 - 6x_2$$

having characteristic equation

$$s^2 + 6s + 5 = 0$$

so that

$$\tau = \frac{1}{\sqrt{5}} \qquad \zeta = \frac{3}{\sqrt{5}}$$

has a critical point such as that of Fig. 32.2a. Here the trajectories enter the critical point directly, without oscillation. This type of critical point is called a node. For comparison, a typical focus is sketched in Fig. 32.2b. In fact, other types of behavior may be exhibited by critical points of a second-order system, depending on the nature of the roots of the characteristic equation. These are summarized for *linear* systems in Table 32.1 and sketched in Fig. 32.2. The distinction between

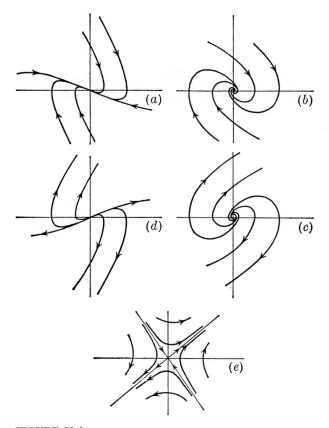

FIGURE 32-2
Second-order critical points: (a) stable node, (b) stable focus, (c) unstable focus, (d) unstable node, (e) saddle point.

stable and unstable nodes or foci is made to indicate that the trajectories move toward the stable type of critical point and away from the unstable point. The saddle point arises when the roots of the characteristic equation are real and have opposite sign. In this case there are only two trajectories that enter the critical point, and after entering, the trajectories may leave the critical point (permanently)

TABLE 32.1
Classification of critical points

Type of critical point	Characteristic equation	Pertinent values of ζ	Nature of roots	Sign of roots
Stable node	$\tau^2 s^2 + 2\zeta\tau s + 1 = 0$	$\zeta > 1$	Real	Both −
Stable focus	$\tau^2 s^2 + 2\zeta\tau s + 1 = 0$	$0 < \zeta < 1$	Complex	Real parts both −
Unstable focus	$\tau^2 s^2 + 2\zeta\tau s + 1 = 0$	$-1 < \zeta < 0$	Complex	Real parts both +
Unstable node	$\tau^2 s^2 + 2\zeta\tau s + 1 = 0$	$\zeta < -1$	Real	Both +
Saddle point	$\tau^2 s^2 + 2\zeta\tau s - 1 = 0$	All	Real	One +, one −

on either of two other trajectories. No other trajectory can enter the critical point, although some approach it very closely.

This categorization of critical points according to the particular linear system is often of value in the analysis of nonlinear systems. The reason for this is that, in a sufficiently small vicinity of a critical point, a nonlinear system behaves approximately linearly. Thus, the system of Eq. (31.10) for the pendulum is nonlinear. It has two physically distinguishable steady states, corresponding to the pendulum pointing up or down. The nonlinear term $\sin \theta$ may be linearized around each steady state. Near the steady state at $\theta = 0$,

$$\sin \theta \approx \theta$$

and near the steady state at $\theta = \pi$, a Taylor series yields

$$\sin \theta = -(\theta - \pi)$$

Therefore, near $\theta = 0$, Eqs. (31.10) are closely approximated by the linear equations

$$\frac{d\theta}{dt} = \dot{\theta}$$

$$\frac{d\dot{\theta}}{dt} = -\omega_n^2 \theta - D\dot{\theta} \tag{32.4}$$

and near $\theta = \pi$, by

$$\frac{dx}{dt} = \dot{x}$$

$$\frac{d\dot{x}}{dt} = \omega_n^2 x - D\dot{x} \tag{32.5}$$

where $x = \theta - \pi$. These linearized versions of Eqs. (31.10) can be easily solved to determine the nature of the *linear approximations* to the critical points. Thus, the characteristic equation for Eqs. (32.4) is

$$s^2 + DS + \omega_n^2 = 0 \tag{32.6}$$

while that for Eqs. (32.5) is

$$s^2 + DS - \omega_n^2 = 0 \tag{32.7}$$

As shown in Table 32.1, Eq. (32.6) yields a stable critical point, which may be a node or focus depending on the degree of damping. (Note that, as the damping is increased, the behavior changes from focus to node, or from oscillatory to nonoscillatory.) On the other hand, Eq. (32.7) indicates a saddle point for the motion near $\theta = \pi$.

These conclusions apply strictly only to the linearized phase equations, Eqs. (32.4) and (32.5). To compare them with the behavior of the true system of Eqs. (31.10), the actual phase diagram is sketched for a lightly damped case in Fig. 32.3. For simplicity, this diagram is extended beyond the range $0 \leq \theta \leq 2\pi$ even though this is the only region of physical significance. Actually, the section for $0 \leq \theta \leq 2\pi$ should be cut out and rolled into a cylinder so that the lines corresponding to $\theta = 0$ and $\theta = 2\pi$ coincide. This phase cylinder would more realistically represent the motion of the pendulum. As seen from Fig. 32.3, the point at $\theta = \pi$ is, indeed, a saddle point and the point $\theta = 0$ (or 2π) is a stable focus. If the system were more heavily damped, this latter point would be a stable node.

A greater understanding of the saddle point may now be obtained by analyzing the $\theta = \pi$ point in terms of what we know to be the physical behavior of the pendulum at this point. That is, the point may be approached from either of two directions. When the pendulum is at the point, an infinitesimal disturbance will cause it to fall in either of two directions. Other trajectories narrowly miss this point, indicating that just the right initial velocity must be imparted to the pendulum at a given initial point to cause it to stop in the $\theta = \pi$ position.

In summary, it can be concluded that in this case the linearized equations give valuable, accurate information about the behavior of the nonlinear system in the vicinity of the critical points. Because the linearized equations are more easily solved, it is always desirable to be able to relate the behavior of the actual system to the behavior of the linearized solutions in the vicinity of the operating point. In fact, in our previous work on control systems, we have assumed for nonlinear systems that design of a stable control system based on the linearized equations was adequate to ensure stable operation of the actual system. The basis for this assumption is given by the following theorem of Liapunov (see Letov, 1961).

Let the nonlinear equations of a motion be linearized by expansion in deviation variables around a particular critical point. If the linearized solution for the deviation variables is stable, the actual motion will be stable in some vicinity of the critical point. If the linearized solution is neutrally stable (i.e., its characteristic equation has roots on the imaginary axis), no statement can be made about the actual motion. If the linearized solution is unstable, then the actual motion will be unstable.

It is necessary to define what is meant by stability and instability of the actual nonlinear motion in the vicinity of the critical point. Although stability in nonlinear systems is a complex subject, for our purposes it will suffice to state that a stable nonlinear motion in the vicinity of a critical point is one for which all phase-plane trajectories in this vicinity travel toward and end at the critical point. An unstable motion is one for which trajectories move away from the critical point. This would mean that, while theoretically the state of the system may remain at the critical point indefinitely, any slight disturbance causes the unstable system

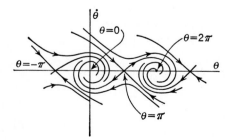

FIGURE 32-3
Phase portrait of lightly damped pendulum.

to move away from the critical point. These conclusions agree with our physical understanding of the pendulum motion, since the steady condition at $\theta = \pi$ is easily destroyed.

It is because of Liapunov's theorem that linear control theory is so successful in control system design. One really hopes to control the system so that it remains permanently in the vicinity of a particular point (i.e., a steady state). However, when serious upsets occur in an automatically controlled plant, moving it far from steady state, it is often necessary to return the plant to manual control until conditions are again close to steady state. This is because the controllers are designed for satisfactory operation in the linear range only. One of the great drawbacks of linear control theory is the fact that stability of the linearized equations guarantees stability of the nonlinear system only in *some* vicinity of the particular critical point. No information about the size of this vicinity or about the behavior outside this vicinity is obtained. If the linear vicinity is extremely small, then unknown to the designer who has used linear methods, almost any plant disturbance of practical size may result in control system failure. An example of this behavior will be given later.

Limit Cycles

The first major difference between linear and nonlinear motions is the possible existence of more than one critical point in the latter type. The second is the possible existence of limit cycles.

A *limit cycle* is defined as a periodic oscillation whose amplitude and frequency depend only on the properties of the system and not on the initial state of the system (provided the initial state lies in a certain non-trivial region of the phase space). In the phase plane, *stable* limit cycles are recognized as closed curves which are approached asymptotically by all nearby trajectories. *Unstable* limit cycles are closed curves from which all nearby trajectories diverge. An example of a stable limit cycle is the "steady-state" behavior of a home heating system when controlled by a thermostat. A periodic oscillation in house temperature is always reached, and the amplitude and frequency of the oscillation are independent of the temperature that existed in the house at the time that the furnace was started. Unstable limit cycles can never be realized physically for any system by definition. However, as will be seen later, they divide the phase plane into regions of totally different dynamic behavior and hence are of considerable importance.

It is important to distinguish between limit cycles and other closed curves which may occur. The linear system

$$\tau^2 \frac{d^2 x}{dt^2} + x = 0$$

has phase-space solution

$$x^2 + \tau^2 (\dot{x})^2 = C^2 \tag{32.8}$$

where $\dot{x} = dx/dt$ and the constant C depends on initial conditions. Equation (32.8) defines a family of concentric ellipses in the phase plane. However, *these are not limit cycles,* because the closed curve which is followed by the system depends on the initial state of the system through the constant C. In the next section, we shall study some limit cycles occurring in typical control systems.

OTHER ASPECTS. We have presented only those aspects of phase-plane analysis that will be of use in the examples to follow. This can be considered only as a brief introduction to the subject, and the interested reader is referred to the references already cited for more information. Among the important subjects that have been omitted are graphical methods for determination of time along a trajectory, various aspects of phase-plane topology, and the mathematical aspects of stability.

EXAMPLES OF PHASE-PLANE ANALYSIS

In this section, we shall consider two different examples of the use of the phase plane to analyze nonlinear control systems. The first is a simple on-off control system for a stirred-tank heater. The second is the chemical reactor of Chap. 31. In both cases, the systems are second-order and autonomous, so that they are ideal situations for use of the phase plane.

On-Off Control of Stirred-tank Heater

The use of on-off control offers significant economic advantages over proportional control or other more sophisticated modes of control. The control mechanism is simply a relay that turns on or off depending on the value of the measured variable. The disadvantage is usually that the quality of control is inferior to that realized with proportional control.

Consider the stirred-tank heater of Fig. 32.4. Water is being heated to a controlled temperature by mixing with steam. It is assumed for the analysis that the cold-water input rate is constant. Heated water overflows into an outlet pipe at the top of the tank, so that no accumulation of mass occurs in the tank. Most of the steam is added, at a fixed flow rate, from the main steam supply. However, this amount of steam is set at a value somewhat less than the amount required to heat the cold water to the desired temperature. An additional amount of steam may be added whenever the solenoid valve is opened. When this additional steam is admitted, the sum of the two steam inputs is enough to heat the water to a temperature somewhat in excess of the desired temperature. A temperature-measuring device such as a thermocouple or vapor-pressure bulb transmits the tank temperature to the relay. When this temperature is below the set point, the relay closes, which opens the solenoid valve, thus admitting more steam. Eventually, the additional steam will result in the temperature exceeding the set point, the relay will open, the valve will close cutting off the additional steam, and the temperature will fall again.

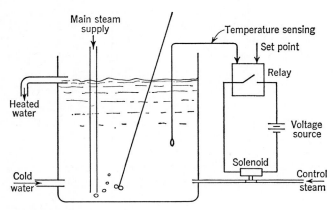

FIGURE 32-4
On-off control of stirred-tank heater.

It is apparent that an oscillating control will be achieved. In fact, from the discussion in the previous section, we recognize that a limit cycle will occur. We consider now a numerical example of this type of control system.

Water at 40°F, at a rate of 100 lb/min, is to be heated to 150°F. The main steam supply is to be set so that it will heat this much water to 125°F, while additional steam, through the controlled solenoid valve, is available to heat the water another 50°F. This means that the steady-state temperatures with the solenoid closed and open, 125 to 175°F, are equally spaced about the set point. Heat losses to the surroundings are negligible. The volume of the tank is 1.6 ft³. The relay control system has a vapor-pressure bulb for measurement of temperature. This measuring system has a time constant of 30 sec. The solenoid valve is very rapid in response.

We first analyze this system considering the relay to behave ideally. This means that it opens precisely at the instant the temperature exceeds the set point and closes similarly. Later, we shall correct this to conform more closely to the behavior of actual relays.

If the tank is perfectly stirred, it is a first-order system with a time constant of

$$\tau = \frac{pV}{\omega} = \frac{(62)(1.6)}{100} = 1.0 \text{ min}$$

and its transfer function relating changes in the steam input rate to temperature is

$$G_p(s) = \frac{10}{s + 1}$$

where 10 (°F) (min)/(lb) is the change in steady-state temperature per unit change in steady-state steam flow. The necessary fixed and controlled steam rates are (using 1,000 Btu/lb for latent heat)

$$Q_{\text{fixed}} = \frac{(125 - 40)(100)}{1,000} = 8.5 \text{ lb/min}$$

$$Q_{\text{controlled}} = \frac{(175 - 125)(100)}{1,000} = 5.0 \text{ lb/min}$$

The amount of steam that would be necessary to maintain the water at a steady-state temperature of 150°F is

$$Q_s = \frac{(150 - 40)(100)}{1,000} = 11.0 \text{ lb/min}$$

Hence, in terms of deviation variables, the controller output may be taken as ±2.5 lb/min of steam.

A block diagram may now be constructed for this system, as shown in Fig. 32.5. This diagram uses deviations from 150°F as temperature variables, so the set point is taken as zero. The action of the relay is symbolized by the input-output relations, indicating that +2.5 lb/min of steam are admitted when the error is positive and −2.5 lb/min when the error is negative, again in deviation variables. The transduction from the vapor-pressure bulb to a temperature reading is included implicitly in Fig. 32.5 in the comparator. The comparator is physically a device that balances the pressure generated by the bulb against a mechanical tension caused by positioning the set point. It need not be explicitly shown because its dynamics are very fast.

It is convenient to use a dimensionless version of Fig. 32.5. This is provided in Fig. 32.6, where the changes

$$M' = \frac{M}{2.5}$$

$$\epsilon' = \frac{\epsilon}{25}$$

$$C' = \frac{C}{25}$$

$$B' = \frac{B}{25}$$

have been made.

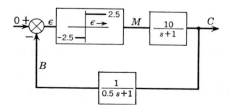

FIGURE 32-5
Block diagram for system of Fig. 32.4.

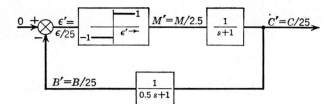

FIGURE 32-6
Dimensionless block diagram
for system of Fig. 32.4.

The usual methods of linear control theory are not applicable to the block diagram of Fig. 32.6 The relay does not obey the principle of superposition in its input-output relation. It is necessary to revert to the differential equations describing the control loop. These are

$$M' = \frac{dC'}{dt} + C' \tag{32.9}$$

$$C' = \frac{1}{2}\frac{dB'}{dt} + B' \tag{32.10}$$

$$\epsilon' = -B' \tag{32.11}$$

In addition we have

$$M' = \begin{cases} 1 & \epsilon' > 0 \\ -1 & \epsilon' < 0 \end{cases} \tag{32.12}$$

Combination of Eqs. (32.9) to (32.12) yields

$$\frac{1}{2}\frac{d^2\epsilon'}{dt^2} + \frac{3}{2}\frac{d\epsilon'}{dt} + \epsilon' = \begin{cases} -1 & \epsilon' > 0 \\ 1 & \epsilon' < 0 \end{cases} \tag{32.13}$$

Equation (32.13) can be rewritten in phase notation as

$$\frac{d\epsilon'}{dt} = \dot{\epsilon}'$$
$$\frac{d\dot{\epsilon}'}{dt} = \begin{cases} -(3\dot{\epsilon}' + 2\epsilon' + 2) & \epsilon' > 0 \\ -(3\dot{\epsilon}' + 2\epsilon' - 2) & \epsilon' < 0 \end{cases} \tag{32.14}$$

Equation (32.14) breaks up into two regions, the region for which $\epsilon' > 0$ will be referred to as R, and that for which $\epsilon' < 0$ as L. The critical point for R occurs at

$$\epsilon' = -1 \qquad \dot{\epsilon}' = 0$$

and that for L at

$$\epsilon' = 1 \qquad \dot{\epsilon}' = 0$$

Note that each critical point is outside the region to which it pertains. In region R, the isocline equation is

$$-\frac{2 + 2\epsilon' + 3\dot{\epsilon}'}{\dot{\epsilon}'} = S_R$$

or

$$\dot{\epsilon}' = \frac{-2(\epsilon' + 1)}{S_R + 3} \tag{32.15}$$

The corresponding isocline equation in L is

$$\dot{\epsilon}' = \frac{-2(\epsilon' - 1)}{S_L + 3} \tag{32.16}$$

The isoclines in R, which is the right half of the $\dot{\epsilon}'\epsilon'$ plane, radiate from the R critical point $(-1,0)$ and have slopes $-2/(S_R + 3)$. The isoclines in L radiate from the critical point $(1,0)$ and have slopes $-2/(S_L + 3)$. These isoclines are indicated in Fig. 32.7. Note that, in this figure, the ϵ' scale has been expanded by a factor of 10 to magnify the behavior near the origin.

A typical trajectory has been constructed, using the method of isoclines. When the trajectory crosses from one region to the other on the $\dot{\epsilon}'$ axis, the applicable isoclines also change. It can be seen from Fig. 32.7 that the trajectory approaches the origin. Since the trajectories must be vertical as they cross the ϵ' axis, the final state is a limit cycle of zero amplitude and infinite frequency about the origin. In other words, the relay alternately opens and closes at very high frequency, a condition known as *chattering*.

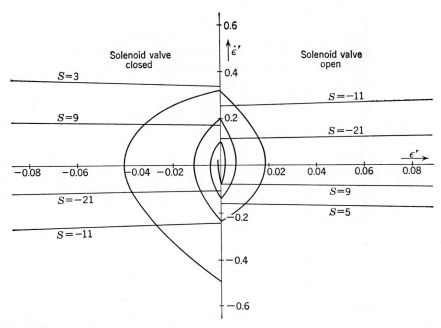

FIGURE 32-7
Phase-plane trajectory for on-off control of system of Fig. 32.4.

Physically, this condition will never be realized because the dynamics of the solenoid valve and the relay itself would become important. Instead, the final condition will be a limit cycle of high, rather than infinite, frequency and low, rather than zero, amplitude.

However, the basic idealization which has led us to this suspect conclusion is in the behavior of the relay. True relays have input-output characteristics more similar to that shown in Fig. 32.8. There is a dead band around the set point, of width $2\epsilon_0'$, over which the relay is insensitive to changes in the error signal. Anyone who has made fine adjustments in the setting of a home thermostat has observed this behavior.

Consider as an example the case for $\epsilon_0' = 0.01$. The effect of this dead zone is to change the dividing line between R and L to that shown in Fig. 32.9. The new dividing line has the equation:

$$\epsilon' = \begin{cases} 0.01 & \dot{\epsilon}' > 0 \\ -0.01 & \dot{\epsilon}' < 0 \end{cases}$$

Now, as shown in Fig. 32.9, all trajectories approach a limit cycle, for which the error amplitude is approximately 0.03. The frequency is finite and is obtained by computing the time around the limit cycle. Although we have not presented here the graphical methods for determining this time, it can always be calculated by noting from the first of Eqs. (32.14) that

$$t = \int dt = \int \frac{d\epsilon'}{\dot{\epsilon}'} \tag{32.17}$$

Thus, time around the limit cycle can be computed by graphical evaluation of the integral in Eq. (32.17). The only difficulty is near the ϵ' axis, where $\dot{\epsilon}'$ goes to zero. To circumvent this, we may use the second of Eqs. (32.14)

$$t = -\int \frac{d\dot{\epsilon}'}{3\dot{\epsilon}' + 2\epsilon' \pm 2}$$

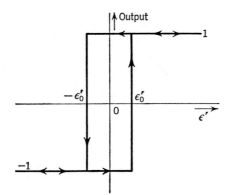

FIGURE 32-8
Characteristics of true relay with dead zone.

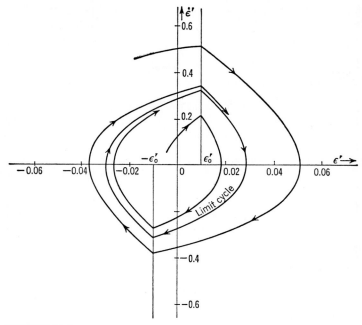

FIGURE 32-9
Phase plane for system of Fig. 32.4 using relay with characteristics of Fig. 32.8.

over a small segment of the trajectory as it crosses the ϵ' axis. The result of this graphical calculation is $\omega = 9.2$ rad/min.

The frequency thus computed for the error signal is, for obvious physical reasons, the same as the frequency of the controlled signal, C'. However, the amplitude of C', which is of more direct interest, is not the same as the amplitude of ϵ'. It may be found in this case by noting from Eqs. (32.10) and (32.11) that

$$C' = -\tfrac{1}{2}\dot{\epsilon}' - \epsilon'$$

It is therefore clear from Fig. 32.9 that C' attains a maximum value near the switching points where

$$C' \approx \pm 0.17$$

Reverting to the original variables, it follows that the water temperature will oscillate with an amplitude of

$$(0.17)(25) = 4.25°F$$

The effect of a small dead zone, $2\epsilon_0 = 2(.01)(25) = 0.5°F$, is thus quite significant.

In practice, the width of this dead zone is usually an adjustable design parameter. This width is always chosen as a compromise. The wider it is made, the lower will be the limit-cycle frequency, thus saving excessive switching or chatter. However, the limit-cycle amplitude increases with dead-zone width, decreasing the quality of control.

The Exothermic Chemical Reactor

We now wish to consider the phase-plane behavior of the chemical reactor of Chap. 31. This study is based on the paper by Aris and Amundson (1958). For convenience, the dynamic equations are reproduced here:

$$\frac{dx_A}{dt} = \frac{F}{V}(x_{A0} - x_A) - k e^{-E/RT} x_A$$

$$\frac{dT}{dt} = \frac{F}{V}(T_0 - T) + \frac{k(\Delta H)e^{-E/RT}}{C_p} x_A - \frac{Q(T)}{\rho V C_p} \tag{31.17}$$

Defining the dimensionless variables

$$\tau = \frac{Ft}{V} \qquad y = \frac{x_A}{x_{A0}} \qquad \theta = \frac{C_p T}{x_{A0}(\Delta H)} \qquad \theta_0 = \frac{C_p T_0}{x_{A0}(\Delta H)}$$

these equations become

$$\frac{dy}{d\tau} = 1 - y - r(y, \theta)$$

$$\frac{d\theta}{d\tau} = \theta_0 - \theta + r(y, \theta) - q(\theta) \tag{32.19}$$

where $r(y, \theta) = \dfrac{kVy}{F} e^{[-EC_p/Rx_{A0}(\Delta H)\theta]}$

$$q(\theta) = \frac{Q(T)}{F\rho x_{A0}(\Delta H)}$$

As a control heat-removal function $q(\theta)$, Aris and Amundson chose the form

$$q(\theta) = U(\theta - \theta_c)[1 + K_c(\theta - \theta_s)] \tag{32.20}$$

where θ_c is the dimensionless mean temperature of water in the cooling coil. This indicates that the heat removal is always proportional to the difference between the reactor temperature and mean cooling-water temperature. In addition, the term in brackets indicates that proportional control on the cooling-water flow rate is present. The flow rate is increased by an amount proportional to the difference between the actual reactor temperature θ and the desired steady-state temperature θ_s. This increase in cooling-water flow rate is assumed for convenience to cause an approximately proportional increase in heat removal. The constant U is a dimensionless analog of $U_0 A$, the overall heat-transfer rate.

As a specific numerical example, Aris and Amundson selected the following values for constants:

$$\frac{kV}{F} = e^{25}$$

$$\frac{EC_p}{Rx_{A_0}(\Delta H)} = 50$$

$$\theta_s = 2$$

$$\theta_0 = \theta_c = 1.75$$

$$U = 1$$

Under these conditions, Eqs. (32.19) become

$$\frac{dy}{d\tau} = 1 - y - ye^{50(1/2 - 1/\theta)}$$

$$\frac{d\theta}{d\tau} = 1.75 - \theta + ye^{50(1/2 - 1/\theta)} - (\theta - 1.75)[1 + K_c(\theta - 2)]$$

(32.21)

It can be seen that there is a critical point of Eqs. (32.21) at

$$y = \tfrac{1}{2} = y_s$$

$$\theta = 2 = \theta_s$$

and this is the location at which control is desired. This point has the correct steady-state temperature and a 50 percent conversion of reactant. In addition, there may be two more critical points of Eq. (32.21) depending on the proportional control constant K_c, as will be discussed below.

Since we are primarily interested in control about θ_s, we make use of Liapunov's theorem on local stability, presented earlier. Linearizing Eq. (32.21) in deviation variables $\theta - \theta_s$ and $y - y_s$ by using Taylor's series yields

$$\frac{d(y - y_s)}{d\tau} = -2(y - y_s) - 6.25(\theta - \theta_s)$$

$$\frac{d(\theta - \theta_s)}{d\tau} = (y - y_s) + \left(4.25 - \frac{K_c}{4}\right)(\theta - \theta_s)$$

(32.22)

where $y_s = \tfrac{1}{2}$. As we have seen before, the solution to this linear system is

$$y - y_s = c_1 e^{s_1 t} + c_2 e^{s_2 t}$$

$$\theta - \theta_s = c_3 e^{s_1 t} + c_4 e^{s_2 t}$$

where, in this case, s_1 and s_2 are the roots of [see Eq. (31.6) and the steps following this equation]

$$s^2 + \frac{K_c - 9}{4} s + \frac{2K_c - 9}{4} = 0$$

(32.23)

According to the Routh criteria, all coefficients in this characteristic equation must be positive in order that the real parts of the roots s_1 and s_2 be negative. Hence, we can see immediately from Eq. (32.23) that, in order to achieve a stable node or focus, it is necessary that $K_c > 9$.

However, Aris and Amundson obtained the phase plane for (among other values) a value of K_c slightly greater than 9. This was accomplished by numerical solution of Eqs. (32.21). It was found that, in the vicinity of the steady-state point, the situation is as depicted in Fig. 32.10. There are two limit cycles surrounding the stable focus critical point. The inner limit cycle is unstable, and the outer limit cycle is stable, according to the definitions given earlier. It may be seen that any disturbance (or initial condition) which moves the system no further from the critical point than the unstable limit cycle can be controlled. That is, the control system will eventually bring the system back to steady state. However, once the system is forced outside this limit cycle, it will eventually spiral out to the stable limit cycle. Control cannot be restored, and the reactor temperature and concentration oscillate continuously. This example illustrates very well the limitations of linear control theory. All that the linear investigation could reveal is that, for $K_c > 9$, the system will be stable in some vicinity of the control point. The phase-plane analysis shows that, for K_c slightly greater than 9, this vicinity is inside the unstable limit cycle of Fig. 32.10. If K_c is increased further, the two limit cycles disappear and good control can be achieved. This example points out the importance of unstable limit cycles. Although a physical system can never follow an unstable limit cycle, the limit cycle divides the phase plane into distinct dynamic regions for the physical system.

Other values of K_c were analyzed by Aris and Amundson. For low values of K_c, there are two other critical points besides the control point. For example, for $K_c = 0.8$, there are critical points at

$$y = 0.95 \qquad \theta = 1.77$$

and

$$y = 0.15 \qquad \theta = 2.15$$

Linear analysis shows that both these are stable, but for $K_c < 9$, the control point ($y = 0.5$, $\theta = 2$) is not. Phase-plane analysis shows that, if the reactor is started at high temperatures, it will come to steady state at the high-temperature critical point and vice versa. Starting the reactor at the desired control point will be of no

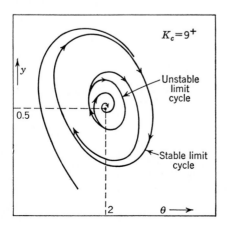

FIGURE 32-10
Stable and unstable limit cycles in exothermic chemical reactor.

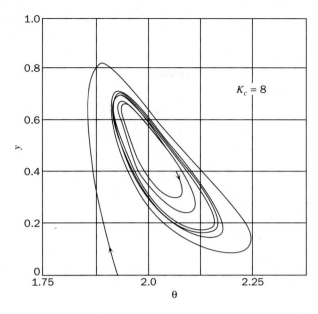

FIGURE 32-11
Phase-plane portrait of the control of a chemical reactor (limit cycle forms).

avail, as it will leave and go to one of the other steady-state points, depending on the direction of the initial disturbance. For high values of K_c, there is only one critical point, which is at the control point. Phase-plane analysis shows that K_c must exceed approximately 30 before rapid return to steady state at the desired control point, following all disturbances, is achieved. Some phase-plane portraits for this system that were obtained by means of a computer are shown in Figs. 32.11 to 32.13.

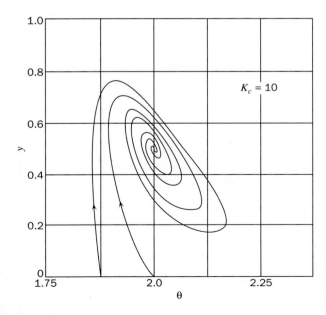

FIGURE 32-12
Phase-plane portrait of the control of a chemical reactor (no limit cycle forms).

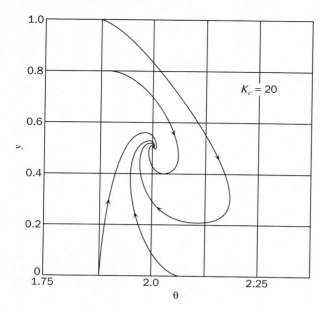

FIGURE 32-13
Phase-plane of the control of a chemical reactor (no limit cycle forms).

This discussion is only a rather brief introduction to the extensive work by Aris and Amundson. The reader is strongly urged to consult the original paper for a more comprehensive treatment of the problem.

SUMMARY

We have seen that phase-plane analysis can be used for two typical nonlinear control problems. The results of this analysis give extensive information about the control system behavior. The responses to various disturbances can be visualized by sketching only a few trajectories.

On the other hand, the method is effectively limited to second-order systems. Furthermore, analysis is considerably more laborious than the linear analysis, and a decision regarding the value of the additional information must be made.

PROBLEMS

32.1. For the system shown in Fig. P32.1, obtain equations for plotting the isoclines in the e versus $\dot{e}$ phase plane. Plot a few isoclines and sketch carefully the trajectory from the initial point $e = 2$ and $\dot{e} = 0$.

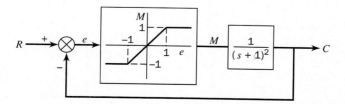

FIGURE P32-1

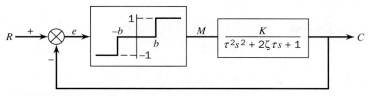

FIGURE P32-2

32.2. For the system shown in Fig. P32.2, plot isoclines for $S = 0, 1, -3$, and ∞ on the phase plane having coordinates $e, \dot{e}$. Use $\tau = 1$, $\zeta = 1/2$, $K = 1$, $b = 0.25$. Let $R = 0$. Sketch the trajectory starting at $e = 1$, $\dot{e} = 0$.

32.3. Consider the phase-plane equations

$$\dot{x}_1 = x_2$$

$$\dot{x}_2 = -x_1 + \tfrac{1}{10}x_2 - x_1^2 - \tfrac{10}{3}x_2^3$$

(a) Determine the type of critical point at $x_1 = -1$, $x_2 = 0$.
(b) If there are any other critical points, find them.

32.4. The system shown in Fig. P32.4 is to be controlled by an ideal on-off relay.

(a) From the block diagram, write the differential equations for phase-plane description of the physical system in the form:

$$\dot{x}_1 = \text{ftn}(x_1, x_2)$$

$$\dot{x}_2 = \text{ftn}(x_1, x_2)$$

where $x_1 = c$ and $x_2 = \dot{c}$.

(b) Plot on the phase plane (x_1, x_2) a few isoclines. Include isoclines for $S = 0$, $1, \infty$. Show clearly the switching line where the forcing changes from one sign to the other.

(c) Make a rough sketch of the trajectory that starts at $x_1 = 2$, $x_2 = 0$ and extend it only to the switching line.

(d) Calculate accurately the values of x_1 and x_2 for the trajectory of part (c) for $t = 0, 0.5, 1, 1.5$, and 2.

(e) Determine the values of x_1 and x_2 where the first switch occurs.

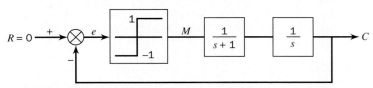

FIGURE P32-4

32.5. Calculate the period of the limit cycle in Fig. 32.9.

CHAPTER

33

THE DESCRIBING FUNCTION TECHNIQUE

In Chap. 32, an on-off temperature-control system was studied in the phase plane. This work led to information about the limit cycle of the system as well as about the manner in which trajectories approached the limit cycle. Very often, this latter information about the transient approach to the limit cycle is unnecessary. Of primary interest to the designer are the amplitude and frequency of the limit cycle. The describing function method facilitates rapid, accurate estimates of these quantities without construction of the phase plane.

In this chapter we shall study application of the describing function method to the analysis of the on-off controller for the temperature bath of Chap. 32. The treatment will be introductory only and largely confined to a single example. The purpose is to indicate the existence of the method and to show how it complements the phase-plane technique. The reader desirous of a more extensive treatment is referred to the text by Graham and McRuer (1961).

HARMONIC ANALYSIS

Consider the block diagram for the on-off control of the stirred-tank heater of Chap. 32, shown in dimensionless form in Fig. 33.1. In the following analysis, we omit the primes from the variables of Fig. 33.1. Our objective is to find the amplitude and frequency of the limit cycle that occurs in the control loop. The describing function method assumes that the error signal, in the limit cycle condition, is sinusoidal:

$$\epsilon = A \sin \omega t \tag{33.1}$$

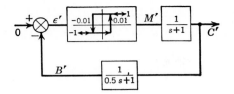

FIGURE 33-1
Block diagram for control of stirred-tank heater
using relay with dead zone.

A glance at Fig. 32.9 shows that the error signal is not actually sinusoidal, since a sinusoidal signal appears as an ellipse in the phase plane. However, the difference between the actual limit cycle and an ellipse is not great, particularly if only the amplitude and frequency are of interest.

If the error signal is sinusoidal, the relay output M can be derived from Fig. 33.2, where it can be seen from the input-output relations that $M(t)$ is a square wave that lags $\epsilon(t)$ by a time $(1/\omega)\sin^{-1}(\epsilon_0/A)$. The time lag is due to the dead zone in the relay. Thus,

$$M(t) = S\left(t - \frac{1}{\omega}\sin^{-1}\frac{\epsilon_0}{A}, \frac{2\pi}{\omega}\right) \tag{33.2}$$

where

$$S(t,P) = \begin{cases} 1 & 0 < t < P/2 \\ -1 & P/2 < t < P \\ S(t+P,P) & \text{all } t \end{cases} \tag{33.3}$$

is the undelayed unit square wave of period P shown in Fig. 33.3.

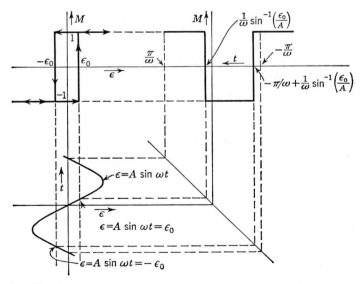

FIGURE 33-2
Result of application of sinusoidal error signal to relay with dead zone.

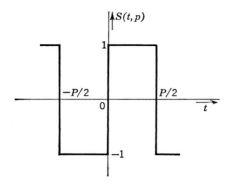

FIGURE 33-3
The unit square wave $S(t, P)$.

As is well known from Fourier series analysis, (see Churchill and Brown, 1986), $S(t, 2\pi/\omega)$ may be expanded in a series of sine waves to give

$$S\left(t, \frac{2\pi}{\omega}\right) = \frac{4}{\pi}\left(\sin \omega t + \tfrac{1}{3} \sin 3\omega t + \tfrac{1}{5} \sin 5\omega t + \cdots\right) \qquad (33.4)$$

Hence, by Eq. (33.2)

$$M(t) = \frac{4}{\pi}\left[\sin\left(\omega t - \sin^{-1}\frac{\epsilon_0}{A}\right) + \tfrac{1}{3}\sin\left(3\omega t - 3\sin^{-1}\frac{\epsilon_0}{A}\right)\right.$$
$$\left. + \tfrac{1}{5}\sin\left(5\omega t - 5\sin^{-1}\frac{\epsilon_0}{A}\right) + \cdots\right] \qquad (33.5)$$

According to Eq. (33.5), $M(t)$ contains a fundamental and odd harmonics. Let us consider what happens to these components of M as they pass around the control loop. Assuming that ω is sufficiently large, the harmonics are much more heavily attenuated by the two first-order elements than is the fundamental, because the harmonic frequencies are higher. For example, if ω is 9 rad/min, the relative attenuation of the fundamental and third harmonic between M and B is expressed by the quotient

$$\frac{\dfrac{1}{\sqrt{1 + (27)^2}\,\sqrt{1 + (27/2)^2}}}{\dfrac{1}{\sqrt{1 + (9)^2}\,\sqrt{1 + (9/2)^2}}} = 0.11$$

Since the initial amplitude of the third harmonic in $M(t)$ is one-third of the fundamental, it is clear that the amplitude of the third harmonic will be less than 4 percent of the amplitude of the fundamental in $B(t)$. The amplitudes of the higher harmonics will be even less. To all intents and purposes, $B(t)$ is sinusoidal and, hence, so is $\epsilon(t)$. Furthermore, the presence of harmonics in $M(t)$ may be ignored, and the approximation

$$M(t) \cong \frac{4}{\pi} \sin\left(\omega t - \sin^{-1}\frac{\epsilon_0}{A}\right) \tag{33.6}$$

is acceptable because the higher harmonics are filtered out by the rest of the loop. In order for a limit cycle to be maintained, it is necessary that

$$B(t) = -\epsilon(t) = -A \sin \omega t \tag{33.7}$$

However, if $M(t)$ is given by Eq. (33.6), $B(t)$ can be calculated by frequency response. The AR between M and B is

$$\left|\frac{B}{M}\right| = \frac{1}{\sqrt{1 + (\omega)^2}\sqrt{1 + (\omega/2)^2}} \tag{33.8}$$

and the phase difference between B and M is

$$\angle B - \angle M = -\tan^{-1}\omega - \tan^{-1}\frac{\omega}{2} \tag{33.9}$$

According to Eq. (33.7), the overall amplitude ratio between B and ϵ must be unity and the overall phase lag 180°. Also, according to Eq. (33.6), the AR between ϵ and M is $4/\pi A$ and the phase lag is $\sin^{-1}(0.01/A)$. Combining these facts results in

$$\frac{4}{\pi A} \frac{1}{\sqrt{1 + (\omega)^2}\sqrt{1 + (\omega/2)^2}} = 1$$

$$-\sin^{-1}\frac{0.01}{A} - \tan^{-1}\omega - \tan^{-1}\frac{\omega}{2} = -180° \tag{33.10}$$

Equations (33.10) are a system of two equations in the unknowns A and ω. Trial-and-error solution yields

$$A = 0.03$$
$$\omega = 9 \text{ rad/min}$$

in excellent agreement with the results of the phase-plane method presented in Chap. 32. The reason for the accuracy of these results is the high attenuation of harmonics provided by the linear elements in the loop. The labor saving of this method over the phase-plane method is apparent.

Of more direct interest is the amplitude of the signal C in the limit cycle. This may now be estimated by frequency response to be

$$|C| = \sqrt{1 + \left(\frac{9}{2}\right)^2}(0.03) = 0.14$$

The true result derived by the phase-plane method is 0.17, so that an error of 18 percent is attributed to the neglect of harmonics in C. The reason for the decreased accuracy in the amplitude of C over that in the amplitude of ϵ is that only one of the linear elements has acted on the squarewave output of the relay before it reaches C. Hence, the harmonics are not fully attenuated in C and the signal C will be less sinusoidal than ϵ. However, for engineering purposes the error in C is probably not excessive.

THE DESCRIBING FUNCTION

Because the basic technique of harmonic analysis often yields accurate results with modest effort, it is profitable to systematize the procedure. To do this, a describing function is defined for the nonlinear loop element. This function assumes a sinusoidal input to the nonlinearity and gives the AR and phase lag of the *fundamental* in the output. Thus, for the relay considered in the last section, the describing function is defined by

$$N = \frac{4}{\pi A} \angle - \sin^{-1} \frac{\epsilon_0}{A}$$

where N is used as the symbol for the describing function.

In general, the loop diagram for a relay control system appears as in Fig. 33.4. As shown previously, the necessary condition for a limit cycle, ignoring harmonics, is

$$|N\|G_pH| = 1$$
$$\angle N + \angle G_pH = -180°$$

$$(33.11)$$

As in the case of the relay, the magnitude and angle of N in general depend on the amplitude A of the input to N. The magnitude and angle of G_pH depend on ω. Equations (33.11) can be rewritten

$$|G_pH(\omega)| = \frac{1}{|N(A)|}$$
$$\angle G_pH(\omega) = -180° - \angle N(A)$$

$$(33.12)$$

Equations (33.12) can be solved graphically on a gain-phase plot. This is a plot of the log of AR versus phase, as shown in Fig. 33.5 for the case treated in the previous section. The linear elements are plotted as $|G_pH|$ versus $\angle G_pH$, with ω plotted as a parameter on the curve. The relay is plotted as $1/|N(A)|$ versus $-180° - \angle N(A)$, with A plotted as a parameter on the curve. According to Eqs. (33.12), a limit cycle occurs at the intersection of the two curves, where the amplitude and frequency can be read from the parametric labeling of A and ω.

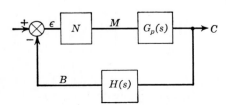

FIGURE 33-4
Typical control loop containing nonlinear element.

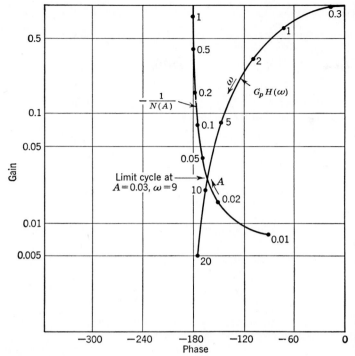

FIGURE 33-5
Gain-phase plot for system of Fig. 33.1.

The advantages of the gain-phase plot are (1) elimination of trial-and-error solution of equations such as Eqs. (33.10) and (2) ease of treatment of complex linear systems G_pH. In addition, the gain-phase plot can be used to estimate the occurrence or nonoccurrence of a limit cycle, according to whether or not an intersection occurs.

SUMMARY

The describing function can be used to good advantage for estimation of amplitude and frequency of limit cycles in systems similar to the one studied here. The success of the method depends on the presence of a sufficient number of linear elements in the loop to filter out the harmonics generated by the nonlinear element. No information about the transient response is obtained. However, the method requires considerably less labor than does the phase-plane method, and the limit cycle amplitude and frequency are often the quantities of primary interest.

It should also be noted that the describing function method is not limited to second-order systems, as is the phase-plane method. In fact, the higher the order of G_pH in Fig. 33.4, the more accurate will be the describing function results.

PROBLEMS

33.1. For the control system shown in Fig. P33.1 determine the frequency and amplitude of the limit cycle if one exists. Use the describing function method.

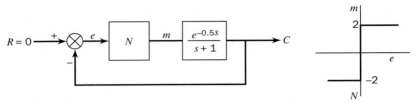

FIGURE P33-1

33.2. (a) For the system shown in Fig. P33.2, $\tau = 1, \zeta = 1/2, K = 1, M = 1, b = 0.25$ and $R = 0$.

(b) Show that the describing function is

$$N = \frac{4M}{A}\sqrt{1 - \left(\frac{b}{A}\right)^2} \qquad \angle 0$$

where A is the amplitude of the sine wave entering the nonlinearity.

(c) For $K = 2$, does a limit cycle exist? If so, describe it.

(d) If a transport lag e^{-s} is introduced in the feedback loop, determine if a limit cycle exists for $K = 2$.

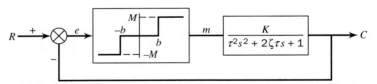

FIGURE P33-2

33.3. The stirred-tank system shown in Fig. P33.3 produces an aqueous solution of salt by use of a solenoid valve that switches from one reagent tank to the other as described below. The reagent tanks contain concentrated solutions of salt. When the measured concentration is above the set point, the control reagent of lower concentration enters the mixing tank at a constant flow rate of 0.01 liter/min. When the measured concentration is less than the set point, the control reagent of higher concentration enters the mixing tank at a constant rate of 0.01 liter/min. The hold-up volume of the tank is 2 liters, the transport lag between the tank and measuring element is 1.2 min, and the set point is 2 g salt/liter.

(a) Obtain a block diagram, in terms of deviation variables, for this control system.

(b) By means of the describing function method, determine the characteristics of the limit cycle (frequency and amplitude), if one exists.

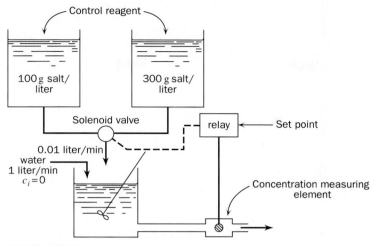

FIGURE P33-3

33.4. For the control system shown in Fig. P33.4, determine if a limit cycle exists for $K = 1$, 2, and 3. If a limit cycle exists, describe it in terms of amplitude and frequency. For the nonlinearity shown,

$$N = \frac{2}{\pi}\left[\sin^{-1}\left(\frac{1}{A}\right) + \frac{1}{A}\sqrt{1 - \frac{1}{A^2}}\right] \qquad \angle 0 \text{ for } A \geq 1$$

$$N = 1 \text{ for } A < 1$$

A is the amplitude of the sine wave entering the nonlinearity.

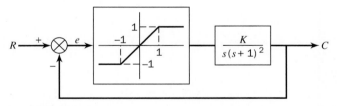

FIGURE P33-4

PART
IX

COMPUTERS IN PROCESS CONTROL

DIGITAL
COMPUTER
SIMULATION
OF CONTROL
SYSTEMS

The purpose of this chapter is to describe some of the methods for obtaining the transient response of a control system from a set of differential equations or transfer functions. Inversion of a high-order transfer function can be a time-consuming task. If a control system includes a nonlinearity or a transport lag, obtaining the response as an analytical expression is often impossible. The appearance of analog computers and digital computers after World War II made the task of solving the dynamic response of control systems much easier.

During the period from the mid-fifties to the mid-seventies, the analog computer was widely used to obtain the response of control systems. During that time, digital computing was very costly and slow compared to the situation today. There was little software available; at the beginning, one had to program the solution to a problem using machine code. The basic elements of the analog computer consisted of integrators, summers, gain potentiometers, and some nonlinear devices, such as multipliers. By connecting these computing elements together with wires, one could obtain the transient response to a rather large-scale control problem in the form of a voltage that varied with time.

As the cost of digital computing decreased and its speed of operation increased, the analog computer was gradually replaced with the digital computer. This change was especially noticed with the availability of the personal computer.

One advantage of the analog computer was that the flow of voltage signals through the computing elements closely resembled the flow of signals in the block diagram of the control system; the analog computer diagram and the block diagram of the control system looked nearly the same. In fact, this advantage has been retained in some of the digital computer simulation software that has been developed to solve control problems.

The basic operation needed to solve control problems by either an analog computer or a digital computer is integration. The integration device, in the case of an analog computer or the simulation software in the case of a digital computer, is called an integrator. Some of the symbols used to represent integration are shown in Fig. 34.1. The operation performed by the integrator is

$$y = \int_0^t x \, dt + y(0) \tag{34.1}$$

where $y(0)$ is the initial value of y at $t = 0$.

The symbol shown in Fig. 34.1a is used in analog computing where sign inversion occurs. The symbol shown in Fig. 34.1b is used in block diagrams for state-space problems. The symbol in Fig. 34.1c is used in digital computer simulation software. Since the focus of this chapter is on the digital computer, the method of achieving integration by means of an analog computer will not be considered further. The reader may consult Coughanowr and Koppel (1965) or other sources for this topic.

In the branch of mathematics called numerical analysis, many routines or algorithms to perform integration have been developed. Perhaps the simplest method, which is often discussed in a course in calculus or differential equations, is the Euler method. The Euler method is easy to understand, but it has a large truncation error that makes it too inaccurate for general use. Many methods of numerical integration have been devised that are far more accurate than the Euler method; one of these is the Runge-Kutta method. In this chapter, only the fourth-order Runge-Kutta method will be used. This method is often used to solve sets of first-order differential equations.

Runge-Kutta Integration

The Runge-Kutta method for solving a differential equation is often called a "marching" solution because the calculation starts at an initial value of the independent variable t and moves forward one integration step at a time.

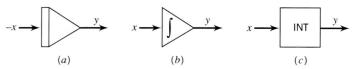

$$(a) \qquad\qquad (b) \qquad\qquad (c)$$

FIGURE 34-1
Symbols used to represent integrators.

Consider the first-order differential equation

$$\frac{dy}{dt} = f(y, t) \tag{34.2}$$

for which $y = y_0$ at $t = t_0$. In control problems, the initial time t_0 is usually taken as zero. When the dependent variable y is defined in terms of a deviation variable, which is usually the case in control problems, the value of y at t_0 is also zero. The Runge-Kutta method divides the independent variable t into increments of equal length Δt as shown in Fig. 34.2.

The fourth-order Runge-Kutta method uses the following equations:

$$k_1 = f(y_0, t_0)\Delta t \tag{34.3}$$
$$k_2 = f(y_0 + k_1/2, t_0 + \Delta t/2)\Delta t \tag{34.4}$$
$$k_3 = f(y_0 + k_2/2, t_0 + \Delta t/2)\Delta t \tag{34.5}$$
$$k_4 = f(y_0 + k_3, t_0 + \Delta t)\Delta t \tag{34.6}$$
$$y_1 = y_0 + (k_1 + 2k_2 + 2k_3 + k_4)/6 \tag{34.7}$$
$$t_1 = t_0 + \Delta t \tag{34.8}$$

The equations just listed are applied during the first increment Δt from t_0 to t_1. The values obtained at the end of the first increment (y_1, t_1) are then used as a new set of initial conditions in these equations to obtain a set of values of y and t at the end of the second interval. This procedure of computing y and t at the end of successive intervals generates the solution to the differential equation.

The set of equations [Eqs. (34.3) to (34.8)] used to solve a single first-order differential equation can be applied to each dependent variable in a set of differential equations. Consider the pair of differential equations

$$\frac{dy}{dt} = f_1(y, w, t) \tag{34.9}$$

$$\frac{dw}{dt} = f_2(y, w, t) \tag{34.10}$$

with the initial conditions y_0, w_0, t_0.

The Runge-Kutta equations used to solve for $y(t)$ and $w(t)$ are given below.

$$k_1 = f_1(y_0, w_0, t_0)\Delta t \tag{34.11}$$
$$l_1 = f_2(y_0, w_0, t_0)\Delta t \tag{34.12}$$

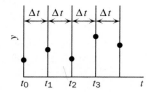

FIGURE 34-2
Dividing the independent variable t into equal increments Δt.

$$k_2 = f_1(y_0 + k_1/2, w_0 + l_1/2, t_0 + \Delta t/2)\Delta t \tag{34.13}$$

$$l_2 = f_2(y_0 + k_1/2, w_0 + l_1/2, t_0 + \Delta t/2)\Delta t \tag{34.14}$$

$$k_3 = f_1(y_0 + k_2/2, w_0 + l_2/2, t_0 + \Delta t/2)\Delta t \tag{34.15}$$

$$l_3 = f_2(y_0 + k_2/2, w_0 + l_2/2, t_0 + \Delta t/2)\Delta t \tag{34.16}$$

$$k_4 = f_1(y_0 + k_3, w_0 + l_3, t_0 + \Delta t)\Delta t \tag{34.17}$$

$$l_4 = f_2(y_0 + k_3, w_0 + l_3, t_0 + \Delta t)\Delta t \tag{34.18}$$

$$y_1 = y_0 + (1/6)(k_1 + 2k_2 + 2k_3 + k_4) \tag{34.19}$$

$$w_1 = w_0 + (1/6)(l_1 + 2l_2 + 2l_3 + l_4) \tag{34.20}$$

$$t_1 = t_0 + \Delta t \tag{34.21}$$

Example 34.1. Simulation of a second-order system. The differential equation describing the dynamics of a second-order system, which was given in Chapter 8, is as follows:

$$\tau^2 \frac{d^2 y}{dt^2} + 2\zeta\tau\frac{dy}{dt} + y = x(t) \tag{34.22}$$

Develop a computer program to solve this problem by use of the Runge-Kutta method, for the following conditions: $\tau = 1$, $\zeta = 0.4$, $x(t) = u(t) = 1$, $y(0) = 0$.

We must first express Eq. (34.22) as two first-order differential equations by letting

$$y_1 = y \tag{34.23}$$

and

$$y_2 = \frac{dy_1}{dt} = \frac{dy}{dt} \tag{34.24}$$

Using these expressions in Eq. (34.22), we obtain

$$\frac{dy_1}{dt} = y_2 \tag{34.25}$$

$$\frac{dy_2}{dt} = -\frac{1}{\tau^2}y_1 - \frac{2\zeta}{\tau}y_2 + \frac{1}{\tau^2}x(t) \tag{34.26}$$

The reader who has studied Chapter 28 will notice that Eqs. (34.25) and (34.26) have been written in terms of the state variables, y_1 and y_2. We are now ready to apply the Runge-Kutta method to these equations.

A computer program written in BASIC is shown in Fig. 34.3. In the program, the functions DY1 and DY2 are defined on lines 20 and 30 and correspond to Eqs. (34.25) and (34.26). After defining the parameters (τ, ζ, and Δt), the Runge-Kutta procedure is started on line 170.

The results from running the program in Fig. 34.3 are shown in Table 34.1. The reader can check these results with the analytical equation given in Chapter 8 [Eq.(8.17)].

Example 34.2. Simulation of a PI control system. In this example, the transient response of the liquid-level control system shown in Fig. 34.4a is to be obtained using a digital computer. The block diagram of this system is shown in Fig. 34.4b. The values of the parameters of the block diagram are as follows.

```
LIST
10 REM RESPONSE OF SECOND ORDER SYSTEM BY RUNGE-KUTTA
20 DEF FNDY1(Y1,Y2) = Y2
30 DEF FNDY2(Y1,Y2) = 1/TAU^2 - Y1/TAU^2 - 2*ZETA*Y2/TAU
40 ZETA = .4
50 TAU = 1
60 DT = .05
70 Y1 = 0
80 Y2 = 0
90 T = 0
100 PRINT "T","Y"
110 K = 0
120 K = K+1
130 IF K = 11,THEN 150
140 GOTO 170
150 PRINT USING "##.####         ";T,Y1
160 K = 1
170 K1 = FNDY1(Y1,Y2)*DT
180 L1 = FNDY2(Y1,Y2)*DT
190 K2 = FNDY1(Y1+.5*K1,Y2+.5*L1)*DT
200 L2 = FNDY2(Y1+.5*K1,Y2+.5*L1)*DT
210 K3 = FNDY1(Y1+.5*K2,Y2+.5*L2)*DT
220 L3 = FNDY2(Y1+.5*K2,Y2+.5*L2)*DT
230 K4 = FNDY1(Y1+K3,Y2+L3)*DT
240 L4 = FNDY2(Y1+K3,Y2+L3)*DT
250 Y1 = Y1 + (K1+2*K2+2*K3+K4)/6
260 Y2 = Y2 + (L1+2*L2+2*L3+L4)/6
270 T = T + DT
280 IF T>10.05 THEN END
290 GOTO 120
300 END
310 RUN
Ok
```

FIGURE 34-3
BASIC program for step response of second-order system of Example 34.1 ($\tau = 1, \zeta = 0.4$).

$$K_c = \text{proportional gain, psi/ft tank level}$$

$$\tau_I = \text{integral time, min}$$

$$K_v = \text{valve constant} = 0.070 \ (\text{ft}^3/\text{min})/\text{psi}$$

$$R_1 = 0.55 \ (\text{ft level})/(\text{ft}^3/\text{min})$$

$$\tau_1 = \text{time constant of tank 1} = 2.0 \ \text{min}$$

$$\tau_2 = \text{time constant of tank 2} = 1.0 \ \text{min}$$

$$\tau_3 = \text{time constant of tank 3} = 1.0 \ \text{min}$$

For convenience in simulating this system, the diagram of Fig. 34.4b has been reduced to that of Fig. 34.5 in which K_v has been combined with the PI control transfer function in one block and the transfer functions for the three tanks have been combined in one block.

TABLE 34.1
Step response of second-order system of Example 34.1

RUN	
T	Y
0.5000	0.1077
1.0000	0.3599
1.5000	0.6582
2.0000	0.9271
2.5000	1.1221
3.0000	1.2281
3.5000	1.2532
4.0000	1.2189
4.5000	1.1517
5.0000	1.0761
5.5000	1.0100
6.0000	0.9637
6.5000	0.9400
7.0000	0.9362
7.5000	0.9465
8.0000	0.9643
8.5000	0.9833
9.0000	0.9995
9.5000	1.0104
10.0000	1.0157
Ok	

To obtain the differential equations for use in the Runge-Kutta method, we proceed as follows. From the controller block, we may write

$$\frac{M(s)}{E(s)} = K_c K_v \left(1 + \frac{1}{\tau_I s}\right) = K_c K_v \frac{(\tau_I s + 1)}{\tau_I s} \tag{34.27}$$

Cross-multiplying gives

$$\tau_I s M(s) = K_c K_v \tau_I s E(s) + K_c K_v E(s)$$

This may be converted to the time domain to give

$$\dot{m} = K_c K_v \dot{e} + [K_c K_v / \tau_I] e \tag{34.28}$$

From the comparator of Fig. 34.5, we have

$$e = 1 - c \tag{34.29}$$

and

$$\dot{e} = -\dot{c} \tag{34.30}$$

Replacing e and $\dot{e}$ in Eq. (34.28) by the expressions in Eqs. (34.29) and (34.30) gives

$$\dot{m} = -K_c K_v \dot{c} + \frac{K_c K_v}{\tau_I}(1 - c) \tag{34.31}$$

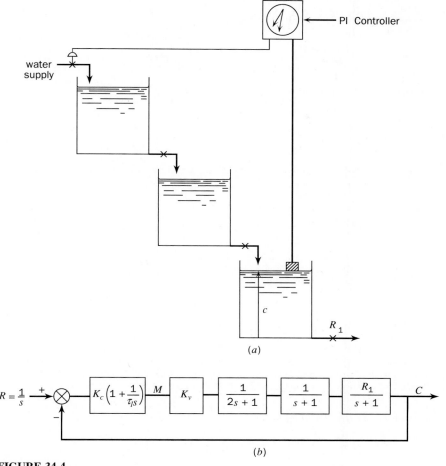

(a)

(b)

FIGURE 34-4
Process for Example 34.2: (a) liquid-level control system, (b) block diagram.

The three tanks are represented by

$$\frac{C(s)}{M(s)} = \frac{0.55}{(2s + 1)(s + 1)^2} \tag{34.32}$$

The differential equation represented by Eq. (34.32) can be formed by cross-multiplying. The result is

$$(2s + 1)(s + 1)^2 C(s) = 0.55M(s)$$

or

$$(2s^3 + 5s^2 + 4s + 1)C(s) = 0.55M(s) \tag{34.33}$$

Recognizing $s^n C(s)$ to be the nth derivative of c in the time domain, Eq. (34.33) can be written as

$$2\,\dddot{c} + 5\,\ddot{c} + 4\dot{c} + c = 0.55m$$

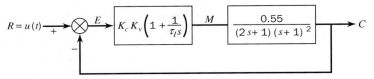

FIGURE 34-5
Reduced diagram of control system for Example 34.2.

or

$$\dddot{c} = -0.5c - 2\dot{c} - 2.5\ddot{c} + 0.275m \tag{34.34}$$

In order to apply the Runge-Kutta method, we must express Eq. (34.34) as three first-order differential equations. The procedure will now be shown.
Let

$$x = c \tag{34.35}$$

$$y = \dot{c} \tag{34.36}$$

$$z = \ddot{c} \tag{34.37}$$

Equation (34.34) can now be written

$$\dot{x} = y \tag{34.38}$$

$$\dot{y} = z \tag{34.39}$$

$$\dot{z} = -0.5x - 2y - 2.5z + 0.275m \tag{34.40}$$

We can now summarize the set of first-order differential equations with initial conditions by listing Eqs. (34.38), (34.39), (34.40), and (34.31). In Eq. (34.31), c and $\dot{c}$ have been replaced by x and y according to Eqs. (34.35) and (34.36).

Summary of differential equations

$$\dot{x} = y \tag{34.38}$$

$$\dot{y} = z \tag{34.39}$$

$$\dot{z} = -0.5x - 2y - 2.5z + 0.275m \tag{34.40}$$

$$\dot{m} = K_c K_v y + (K_c K_v / \tau_I)(1 - x) \tag{34.41}$$

Initial conditions

$$x(0) = 0$$

$$y(0) = 0$$

$$z(0) = 0$$

$$m(0) = K_c K_v$$

Notice that the control problem has been converted to a state-variable representation in which the state variables are x, y, z, and m. The initial conditions for the state variables x, y, and z are all zero, in keeping with the fact that these variables represent deviation variables that are, by definition, zero initially. In this formulation x, y, and z represent level, derivative of level, and second derivative of level, respectively.

A comment is needed to explain the fact that the initial value of m is $K_c K_v$. At time zero, the system is disturbed by a unit-step change in set point. This signal is transmitted through the controller block and causes m to jump to $K_c K_v$ because of the proportional action present in the controller.

The Runge-Kutta method will now be applied to solving Eqs. (34.38) to (34.41). The Runge-Kutta equations given by Eqs. (34.11) through (34.21) must, of course, be extended to handle the four differential equations. A BASIC computer program for this problem is shown in Fig. 34.6. The output from running the program is given in Table 34.2.

Example 34.3. Simulation of a control system with transport lag. Consider the proportional control system in Fig. 34.7 in which a transport lag is located in the feedback path. The equations representing this system are as follows:

$$\dot{y} = -\frac{1}{\tau}y + \frac{1}{\tau}m$$

$$m = K_c e$$

$$e = r - x$$

$$x = y(t - \tau_d)$$

The difference between this problem and the previous ones considered in this chapter is the presence of a transport lag. In the previous digital simulations, only the current value of y was needed and hence stored. In this problem, we must store values of y over the time interval $t - \tau_d$ to t (i.e., over the interval τ_d). Since we

TABLE 34.2
Response of level for Example 34.2

TIME,MIN	LEVEL,FT
0.500	0.033
1.000	0.199
1.500	0.504
2.000	0.884
2.500	1.246
3.000	1.505
3.500	1.608
4.000	1.543
4.500	1.342
5.000	1.067
5.500	0.794
6.000	0.591
6.500	0.503
7.000	0.543
7.500	0.690
8.000	0.899
8.500	1.113
9.000	1.279
9.500	1.359
10.000	1.342
Ok	

```
5   REM CONTROL OF THREE-TANK SYSTEM; USE OF RUNGE-KUTTA;EX. 2
10  DEF FNM(M,X,Y,Z) = -KV*KC*Y + KC*KV*(1-X)/TAUI
20  DEF FNX(M,X,Y,Z) = Y
30  DEF FNY(M,X,Y,Z) = Z
40  DEF FNZ(M,X,Y,Z) = -2.5*Z -2*Y -.5*X + .275*M
60  KV = .07
70  KC = 107.2
80  TAUI = 3.7
90  M = KC*KV
100 X = 0
110 Y = 0
120 Z = 0
130 T = 0
140 DT = .1
141 PRINT "TIME,MIN","LEVEL,FT"
142 K = 0
144 K = K+1
145 IF K=6 THEN 147
146 GOTO 150
147 PRINT USING "##.###          ";T,X
148 K=1
150 M1 = FNM(M,X,Y,Z)*DT
160 X1 = FNX(M,X,Y,Z)*DT
170 Y1 = FNY(M,X,Y,Z)*DT
180 Z1 = FNZ(M,X,Y,Z)*DT
190 M2 = FNM(M+M1*.5,X+X1*.5,Y+Y1*.5,Z+Z1*.5)*DT
200 X2 = FNX(M+M1*.5,X+X1*.5,Y+Y1*.5,Z+Z1*.5)*DT
210 Y2 = FNY(M+M1*.5,X+X1*.5,Y+Y1*.5,Z+Z1*.5)*DT
220 Z2 = FNZ(M+M1*.5,X+X1*.5,Y+Y1*.5,Z+Z1*.5)*DT
230 M3 = FNM(M+M2*.5,X+X2*.5,Y+Y2*.5,Z+Z2*.5)*DT
240 X3 = FNX(M+M2*.5,X+X2*.5,Y+Y2*.5,Z+Z2*.5)*DT
250 Y3 = FNY(M+M2*.5,X+X2*.5,Y+Y2*.5,Z+Z2*.5)*DT
260 Z3 = FNZ(M+M2*.5,X+X2*.5,Y+Y2*.5,Z+Z2*.5)*DT
270 M4 = FNM(M+M3,X+X3,Y+Y3,Z+Z3)*DT
280 X4 = FNX(M+M3,X+X3,Y+Y3,Z+Z3)*DT
290 Y4 = FNY(M+M3,X+X3,Y+Y3,Z+Z3)*DT
300 Z4 = FNZ(M+M3,X+X3,Y+Y3,Z+Z3)*DT
310 M = M+(1/6)*(M1+M2*2+M3*2+M4)
320 X = X+(1/6)*(X1+X2*2+X3*2+X4)
330 Y = Y+(1/6)*(Y1+Y2*2+Y3*2+Y4)
340 Z = Z+(1/6)*(Z1+Z2*2+Z3*2+Z4)
350 T = T+DT
370 IF T>10.1 THEN END
380 GOTO 144
390 END
```

FIGURE 34-6

BASIC computer program for liquid-level control system for Example 34.2 for a step change in set point of 1.0 ft and $K_c = 107$ psi/ft and $\tau_I = 3.7$ min.

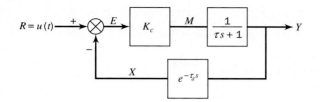

FIGURE 34-7
Block diagram of a proportional control system with transport lag (Example 34.3).

compute y only at discrete times, we must store values of y in an array of computer storage locations, called a stack. The diagram in Fig. 34.8 will help clarify this storage.

The array will be used to store past values of y that were computed at the end of each computation interval. At the end of each computation interval the values of y will be moved one position toward the end of the stack and the value of y just computed will be placed in the first storage location of the stack. By this means, a current value of y will not appear at the end of the stack until it has moved through each storage location. The amount of time the current value of y is delayed will depend on the number of storage locations and Δt. The number of storage locations N is determined by:

$$N = \tau_d / \Delta t$$

Let the values stored in the array be $S(i)$ where i, which represents the array position, will vary from 1 to $N + 1$. The following terms are now defined for the computer program to be developed.

$$Y = y, \text{ present value of } y$$

$$S(i) = \text{stored past values of } y$$

$$S(1) = \text{current value of } y, \text{ obtained at end of most recent computation interval}$$

$$S(N + 1) = X, \text{ the delayed value of } y, \text{ i.e., } X = y(t - \tau_d)$$

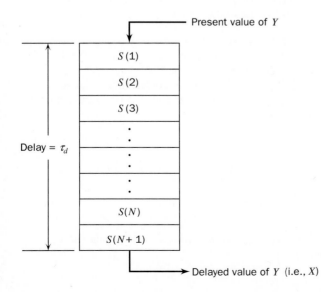

FIGURE 34-8
Array used to obtain transport lag in Example 34.3.

An outline of the procedure for computing y at discrete values of t is as follows.

1. Let the array for storing values of y be of "length" $N + 1$
   where $N = \tau_d / \Delta t$.
2. Initialize the elements of the array to zero.
3. Initialize the time variable, $T = 0$.
4. Set $X = S(N + 1)$.
5. Print T, Y, and X.
6. Start the Runge-Kutta routine to integrate the differential equation over the first computation interval ΔT.
7. Rearrange the contents of the array as shown in Fig. 34.8 by shifting the contents of each storage location by one position. Start shifting from the bottom. In this shifting, the oldest value of Y will be discarded and the value of Y just computed will enter the first cell to become $S(1)$.
8. Store the value of Y just computed into $S(1)$, i.e., $S(1) = Y$.
9. Increment T by ΔT and return to step (4) to repeat another cycle of calculation.

Using the steps just listed, the BASIC computer program shown in Fig. 34.9 was written for the conditions: $\tau = 1.0$, $\tau_d = 0.2$, $R = u(t)$, $K_c = 8.4$, and $\Delta t = 0.02$. For these conditions,

$$N = \tau_d / \Delta t = 0.2/0.02 = 10$$

The output from the computer program is shown in Table 34.3.

The computer program for simulating a transport lag that has just been presented is quite primitive compared to those provided in commercial software in which the delayed function is not held constant during the time step Δt, but is allowed to vary by use of an interpolation scheme. Some of the simulation software packages that provide the simulation of transport lag (e.g., TUTSIM, ACSL, and CC) are listed at the end of this chapter.

Example 34.4. Simulation of PID control. The presence of derivative action in a control algorithm, such as PID control, gives some difficulty in the writing of a program for digital computer simulation. Consider the PID control of a first-order system as shown in Fig. 34.10. To obtain a set of first-order differential equations for use with the Runge-Kutta method, we proceed as follows. From the controller block, we obtain

$$\frac{M(s)}{E(s)} = K_c \left(1 + \frac{1}{\tau_I s} + \tau_d s \right) = \frac{K_c}{\tau_I s} (\tau_d \tau_I s^2 + \tau_I s + 1) \qquad (34.42)$$

Cross-multiplying this expression, solving for sM, and writing the result in the time domain give

$$\dot{m} = K_c \dot{e} + (K_c/\tau_I)e + K_c \tau_d \ddot{e} \qquad (34.43)$$

This expression is not in the form in which the right side is free of derivatives of the variables. To obtain the correct form, we proceed as follows. Since $R = 0$ for this problem, we may write

```
10 REM            FIRST ORDER SYSTEM WITH TRANSPORT LAG
15 DIM S(11)
20 DEF FNDY(Y) = -Y+KC-KC*X
30 KC=8.399999  .
40 Y = 0
50 T = 0
60 DT=.02
70 FOR I = 1 TO 11
80 S(I) = 0
90 NEXT I
100 PRINT "T","Y","X"
110 X=S(11)
120 K1=FNDY(Y)*DT
130 K2=FNDY(Y+.5*K1)*DT
140 K3=FNDY(Y+.5*K2)*DT
150 K4=FNDY(Y+K3)*DT
160 Y=Y+(K1+2*K2+2*K3+K4)/6
170 T=T+DT
180 PRINT USING "#.###          ";T,Y,X
190 K=10
200 FOR I = 1 TO 10
210 S(K+1)=S(K)
220 K=K-1
230 NEXT I
240 S(1)=Y
250 IF T > 1.001 THEN END
260 GOTO 110
270 END
```

FIGURE 34-9
BASIC computer program for control of a first-order system with transport lag (Example 34.3).

$$e = -y \tag{34.44}$$

$$\dot{e} = -\dot{y} \tag{34.45}$$

$$\ddot{e} = -\ddot{y} \tag{34.46}$$

From Fig. 34.10, we may write

$$Y(s) = \frac{1}{\tau s + 1}[M(s) + U(s)]$$

In the time domain, this equation becomes

$$\dot{y} = (1/\tau)[u(t) + m - y] \tag{34.47}$$

where $U(s) = 1/s$ has been written as $u(t)$ (a unit step) in the time domain. Taking the derivative of both sides of Eq. (34.47) gives

$$\ddot{y} = (1/\tau)[\delta(t) + \dot{m} - \dot{y}] \tag{34.48}$$

where use has been made of the fact that the derivative of a unit-step function is a unit-impulse function (see Chap. 4). Combining Eqs. (34.44), (34.45), and (34.46) with Eqs. (34.47) and (34.48) gives

TABLE 34.3
Computer output for control of first-order system with transport lag (Example 34.3)

T	Y	X
0.020	0.166	0.000
0.040	0.329	0.000
0.060	0.489	0.000
0.080	0.646	0.000
0.100	0.799	0.000
0.120	0.950	0.000
0.140	1.097	0.000
0.160	1.242	0.000
0.180	1.384	0.000
0.200	1.523	0.000
0.220	1.659	0.000
0.240	1.765	0.166
0.260	1.841	0.329
0.280	1.890	0.489
0.300	1.911	0.646
0.320	1.907	0.799
0.340	1.877	0.950
0.360	1.824	1.097
0.380	1.748	1.242
0.400	1.649	1.384
0.420	1.530	1.523
0.440	1.390	1.659
0.460	1.235	1.765
0.480	1.071	1.841
0.500	0.901	1.890
0.520	0.732	1.911
0.540	0.567	1.907
0.560	0.410	1.877
0.580	0.264	1.824
0.600	0.135	1.748
0.620	0.024	1.649
0.640	-.064	1.530
0.660	-.128	1.390
0.680	-.165	1.235
0.700	-.173	1.071
0.720	-.153	0.901
0.740	-.106	0.732
0.760	-.031	0.567
0.780	0.067	0.410
0.800	0.188	0.264
0.820	0.329	0.135
0.840	0.484	0.024
0.860	0.652	-.064
0.880	0.827	-.128
0.900	1.004	-.165
0.920	1.179	-.173
0.940	1.348	-.153
0.960	1.505	-.106
0.980	1.647	-.031
1.000	1.769	0.067
1.020	1.869	0.188

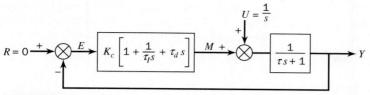

FIGURE 34-10
PID control of a first-order process (Example 34.4).

$$\dot{e} = -\dot{y} = -(1/\tau)[u(t) + m - y] \tag{34.49}$$

$$\ddot{e} = -\ddot{y} = -(1/\tau)\{\delta(t) + \dot{m} - (1/\tau)[u(t) + m - y]\} \tag{34.50}$$

Substituting the expressions for e, $\dot{e}$, and $\ddot{e}$ from Eqs. (34.44), (34.49), and (34.50) into Eq. (34.43) gives after simplification

$$\dot{m} = C[-\tau_d\tau\delta(t) + A + By + Am] \tag{34.51}$$

where $A = \tau_d - \tau$

$$B = \tau - \frac{\tau^2}{\tau_I} - \tau_d$$

$$C = \frac{K_c}{(\tau + K_c\tau_d)\tau}$$

The right side of Eq. (34.51) contains the forcing term $-C\tau_d\tau\delta(t)$. If Eq. (34.51) were integrated, this term would contribute a constant value of $-C\tau_d\tau$. The reason for this is that the integration of a unit impulse is a unit step, thus

$$\int_0^t \delta(t)dt = u(t) = 1$$

We may now write Eq. (34.51) in the form

$$\dot{m} = C(A + By + Am)$$

with

$$m(0) = -C\tau\tau_d = -K_c\tau_d/(\tau + K_c\tau_d)$$

The differential equations to be solved by the Runge-Kutta method now can be summarized

$$\dot{y} = (1/\tau)(1 + m - y) \tag{34.52}$$

$$\dot{m} = CA + CBy + CAm \tag{34.53}$$

with

$$y(0) = 0$$

$$m(0) = -(K_c\tau_d)/(\tau + K_c\tau_d)$$

Solving Eqs. (34.52) and (34.53) with the initial conditions given will produce a response for the control system of Fig. 34.10. The procedure for programming Eqs. (34.52) and (34.53) by use of the Runge-Kutta method is straightforward and will not be done here.

SIMULATION SOFTWARE

In the first part of this chapter, we have seen how one can write a digital computer program for the solution of a control problem. Even for the simple examples presented, there is considerable work in writing and debugging the program. A number of software programs have been written to solve dynamic problems, including process control problems. One of the earliest was CSMP developed by IBM. More recent programs include ACSL, TUTSIM, Simnon, and CC. The sources of these simulation programs are given in the list of references at the end of the chapter. Some of these programs provide blocks that simulate the basic transfer functions of process control such as integrator, first-order, second-order, lead-lag, and transport lag.

TUTSIM Simulation

One of these simulation programs to be discussed here is TUTSIM, which is distributed in North America by Tutsim Products (formerly Applied i) in Palo Alto, California. This program provides about eighty blocks, such as summer, integrator, gain, and transport lag. The use of this software is similar to the use of the analog computer in that computing blocks are selected and connected to one another in a manner similar to the connecting of analog computing elements by wires. The connection of the blocks, which is done with computer code in the software, is sometimes referred to as "softwiring." For the purpose of illustrating the solution of control problems by simulation software, only a few of the many blocks available in TUTSIM will be considered. A complete manual on TUTSIM is available from the distributor. (See references at the end of this chapter.) For use in our first example, the following TUTSIM blocks will be described: Summer, Integrator, Gain, and Pulse.

   SUM. The summer block, designated as SUM, is shown in Fig. 34.11a. The output U is the sum of the inputs. The sign of the inputs can be designated as plus or minus.
   INT. One of two types of integrator blocks available in TUTSIM, designated as INT, is shown in Fig. 34.11b. For this block, the output U is the sum of the inputs integrated with respect to the independent variable t. The initial condition of U, designated as l_c, is a parameter that may be assigned. The inputs can be labeled plus or minus.
   GAI. The gain block, designated as GAI, is shown in Fig. 34.11c. This block multiplies the sum of inputs by a gain P. The inputs can be labeled plus or minus.

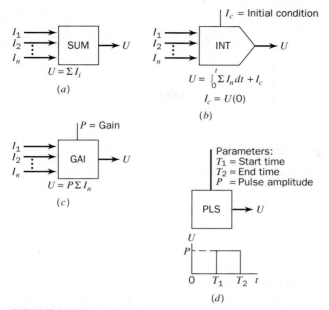

FIGURE 34-11
Some TUTSIM blocks: (*a*) Summer, (*b*) Integrator, (*c*) Gain, (*d*) Pulse.

PLS. The pulse function block, designated as PLS, is shown in Fig. 34.11*d*. This block provides a pulse of magnitude P starting at T_1 and ending at T_2. If T_1 is taken as 0.0 and T_2 is equal to or greater than the length of the run, one obtains a step function of magnitude P.

In the next example, we shall learn how one uses the TUTSIM blocks to solve a control problem.

Example 34.5. Simulation of proportional control with TUTSIM. Consider the control system shown in Fig. 34.12. To simulate this system by TUTSIM, we shall use the blocks shown in Fig. 34.11. To simulate the first-order lag, we first write

$$\frac{Y(s)}{M(s)} = \frac{1}{\tau s + 1}$$

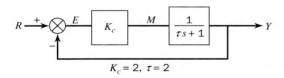

$K_c = 2, \ \tau = 2$

FIGURE 34-12
Block diagram for control system of Example 34.5.

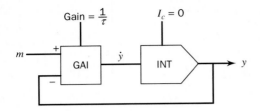

FIGURE 34-13
TUTSIM block diagram for a first-order system, $1/(\tau s + 1)$.

Cross-multiplication of this expression and solving for $sY(s)$ gives

$$sY(s) = \frac{-Y(s)}{\tau} + \frac{M(s)}{\tau}$$

The equivalent expression in the time domain is

$$\dot{y} = -\frac{y}{\tau} + \frac{m}{\tau} \tag{34.54}$$

We can obtain a TUTSIM diagram for this equation by using a gain block and an integrator block as shown in Fig. 34.13. In this figure, the gain block multiplies m by $1/\tau$ and $(-y)$ by $1/\tau$. The sum of these two signals is then integrated by the integrator. The operations performed by the software blocks of Fig. 34.13 match those in Eq. (34.54).

   With a block diagram for a first-order lag available (Fig. 34.13), we can now simulate the control system of Fig. 34.12. The result is shown in Fig. 34.14. In this figure, a gain block, no. 2, combines the function of the comparator and the proportional controller.

Set-up of model with TUTSIM software. The method for setting up a model with TUTSIM software is straightforward and the diagram of Fig. 34.12 will be used to illustrate the setup. After translating the control problem of Fig. 34.12 into a TUTSIM simulation diagram of Fig. 34.14, one enters into the computer through keyboard commands the following blocks of data:

   Model structure
   Model parameters
   Plotblocks and ranges
   Timing data

Each block of data will be described below in terms of Fig. 34.14.

Model structure. The model structure lists the types of computing blocks needed to solve the problem; a number is assigned arbitrarily to each block. The sources to the input of each block, with appropriate sign, are also listed. The format for the model structure data is

Format : Block number, Type, input 1, input2, . . .

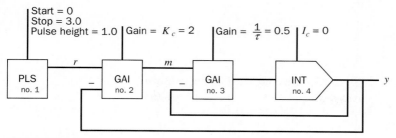

FIGURE 34-14
TUTSIM block diagram for Example 34.5.

For Fig. 34.14, the data to be entered into the computer are

1, PLS (Block 1 is a pulse generator, there are no inputs)

2, GAI, 1, −4 (Block 2 is a gain block, inputs are from blocks 1 and −4)

3, GAI, 2, −4 (Block 3 is a gain block, inputs are from blocks 2 and −4)

4, INT, 3 (Block 4 is an integrator, input is from block 3)

Model parameters. The model parameters for each block to be entered into the computer are entered with the following format.

Format : Block Number, parameter 1, parameter 2, ...

The number of parameters entered depends on the specific block. For Fig. 34.14, the parameter data take the form

1,0.0,3.0,1.0 (The pulse starts at 0, ends at 3.0, and is of magnitude 1.0)

2, 2.0 (The gain is 2.0)

3, 0.5 (The gain is 0.5)

4, 0.0 (The initial condition on the integrator is 0)

Plotblocks and ranges. In TUTSIM, one can plot up to four dependent variables, designated as Y1, Y2, Y3, and Y4, with a range specified for each variable. These variables appear on the vertical axis of the plot. One must also select the independent variable (usually time) on the horizontal axis, designated as HORIZ. Block 0 is reserved for a block that produces time. The choice of the variables and ranges is determined by the programmer. The format for this data is

Format : Block number, minimum, maximum

For Fig. 34.14, the following choices are made

HORIZ: 0, 0.0, 3.0 (Run varies from $t = 0$ to $t = 3.0$)

Y1: 4, 0.0, 1.0 (Y1 represents output from block 4; range is from 0 to 1)

Y2:   (Y2 not used)

Y3:   (Y3 not used)

Y4:   (Y4 not used)

Outputs Y2, Y3, and Y4 were not used in this example.

Timing Data. The timing data selects the step size and the length of the run, the format is

Format : Delta time, Final time

The step size of delta time must be chosen to fit the nature of the transient. The simulation starts at $t = 0.0$ and ends when t is equal to the final time. The ratio of final time/delta time determines the number of steps taken in the calculation and determines the number of points plotted. If this ratio is too small, numerical instability may occur because of inexact integrations. The TUTSIM manual recommends a ratio that is in the range of 500 to 5000.

A summary of the data used in setting up the model can be listed by keyboard command. For the system under consideration, this summary is shown in Fig. 34.15.

After the model is set up, it can be exercised by simple keyboard commands to produce numerical output and plots. The results for Fig. 34.14 using the data chosen is shown in Table 34.4 and Fig. 34.16.

After the model is set up, one can change easily the structure, the parameters, the choice of plots, and the timing by keyboard commands.

```
Model File: PCON.SIM
Date:      2 /     26 /   1989
Time:     10 :     24
Timing:    0.0010000  ,DELTA  ;   4.0000       ,RANGE
PlotBlocks and Scales:
Format:
      BlockNo,  Plot-MINimum,  Plot-MAXimum;  Comment
Horz:     0 ,     0.0000    ,     3.0000    ; Time
  Y1:     4 ,     0.0000    ,     1.0000    ;
  Y2:        ,              ,              ;
  Y3:        ,              ,              ;
  Y4:        ,              ,              ;


     0.0000           1 PLS
     3.0000
     1.0000
     2.0000           2 GAI      1     -4
     0.5000000        3 GAI      2     -4
     0.0000           4 INT      3
```

FIGURE 34-15
Summary of TUTSIM model for Example 34.5.

TABLE 34.4
Numerical output for Example 34.5, using TUTSIM

```
Model File: PCON.SIM
Date:       2 /     26 /  1989
Time:      10 :     25
Timing:    0.0010000  ,DELTA  ;   4.000       ,RANGE
PlotBlocks and Scales:
Format:
      BlockNo,  Plot-MINimum,  Plot-MAXimum;  Comment
Horz:      0 ,   0.0000     ,    3.0000     ; Time
  Y1:      4 ,   0.0000     ,    1.0000     ;
  Y2:           ,              ,            ;
  Y3:           ,              ,            ;
  Y4:           ,              ,            ;


      0.0000          0.0000
      0.2000          0.1727880
      0.4000          0.3007920
      0.6000          0.3956200
      0.8000          0.4658700
      1.0000          0.5179130
      1.2000          0.5564670
      1.4000          0.5850290
      1.6000          0.6061880
      1.8000          0.6218630
      2.0000          0.6334750
      2.2000          0.6420780
      2.4000          0.6484510
      2.5999          0.6531720
      2.7999          0.6566700
      2.9999          0.6592610
```

OTHER TUTSIM BLOCKS. The TUTSIM software can solve far more complicated systems than the one shown in Fig. 34.14. Many of the computer blocks were developed specifically for process control calculations, such as blocks that simulate a first-order lag, a second-order lag, a transport lag, a lead-lag transfer function, and a PID controller. There are also blocks (referred

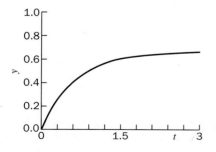

FIGURE 34-16
Plot of response from TUTSIM model for Example 34.5.

to as Z blocks) that are used to simulate sampled-data systems. There are also thermodynamic property blocks that provide the thermodynamic properties (enthalpy, temperature, heat capacity, etc.) of air, steam, and other substances. To comprehend the full range of the TUTSIM software, one should have access to the TUTSIM user's manual.

SUMMARY

Obtaining the response of a control system analytically can be very difficult, if not impossible, for high-order systems or for systems containing nonlinearities and transport lags. To study the effect of control strategies and controller parameters on the response of a complex control system, one must often use a computer simulation. Fortunately, we have today a number of simulation software packages for obtaining the response of control systems. The Appendix lists some of these software packages, along with their features and sources. If one does not have such simulation software, it is necessary to write one's own computer program. Even if one has such simulation software, it is still instructive to learn how to write some computer programs for the purpose of understanding the problems and limitations associated with commercially available simulation software. Examples of such problems are selecting the step size of the independent variable (Δt), establishing initial conditions, and providing sufficient storage for the simulation of transport lag.

In the first part of this chapter, the methods for writing computer programs were presented by means of four examples of control system. A computer routine for integration is at the center of any computer program for dynamic simulation. The fourth-order Runge-Kutta method was selected because of its accuracy and wide use. The literature on numerical analysis, of course, covers many other methods of integration, some of which are needed for difficult cases. The first step in obtaining a computer simulation based on numerical methods is to reduce the block diagram containing the transfer functions of the components to a set of first-order differential equations. This step is equivalent to obtaining a state-space model of the control system (see Chap. 28). Since the Runge-Kutta method is a marching solution, which starts at time zero and moves one time step with each cycle of computation, the selection of initial conditions must be considered. In some cases, the initial conditions are obvious, in other cases (e.g., the presence of derivative action in a controller), selection of the initial conditions are more subtle. The simulation of a transport lag is of great importance in the simulation of control systems. In the example involving a transport lag, a simple method using a stack (an array of computer storage locations) was used and it was shown that the size of the stack depends on the ratio of the transport lag to the step size ($\tau_d/\Delta t$).

In the second part of this chapter, the use of commercially available software for solving control problems was discussed. The program TUTSIM was presented and applied to some simple control problems. The use of TUTSIM is closely related to the block diagram of a control system in that a specific computer calculation block is used for each transfer function in the block diagram. Much of the other simulation software requires that the model be represented as a state-space model (i.e., a set of first-order differential equations.)

Since so many commercial simulation software packages for solving control problems are now being developed and distributed it is impossible to cover all of them in a book of this type. A listing is given in the appendix; however, the list is by no means complete and an entry in the list should not be considered as a recommended product. The field of computer applications changes so rapidly that the reader must keep abreast of developments in this area through technical journals and contacts with professional colleagues.

APPENDIX 34A
COMPUTER SOFTWARE
FOR PROCESS CONTROL

Many computer software packages are available for use in solving problems in process dynamics and control. A short selection of software packages that are useful for simulation of control systems is listed here. A more extensive list (17 packages) is given in Seborg, Edgar, and Mellichamp (1989). Since software changes occur frequently in terms of version and cost, the reader is advised to write to the vendor for the most recent information.

TUTSIM. TUTSIM is a computer simulation program that provides a numerical and graphic representation of linear or piecewise linear systems. It can also handle nonlinear functions. A problem is solved by constructing a TUTSIM model consisting of interconnected blocks that match the block diagram of the control system. The block diagram for the model resembles an analog computer diagram, but all the computations are done numerically by the digital computer. Once a TUTSIM model has been created, it is very easy to change its structure and parameters. The output on the screen of the monitor can display four process variables versus time. The blocks for continuous control include the usual ones, such as first-order, second-order, lead-lag, and transport lag. TUTSIM also provides "Z" blocks for use in sampled-data systems. TUTSIM, which contains about eighty different blocks, is available in three versions:

    short version, 15 blocks allowed per model, $35
    collegiate version, 35 blocks allowed per model, $150
    professional version, 999 blocks allowed per model, $500

There are special prices for academic use. A student version is available for $28.00. TUTSIM was described in detail in this chapter. TUTSIM was adapted from Twente University of Technology in the Netherlands. TUTSIM is available for the IBM-PC computers and their compatibles. Information can be obtained from:

TUTSIM Products
200 California Avenue, Suite 212
Palo Alto, CA 94306

PROGRAM CC. Program CC is a simulation software package for the analysis and design of linear control systems. Linear systems are represented in the program by transfer functions that can be continuous (Laplace transforms) or sampled-data (Z-transforms). The linear systems can also be represented by state-space equations. Approximately sixty commands are available and include transient response, root-locus plots, frequency analysis (Bode, Nyquist, Nichols), and conversion between transfer function and state-space systems. CC software is available for the IBM-PC computers and their compatibles. A student version, which is a subset of a much larger package, is available at the suggested price of $95.00. Information on CC can be obtained from:

Systems Technology, Inc.
13766 S. Hawthorne Blvd.
Hawthorne, CA 90250

ACSL. ACSL (Advanced Continuous Simulation Language) simulates the dynamic response of physical systems through the integration of differential equations. This software is used for very complex systems containing a large number of independent variables (states). A variety of integration algorithms (Adams-Moulton, Euler, Runge-Kutta, etc.) can be selected. The software has a discrete section for simulating sampled-data systems. Many plot commands are available for displaying the output in a variety of graphical forms. This software is available at a discount to educational institutions. ACSL is available for mainframe computers and also for IBM-AT computers and their compatibles. Information can be obtained from:

Mitchell and Gauthier Associates
73 Junction Square Drive
Concord, MA 01742-9990

SIMNON. Simnon is a general-purpose software for simulation of linear and nonlinear systems, operating in continuous time or discrete (e.g., sampled-data) time. Models containing up to three hundred states can be simulated. The software is available for a VAX computer or for IBM-AT computers and their compatibles. The IBM version of the software is available to educational institutions for $350. This software, which was developed in the Department of Automatic Control at the Lund Institute of Technology in Lund, Sweden, is available in North America from

Engineering Software Concepts
436 Palo Alto Ave.
P.O. Box 66
Palo Alto, CA 94301

PROBLEMS

34.1. The following differential equation is to be solved by digital computation:

$$\frac{dy}{dt} = 2t - 1.5y \qquad y(0) = 0$$

A portion of a computer program, which uses the Runge-Kutta method, is shown below:

    25 DY (T,Y) = 2*T − 1.5*Y
    26 DT = 0.1
    27 Y = 0.0
    28 T = 0.0
    29 K1 = DY (T,Y)*DT
    30 K2 = DY (T + DT*.5, Y + K1*.5)*DT
    31 K3 = DY (T + DT*.5, Y + K2*.5)*DT
    32 K4 = DY (T + DT*.5, Y + K3*.5)*DT
    33 Y = Y + (K1 + 2*K2 + 2*K3 + K4)*DT/6
    34 T = T + Δ T
    35 PRINT T,Y
    36 IF (T.LT.2.) GO TO 28
    37 STOP
    38 END

(a) Indicate any errors you find in this program by noting the statement number of the line where it appears; also describe the error and correct it if you can.

(b) Do one cycle of calculation by hand using the Runge-Kutta method and obtain the value of $K1$, $K2$, $K3$, and $K4$ for use in getting Y at $t = 0.1$.

(c) Also obtain Y at $t = 0.1$ by using the Runge-Kutta method.

34.2. The control system shown in Fig. P34.2 is to be simulated by digital computation.

A portion of a computer program, which uses the Runge-Kutta method, is shown below.

    24 DY (T,Y) = (1 + KC)*DT − KC*Y
    25 KC = 2.0
    26 DT = 0.1
    27 Y = 1.0
    28 T = 0.0
    29 K1 = DY(T,Y)*DT
    30 K2 = DY(T + DT, Y + K1)*DT
    31 K3 = DY(T + DT*.5, Y + K2*.5)*DT
    32 K4 = DY(T + DT*.5, Y + K3*.5)*DT
    33 Y = Y + (K1 + 2*K2 + 2*K3+K4)*DT

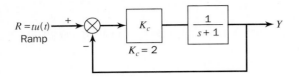

$R = tu(t)$ — Ramp

K_c

$\dfrac{1}{s+1}$

Y

$K_c = 2$

```
34 T = T +ΔT
35 PRINT T,Y
36 IF (T.LT.2.) GO TO 27
37 STOP
38 END
```

(a) Indicate any errors you find in this program by noting the statement number of the line where it appears; also describe the error and correct it it you can.

(b) After correcting the program, do one cycle of calculation by hand using the Runge-Kutta method and obtain the value of $K1$, $K2$, $K3$, and $K4$ for use in getting Y at $t = 0.1$.

34.3. The step response of the following differential equation is to be obtained numerically with the aid of a digital computer.

$$\frac{d^2y}{dt^2} + 0.8\frac{dy}{dt} + y = 1$$

$dy/dt = 0 \quad$ and $\quad y = 0$ at $t = 0$

Integration step sizes (Δt) of 0.1, 0.5, and 1.0 are to be used.

(a) Which of the step sizes will give a numerical solution closest to the analytical solution?

(b) Which step size will require the least computation time?

(c) If it is possible to get an impulse response for the above differential equation, show how you would provide for it in solving the differential equation by the computer.

34.4. Write a computer program in BASIC to simulate the response of the PID control system of Example 34.4 for a unit-step change in load ($U = 1/s$) for the case of $K_c = 2.0$, $\tau_I = 1$, $\tau_d = 1$, and $\tau = 2$.

MICROPROCESSOR-BASED CONTROLLERS AND DISTRIBUTED CONTROL

In this chapter, some of the highlights of modern industrial microprocessor-based controllers and distributed control systems will be presented. A microprocessor-based controller is essentially a digital computer programmed to perform the function of a process controller. For our purpose, the term microprocessor is synonymous with computer and we could refer to a microprocessor-based controller as a computer-based controller. The number of features of these modern controllers is far too great to cover in one chapter. The best way for the reader to acquire some experience with modern controllers is through laboratory and plant use and by attending some of the short courses offered by the major suppliers of the equipment.

HISTORICAL BACKGROUND

During the past fifty years tremendous development has occurred in process control hardware. The three phases of development are pneumatic control, electronic control, and microprocessor-based control. During the 1940s, the predominant controller was pneumatic, meaning that signals to and from the controller and within the controller mechanism were air-pressure signals that usually varied from 3 to 15 psig. The development of the high-gain operational electronic amplifier during the second world war led to the development of the electronic controller and also the analog computer. The electronic controller mimicked the control functions of the pneumatic controller. It also provided some improvements, such as accurate and reproducible control parameter settings and reduction in size of the instruments. In contrast, the pneumatic controller required frequent calibration of the knobs used to set the various controller parameters (K_c, τ_I, τ_D). The pneumatic controller had interaction among the control modes and had inherent lags that became significant at high-frequency operation. There were frequent debates over the pros and cons of pneumatic and electronic controllers. For example, the pneumatic controller was rugged, simple to install, and required little maintenance. Only a source of air pressure was needed to operate the controller. There was initially great concern about the possibility of explosions with the use of electronic controllers, so the instrument cases for these controllers were purged with steady streams of air when used in plants producing flammable substances. The maintenance of electronic controllers also required highly trained technicians.

In the 1960s, the chemical industry made its first attempt at computer process control. These control systems used large mainframe computers, for which the control programs had to be written from scratch. The first attempts at computer control were met with mixed reactions. In the 1970s, there appeared on the market the first generation of digital control hardware, which was based on the advances in microprocessor-based technology. This equipment was user friendly and all the software accompanied the hardware. The operator did not face the problem of writing computer code to implement the control functions; it was only necessary to learn the instructions needed to configure (set up) the controllers.

HARDWARE COMPONENTS

The hardware requirements for pneumatic, electronic, and microprocessor-based controls are shown in Fig. 35.1. In this figure, all of the components are obtained from a manufacturer of control equipment; several of the components are common to the three systems. In Fig. 35.1a, all the signals are pneumatic (3–15 psig). The energy needed to operate these pneumatic components is a source of clean, dry air at a pressure of about 20 psig. The pressure can vary from 20 psig by about $\pm 10\%$ without adversely affecting the operation of the instruments.

The electronic system shown in Fig. 35.1b requires both electrical and pneumatic power to operate the components. A transducer or converter is needed between the controller and the valve to convert current (4–20 ma) to pressure (3–15 psig).

The components for a microprocessor-based system are shown in Fig. 35.1c. In this case, the control algorithm resides as a computer program in the memory of the computer. The operator communicates with the control system with a keyboard, a monitor, and a printer. The computer can perform many more functions than implementation of the control algorithm as will be discussed later. The recorder of the pneumatic or electronic system is replaced by a monitor screen on which the transients are shown.

In a modern controller both analog and digital signals are processed. The analog signal is the type that represents a continuous variable that varies over a range of values. The digital signal is a binary signal that can be represented by two states (*on, off,* or logic 1, logic 0, etc.). Examples of analog signals are the measurement from a temperature transmitter or the signal sent to a valve. Examples of digital signals are the output to a motor, which causes it to be on or off, or the output to an alarm light causing it to be on or off. The focus

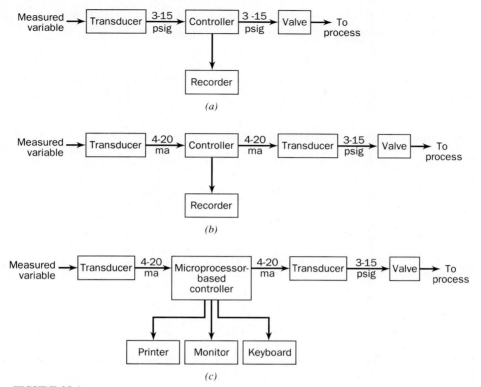

FIGURE 35-1
Controller components for (*a*) pneumatic control, (*b*) electronic control, (*c*) microprocessor-based control.

of this book has been on analog signals that are applicable to continuous control systems. However, there is an important area of control called *batch control,* which is receiving more attention in industry. Batch control, as the name suggests, is the control of processes that are done in a batch operation. Many examples of batch processing occur in the pharmaceutical industry where small amounts of products of high unit cost are produced.

TASKS OF A MICROPROCESSOR-BASED CONTROLLER

The primary task of a microprocessor-based controller is implementation of a control algorithm; however, the presence of a computer makes it possible to assign a number of peripheral tasks that are useful in process control. Some of these tasks provided in a modern control system are to:

Implement classical and advanced control algorithms
Provide static and dynamic displays on the monitor

Provide process and diagnostic alarms

Provide mathematical functions

Provide data acquisition and storage (archiving)

The software to support all of these tasks is supplied by the manufacturer of the control equipment. We shall now look briefly at the nature of each task.

Implementation of Control Algorithms

The portion of the software that covers this task is organized into large numbers of blocks that can be connected together to solve a specific control problem. A partial listing of the blocks typically provided are as follows:

analog input

analog output

conventional control algorithms (P, PI, PD, PID)

linearization

lead lag

dead time

self-tuning

There are many other blocks that have been omitted from this list because of the limitation of space in this chapter. There are also a number of blocks that process digital (or logic) signals (on/off) such as comparators, selectors, or timers, which are needed in batch control and automatic plant start-up and shut down.

ANALOG INPUT BLOCK. The analog input block is an analog-to-digital device that converts a continuous signal from a transducer, which is in the form of a current or voltage, to a digital signal that can be used in the microprocessor.

ANALOG OUTPUT BLOCK. The analog output block reverses the operation of the analog input block by converting a digital signal, which has been computed in the microprocessor, to a voltage or a current that can be sent out to a transducer in the process in the field. Sometimes this block is called a field output block.

CONTROL BLOCK. The control block is a block for which many parameters can be specified. The manufacturer does not give any information on the method of implementing the control algorithms; however, the reader who has read the section of Chap. 27 on the design of conventional control algorithms [$D(z)$] will have some idea on how the signals are manipulated within the microprocessor to implement the desired control action. The sampling period T is one parameter that cannot be adjusted in a commercial controller; it is fixed by the developer of the software. In most of the operating manuals provided with the control equipment, the sampling time may not be mentioned. Typical values of T in commercial controllers vary from 0.1 to 0.25 sec. A controller operating with such a small T can be considered as a continuous controller for many chemical processes with large time constants. Parameters that can be selected are the controller parameters (K_c, τ_I, τ_D), limits on set point and controller output, and others.

LINEARIZATION BLOCK. The linearization block is used to "straighten out" a nonlinear relation. The most common example of the need for this block is in processing a signal from an orifice plate used to measure flow. The signal (pressure) across an orifice plate is proportional to the square of the flow. To obtain a linear relation between flow rate and signal, the signal is sent through a linearization block, which has been configured to extract the square root of the input signal. The linearization block can also be configured to linearize any nonlinear relation that can be plotted on a coordinate system. This aspect of the linearization block can be useful for linearizing the input-output relation to a valve that is nonlinear in behavior. In Chap. 20, an equal percentage valve was proposed as a device to linearize the relation between flow and valve-top pressure when line loss was large.

LEAD-LAG BLOCK. The lead-lag block simulates the lead-lag transfer function, $K(T_1 s + 1)/(T_2 s + 1)$. The parameters K, T_1, and T_2 can be selected over a wide range of values. If one needs a first-order lag, T_1 can be set to zero. We have seen the need for the lead-lag block in feedforward control in Chap. 18.

DEAD-TIME BLOCK. The dead-time block simulates dead time (or transport lag), $e^{-\tau_d s}$. For this block, τ_d can be selected over a wide range of values. We have seen the need for this block in the Smith predictor control algorithm of Chap. 18. The nature of the computer program needed to simulate the dead-time block was presented as an example in Chap. 34.

Figure 35.2 shows a simple flow example using some of the blocks just described. The blocks are connected together by computer code at a keyboard during the configuration of the control system. This connection of blocks is called *softwiring* since it is done through the use of computer software. The actual connection between the flow transmitter and the analog input block in the controller, which is made with wires, is called *hardwiring*.

SELF-TUNE BLOCK. For years, one of the goals of control engineers has been to develop a device that would automatically tune a controller, on-line, while the process is operating. Until recently, this goal was reached for some special cases by the application of adaptive control theory, a branch of control that is beyond the scope of this book. Recently (mid-1980s) a commercial device became available that uses the normal transients occurring in a controlled process (caused by set-point and load upsets) to update the control parameters of a PID controller. This device is called a self-tuner and is one of the blocks available in the microprocessor-based controller of several hardware manufacturers. When the self-tuner is first applied to a process for which no process identification has been performed, the self-tuner is placed in the pre-tune phase, during which time the

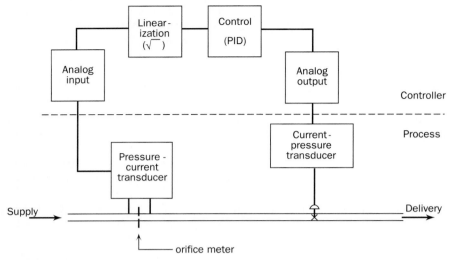

FIGURE 35-2
Example of the use of control blocks to control a flow process.

process is subjected to a pulse while it is operated open-loop. The introduction of the pulse and the analysis of the transient is done automatically by the self-tuner. The outcome of the pre-tune phase of operation is the selection of controller parameters. A conceivable approach to the development of the pre-tune phase of a self-tuner is to monitor an open-loop step response and apply a tuning method similar to the Cohen-Coon tuning method of Chap. 19. After the pre-tune phase, the control system is returned to closed-loop and the self-tuner continues to monitor the transients and make changes in controller parameters when needed. The self-tuning that occurs during closed-loop operation is based on the characteristics of the transients, such as decay ratio, overshoot, period of oscillation, etc. The self-tuning algorithm, being proprietary information, is described in only a general manner in the reference manual that accompanies the control equipment. Since many industrial processes are poorly tuned, the general-purpose self-tuner represents an impressive achievement in the application of a digital computer to control technology. The reader may consult the paper by Krause and Myron (1984) to obtain more information on the development of the EXACT self-tuner of the Foxboro Co. The term EXACT stands for expert automatic control tuning and suggests that the method of tuning is based on a branch of artificial intelligence known as "expert system" design.

Displays

The software in a modern controller has made the strip chart recorder almost unnecessary. Through the use of skilled programming, the transients (or trends) produced in a control system can be displayed on a monitor screen dynamically. As time progresses, the values of selected variables are displayed as a function of time. The segment of time shown on the screen can be selected to be a few minutes to a few hours to show dynamic detail or long-term trends. Transients that occurred in the past can be stored and displayed again.

Many process operators are more comfortable with control instruments that have a faceplate which shows bar graphs or pointers indicating set point, control variable, and output to the valve. In the older instruments, those indicators were obtained by use of mechanical motion or other means. The software provided with a microprocessor-based controller can be used to obtain a dynamic display on the screen that mimics the faceplates of traditional instruments.

Alarms

An extensive amount of the software in modern controllers is devoted to detecting and reporting a problem in the form of an alarm. The alarm takes the form of a visual signal (flashing light), an audible signal (beeping horn), or the actuation of a switch. Examples of the use of switch closures include turning on or off a pump motor or opening or shutting a valve. The alarms are classified as *process alarms* and *diagnostic alarms*. The diagnostic alarm detects a malfunction in the control equipment or the loss of communication. For example, if a wire connecting the output of a temperature transmitter to an analog input block breaks, a diagnostic alarm would go off indicating that the signal to the analog input block is out of range. The manufacturer of the control equipment provides all the software for detecting the problems that trigger diagnostic alarms.

The engineer who configures the control blocks selects the variables that are to trigger process alarms and specifies the alarm limits and the type of annunciation (flashing light, beeping horn, etc.) The alarms can be assigned a priority rating. Those variables in a process that are most critical are given the highest priority; less critical variables are given a lower priority.

Mathematical Functions

The software provides basic mathematical functions such as summation, subtraction, multiplication, and accumulation (i.e., integration). These functions can be used along with other blocks in the design of a control system. A simple example of these functions is the calculation of mass flow rate of a gas from measurements of velocity, pressure, and temperature. These three measurements are combined according to the following relationship, which is based on the ideal gas law:

$$w = vAPM/RT$$

where w = the mass flow rate, mass/time

$\quad\quad v$ = the velocity

$\quad\quad P$ = pressure

$\quad\quad T$ = absolute temperature

$\quad\quad M$ = molecular weight of the gas

$\quad\quad R$ = the gas constant

$\quad\quad A$ = cross-sectional area for flow

The signal from the math block that represents w can then be sent as the control variable to a control block that controls the mass flow of gas.

Data Acquisition and Storage

Long-term storage of the transients can be obtained easily with a digital computer. This task is referred to as *archiving*. The automatic storage of critical process-control variables on disk or tape can be retrieved later to explain process operating difficulties. The computer can also be used to automatically record or log the type and location of an alarm, the time of a process alarm, the time of acknowledgment of an alarm, and the time it was cleared by operator intervention: this information is useful to supervisors in detecting violation of safety regulations or process malfunctions.

SPECIAL FEATURES OF MICROPROCESSOR-BASED CONTROLLERS

In addition to the tasks just described, there are three special features available in modern microprocessor-based controllers that deserve attention. These are limiting, tracking, and anti-reset windup. Each will be discussed separately.

Limiting

In configuring a control system from basic control blocks, one can select lower and upper limits on controller output and set point. These limits are narrower than the limits inherently present in the hardware. Limits are often placed on a controller output for safety reasons or to protect equipment. For example, if one knows the flow rate of a liquid that causes a tank to overflow, one can set the limit on the output of a controller at a value less than the value that causes overflow. The limits on the controller output are active when the controller is in either automatic or manual mode. An example of a limit on set point is the selection of an upper limit on pressure for a steam-heated sterilizer to prevent damage to the equipment.

Tracking

A very useful feature of a microprocessor-based controller is tracking. Although tracking is not needed to successfully control a system, its presence is of great convenience to the process operator. Two examples of tracking are set-point tracking and controller-output tracking.

Set-point tracking is useful when a controller is transferred from manual to automatic. When a process is started up for the first time, a common procedure is to bring the process on-stream in manual mode. In this case the operator adjusts the output of the controller (which goes to the valve) until the process variable comes to a desired steady state. When the tracking feature is not present in the controller, the set point must be manually adjusted until it equals the process variable before the controller is transferred to automatic; the process then continues running in a smooth manner. If the operator adjusts the set point to the process variable after switching to automatic, there may be a temporary disturbance in the process variable. The expression for the disturbance is called a "bump." With set-point tracking, the operator does not need to think about adjusting the set point to the process variable, because it is done automatically. In other words, set-point tracking provides "bumpless" transfer when switching from manual to automatic.

A second example of tracking can be seen in its use for transferring a cascade system from manual to automatic. (The reader should be familiar with the information on cascade control provided in Chap. 18 to understand this example.) To explain the use of tracking in cascade control, reference to Figs. 18.1b and 18.2b will be made. In starting up this system, the primary controller is placed in a stand-by condition and the secondary controller is placed in manual mode. The means for accomplishing this is built into the software of the controller. With the secondary controller in manual mode, its output is adjusted until the temperature of the tank contents (T_0) is at the desired value. Then, with the control system at steady state and T_0 at the desired value, the system is transferred to cascade mode by placing both controllers in automatic. Since the output of the primary controller adjusts the set point of the secondary controller, it is necessary to have the output of the primary controller equal to the jacket temperature (T_j) when the system is transferred to cascade mode. This goal can be achieved by having the output of the primary controller track the jacket temperature while the secondary controller is used in manual mode to adjust the tank temperature to the desired value. For this example, the set point of the primary controller can also automatically track the tank temperature (T_0) before the transfer to cascade mode occurs. In this cascade control example, we have seen tracking used for both the set-point and the controller output.

Anti-Reset Windup

A troublesome problem with a controller having integral action (PI or PID) is the possible occurrence of reset windup. When the error to a controller remains large for a long time, the integral action of the controller builds up a large value of output which often approaches the saturation value of the controller output. This accumulation of output is called *reset windup*. When the process variable returns to the set point, the output of the controller does not immediately return to a value that will hold the process variable at the set point because the controller output has built up (or has been wound up) and must be reduced by the presence of error of opposite sign over some duration of time. Thus the transient for the control variable exhibits a large overshoot that can persist while the output signal is being reduced through integral action being applied to the error of reversed sign.

Reset windup typically occurs during the start-up of a process. To gain some insight into the cause of reset windup, consider the start-up of the liquid-level process shown in Fig. 35.3 in which the level in the third tank is to be controlled by a PI controller. The valve is linear and saturates at 0 and at 0.5 as shown in

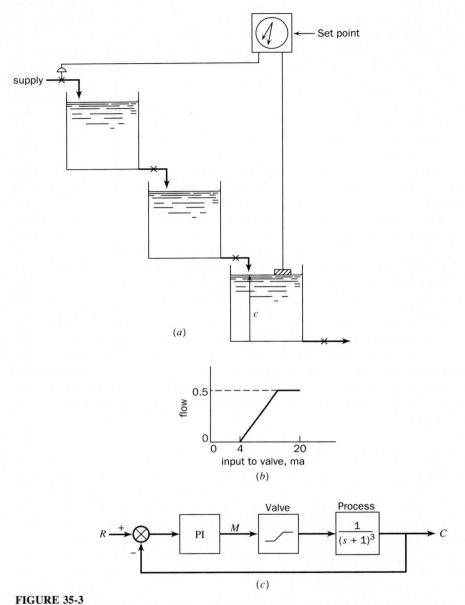

FIGURE 35-3
Plant start-up illustrating reset windup (tanks are initially empty): (*a*) process, (*b*) linear valve with saturation limits, (*c*) block diagram of process.

Fig. 35.3*b*. Upon start-up with the PI controller in automatic mode, the tanks are empty, and the error $(R - C)$ is large and positive. The action of the controller on this error will result in a large output M due to proportional action and a rising contribution to M due to the integral action. The output of the controller will be at its saturation value, which is typically about 10% above the top of the 4 to 20 ma scale (i.e., 22 ma). The large saturated value of M will in turn cause the valve to reach its saturation value, which has been taken as 0.5. During the initial phase of the operation, the tanks are being filled at the maximum rate of flow provided by the upper limit of the control valve. During this filling stage of operation the controller is not exercising any control since the valve is at its limit. As the level rises toward the set point, the large error that existed at start-up gradually diminishes toward zero. If only proportional action were present in the controller, the output of the controller would return quickly to a mid-scale value; however, because of the integral action, the controller output remains high, at its saturation value, long after the process variable first reaches the set point. To reduce the output M, the integral action must be applied to negative error so that the integration can lower the output to mid-scale. This negative error occurs as a result of the tank level remaining *above* the set point for some time after the tank level reaches the set point. Other causes of reset windup and some methods to prevent it are discussed by Shinskey (1979).

The control system shown in Fig. 35.3*c* was simulated for a start-up transient with the tanks initially empty; the transient is shown as Curve I in Fig. 35.4. The large overshoot in tank level after the level reaches the set point is clearly illustrated. Now that the problem of reset windup has been described, we focus our attention on how to reduce or eliminate it. The development that follows on the use of external feedback to eliminate reset windup is based on the work of Shunta and Klein (1979).

A feature of microprocessor-based controllers is the availability of external feedback in the configuration of a PI or PID controller. The block diagram of a PI controller with external feedback is shown in Fig. 35.5. The output of this controller is given by

$$M(t) = K_c e(t) + \frac{K_c}{\tau_I} \int_0^t e(t)dt + \frac{1}{\tau_I} \int_0^t [F(t) - M(t)]dt \qquad (35.1)$$

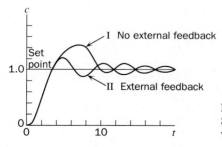

FIGURE 35-4
Start-up transients for system in Fig. 35.3 with and without external feedback.

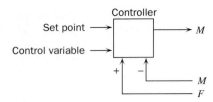

FIGURE 35-5
Controller with external feedback for use in anti-reset windup.

where $M(t)$ = controller output

$e(t)$ = error = setpoint − control variable

$F(t)$ = external feedback signal

If the Laplace transform of both sides of Eq. (35.1) is taken, the result is

$$M(s) = K_c e(s) + \frac{K_c e(s)}{s\tau_I} + \frac{1}{s\tau_I}[F(s) - M(s)] \tag{35.2}$$

If the feedback signal is the controller output, $F(s) = M(s)$, Eq. (35.2) becomes the usual transfer function for a PI controller:

$$M(s) = K_c \left(1 + \frac{1}{\tau_I s}\right) e(s) \tag{35.3}$$

The feedback signal $F(t)$ can be any signal available to the microprocessor-based controller. When $F(t)$ is not equal to $M(t)$, Eq. (35.2) can be solved for $M(s)$ to give

$$M(s) = K_c e(s) + \frac{F(s)}{\tau_I s + 1} \tag{35.4}$$

A controller following this equation provides a signal consisting of proportional action plus first-order tracking of $F(t)$. If $F(t)$ in Eq. (35.1) is taken as the output of the valve (or the output signal of the current-to-pressure transducer that goes to the valve) in our example in Fig. 35.3c, we have the basis for eliminating reset windup. During the filling stage of the tank, the feedback signal $F(t)$ will be constant at the saturation value of the valve output. When the tank level reaches the set point, the error will be zero and the only contribution from the controller output will be the tracked signal represented by the second term on the right side of Eq. (35.4). This value will be less than would be the case if external feedback were not employed. The overall result is that the controller output is less with the external feedback at the time the level first equals the set point and the overshoot is reduced. The transient using external feedback is also shown in Fig. 35.4 as Curve II. Notice that the overshoot is less when external feedback is used. To emphasize the benefit of external feedback for eliminating reset windup, no limits were placed on the output of the controller in the simulation of Fig. 35.3. In practice, there are physical limits on the controller output, and when this is the case, the reduction of overshoot with the use of external feedback may not be so pronounced as shown in Fig. 35.4.

DISTRIBUTED CONTROL

So far we have been concerned in this chapter with the operation of a single controller. Such a controller is referred to as a *stand-alone controller* because it is not communicating with other controllers, but only with the one control loop of which it is a part. Present-day microcomputer-based control systems have the capability of communicating with other controllers through a network, which is called *distributed control*. Figure 35.6 shows one version of the communication linkages that are usually present in a distributed control system. Each manufacturer of distributed control systems has a different way of organizing them.

A distributed control system is intended to be used for a large processing facility that involves as many as fifty to one hundred loops. Examples include a refinery, a brewery, a power plant, and the like. In Fig. 35.6, the modules of control equipment that communicate with each other are as follows.

> Control processor (CP)
> Applications processor (AP)
> Workstation (WS)
> Fieldbus module (FBM)

The first three of these modules communicate with each other through a nodebus or "data highway," as it has been called. The fieldbus modules serve as devices that interface with transducers and valves in the process.

The control processor contains the blocks described earlier (analog input, analog output, control, linearization, etc.) that are connected together by soft-wiring to provide the control algorithm required for each loop. Communication between the control processor and the process (a distance away) in the field takes place in the fieldbus module. Two types of fieldbus modules are available. One type provides a set of analog inputs and a set of analog outputs that send to and receive from the field continuous signals (4–20 ma). The other type of module

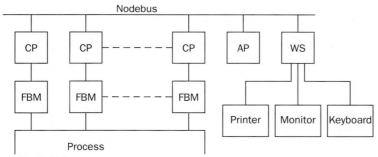

FIGURE 35-6
Typical connections in a distributed control system: CP: control processor, AP: applications processor, WS: workstation, FBM: fieldbus module.

sends to and receives from the field digital signals that often take the form of switch-contact closures.

The application processor is a microprocessor (or computer) in which the programs (or software) are stored for performing the many tasks described earlier and for managing the communication among modules.

The workstation module is connected to a keyboard, a mouse, a monitor, and a printer for use by process operators to interact with the system. At the workstation, the process operator can call up on the screen various displays, change set points and controller parameters, switch from automatic to manual, acknowledge alarms, and perform other tasks needed to operate a control system consisting of many loops. A control system can also be configured as an off-line task at the workstation. After configuration, the configured control system is downloaded to the control processor. If necessary, more than one workstation can be attached to the nodebus in order to provide communication at several locations in a plant. If more than one workstation is used, only one of them should have the authority at a given time to be in charge of the control system.

SUMMARY

During the past 15 years, the computer has greatly changed the nature of industrial process control equipment. The microprocessor has become the heart of control instruments and the computer programs stored in the memory of the hardware have provided many functions besides the basic control algorithm. When the pneumatic controller was the predominant type, one purchased a controller with very specific attributes (e.g., mode of control, type of measured variable, chart speed, etc.). The microprocessor-based control instruments available today contain not only the conventional control algorithms, but many other functions such as simulation of basic transfer functions (e.g., lead-lag and transport lag), display-building, mathematical functions, process and diagnostic alarms and data acquisition. The modern instruments also provide logic functions (comparators, timers, counters, etc.) for use in batch control and plant start-up and shutdown. Recently, self-tuning algorithms have been added to the microprocessor-based instruments.

In this chapter, some of the special features of modern controllers were discussed (e.g., limiting, tracking, and anti-reset windup). Any controller having integral action can cause reset windup under certain conditions when the error persists for a long time. The result of such a phenomenon is a transient that has large overshoot. Manufacturers of control instruments now offer several methods for reducing reset windup; the one presented in this chapter was use of external feedback.

Before computer control appeared, most process loops were served by individual controllers with signals to and from these controllers being collected on a large panel board in a special control room. To obtain communication between the control room and the controllers required much wiring and piping (for pneumatic systems). Today, microprocessor-based control systems have the capability of communicating with other control instruments through networks, called distributed control, with the result that much of the hardwiring used in the older systems is done within the computer. Such internal computer connections are called softwiring because the connections are made through software. A distributed control system can control an entire plant and involve as many as one hundred or more control loops. Since each manufacturer has a different way of organizing a distributed control system, the practicing engineer must obtain the details of a particular system from the manufacturer. Most manufacturers offer a variety of short courses for technicians and engineers on the installation and use of their hardware and software.

BIBLIOGRAPHY

Anton, H. (1984). *Elementary Linear Algebra*, 4th ed., New York: Wiley.

Aris, R., and N. R. Amundson (1958). *Chem. Eng. Sci., 7*, 121–155.

Bennett, C. O., and J. E. Myers (1982). *Momentum, Heat, and Mass Transfer*, 3rd ed., New York: McGraw-Hill.

Bergen, A. R., and J. R. Ragazzini (1954). "Sampled-Data Processing Techniques for Feedback Control Systems." *Trans. AIEE, 73,* part 2, 236.

Carslaw, H. S., and J. C. Jaeger (1959). *Conduction of Heat in Solids*, 2nd ed., Oxford at the Clarendon Press.

Churchill, R. V. (1972). *Operational Mathematics*. 3rd ed., New York: McGraw-Hill.

Churchill, R. V., and J. W. Brown (1986). *Fourier Series and Boundary Value Problems*, 4th ed., New York: McGraw-Hill.

Cohen, G. H., and G. A. Coon (1953). *Trans. ASME, 75,* 827.

Cohen, W. C., and E. F. Johnson (1956). *IEC, 48,* 1031–34.

Coughanowr, D. R., and L. B. Koppel (1965). *Process Systems Analysis and Control*. New York: McGraw-Hill.

Eckman, D. P. (1958). *Automatic Process Control*. New York: Wiley.

Evans, W. R. (1948). "Graphical Analysis of Control Systems." *Trans. AIEE, 67,* 547–551.

Evans, W. R. (1954). *Control-system Dynamics*. New York: McGraw-Hill.

Foxboro Co. (1978). "Principles of Feedforward Control." Audio-visual tape, No. B0150ME, (1hr, 23 min; 3 parts), Foxboro, MA.

Graham, D., and D. McRuer (1961). *Analysis of Nonlinear Control Systems*. New York: Wiley.

Hougen, J. O. (1964). *Experiences and Experiments with Process Dynamics*. CEP Monograph Series, *60*, No. 4.

Hovanessian, S. A., and L. A. Pipes (1969). *Digital Computer Methods in Engineering*. New York: McGraw-Hill.

Iinoya, K., and R. J. Altpeter (1962). "Inverse Response in Process Control." *IEC, 54* (7), 39–43.

Jury, E. I. (1964). *Theory and Application of the Z-Transform*. New York: Wiley.

Krause, T. W., and T. J. Myron (June 1984). "Self-Tuning PID Controller uses Pattern Recognition Approach." Control Eng.

Kuo, B. C. (1987). *Automatic Control Systems*, 5th ed., Englewood Cliffs, NJ: Prentice-Hall.

Lees, S., and J. O. Hougen (1956). *Ind. Eng. Chem., 48,* 1064.

Letov, A. M. (1961). *Stability in Nonlinear Control Systems.* Princeton, NJ: Princeton University Press.

Lopez, A. M., P. W. Murill, and C. L. Smith (1967). "Controller Tuning Relationships Based on Integral Performance Criteria." *Instrumentation Technology, 14*, No. 11, 57.

Mickley, H. S., T. K. Sherwood, and C. E. Reed (1957). *Applied Mathematics in Chemical Engineering*, 2nd ed., New York: McGraw-Hill.

Morari, M., and E. Zafiriou (1989). *Robust Process Control.* Englewood Cliffs, NJ: Prentice Hall.

Mosler, H. A., L. B. Koppel, and D. R. Coughanowr (1966). "Sampled-data Proportional Control of a Class of Stable Processes." *Industrial and Engineering Chemistry Process Design and Development, 5,* 297–309.

Mosler, H. A., L. B. Koppel, and D. R. Coughanowr (1967). "Process Control by Digital Compensation." *AIChE J., 13,* No. 4, 768–78.

Nobbe, L. B. (1961). "Transient Response of a Bubble-Cap Plate Absorber." M. S. Thesis, Purdue University, Indiana.

Oldenbourg, R. C., and H. Sartorius (1948). *The Dynamics of Automatic Controls, Trans.* ASME, p. 78.

Perry, R. H., and C. H. Chilton (1973). *Chemical Engineers' Handbook*, 5th ed., New York: McGraw-Hill.

Ragazzini, J. R., and G. F. Franklin (1958). *Sampled-Data Control Systems.* New York: McGraw-Hill.

Richards, R. J. (1979). "An Introduction to Dynamics and Control." London and New York: Longman.

Routh, E. J. (1905). *Dynamics of a System of Rigid Bodies.* Part II. London: Macmillan & Co., Ltd.

Seborg, D. E., T. F. Edgar, and D. A. Mellichamp (1989). *Process Dynamics and Control.* New York: Wiley.

Shinskey, F. G. (1979). *Process Control Systems.* 2nd ed., New York: McGraw-Hill.

Shunta, J. P., and W. F. Klein (1979). "Microcomputer Digital Control—What it ought to do." *ISA Trans., 18*, No. 1, 63–69.

Smith, C. A., and A. B. Corripio (1985). *Principles and Practice of Automatic Process Control.* New York: Wiley.

Smith, O. J. M. (1957). "Closer Control of Loops with Dead Time." *Chem. Eng. Prog., 53,* 217.

Sokolnikoff, I. S., and R. M. Redheffer (1966). *Mathematics of Physics and Modern Engineering.* New York: McGraw-Hill.

Soucek, H. E., H. E. Howe, and F. T. Mavis (Nov. 12, 1936). *Eng. News Rec.,* 679–680.

Thaler, G. J., and M. P. Pastel (1962). *Analysis and Design of Nonlinear Feedback Control Systems*, New York: McGraw-Hill.

Tou, J. T. (1959). *Digital and Sampled-Data Control Systems.* New York: McGraw-Hill.

Wilts, C. H. (1960). *Principles of Feedback Control.* Reading, MA: Addison-Wesley Publishing Co.

Ziegler, J. G., and N. B. Nichols (1942). "Optimum Settings for Automatic Controllers." *Trans. ASME, 64,* 759.

INDEX